Marx and the Earth

Historical Materialism Book Series

The Historical Materialism Book Series is a major publishing initiative of the radical left. The capitalist crisis of the twenty-first century has been met by a resurgence of interest in critical Marxist theory. At the same time, the publishing institutions committed to Marxism have contracted markedly since the high point of the 1970s. The Historical Materialism Book Series is dedicated to addressing this situation by making available important works of Marxist theory. The aim of the series is to publish important theoretical contributions as the basis for vigorous intellectual debate and exchange on the left.

The peer-reviewed series publishes original monographs, translated texts, and reprints of classics across the bounds of academic disciplinary agendas and across the divisions of the left. The series is particularly concerned to encourage the internationalization of Marxist debate and aims to translate significant studies from beyond the English-speaking world.

For a full list of titles in the Historical Materialism Book Series
available in paperback from Haymarket Books, visit:
https://www.haymarketbooks.org/series_collections/1-historical-materialism

Marx and the Earth

An Anti-Critique

John Bellamy Foster
Paul Burkett

with the editorial assistance of
Ryan Wishart

Haymarket Books
Chicago, IL

First published in 2016 by Brill Academic Publishers, The Netherlands

Published in paperback in 2017 by
Haymarket Books
P.O. Box 180165
Chicago, IL 60618
773-583-7884
www.haymarketbooks.org

ISBN: 978-1-60846-705-1

Trade distribution:
In the US, Consortium Book Sales, www.cbsd.com
In Canada, Publishers Group Canada, www.pgcbooks.ca
In the UK, Turnaround Publisher Services, www.turnaround-uk.com
All other countries, Ingram Publisher Services International, intlsales@perseusbooks.com

Cover design by Jamie Kerry of Belle Étoile Studios and Ragina Johnson.

This book was published with the generous support of Lannan Foundation and the Wallace Action Fund.

Entered into digital printing June 2018.

Library of Congress Cataloging-in-Publication data is available.

Contents

Preface

Ecology as a field of inquiry can be traced back to the beginnings of civilisation. But its modern development dates back largely to the rise of capitalism in the sixteenth and seventeenth centuries, and especially to the emergence of industrial capitalism in the late eighteenth century. The word 'ecology' was first introduced in 1866 in the work of German zoologist Ernst Haeckel, where it was used synonymously at first with what Charles Darwin called the 'economy of nature'.[1] Karl Marx, while not himself employing the term 'ecology', which had little currency in his day, introduced the notion of 'social metabolism' (or socio-ecological metabolism), defining the labour process as the metabolic relation between humans and nature.[2] In this way he provided an ecological perspective that was to underpin his entire critique of political economy. A similar conception of metabolic relations was to underlie the concept of ecosystem as developed by early twentieth-century system ecologists. Given this historical background, it should not be surprising that Marx's approach to the social metabolism and his concept of metabolic rift (or ecological crisis) have increasingly been seen as central to the political-economic critique of the alienation of nature under capitalism, constituting the single most important legacy of social science in this realm.

This understanding of the ecological foundations of Marx's critique was put forward in our previous work, particularly *Marx and Nature* (Burkett) and *Marx's Ecology* (Foster), and is now widely accepted.[3] Yet, there remain a number of criticisms, most of recent origin, levelled at Marx and Engels for their supposed ecological flaws. The present work is designed to address these latest criticisms, which mainly emanate from self-characterised 'ecosocialist' (or what we call 'first-stage ecosocialist') thinkers. Because of the nature of the present work as a response to ecological critiques of Marx and Engels, we have given it the subtitle *An Anti-Critique*. The concept of 'anti-critique' has a clear history and meaning in historical materialism, deriving principally from Rosa Luxemburg's famous *The Accumulation of Capital: An Anti-Critique* (1915) – usually referred to by its subtitle (to avoid confusion with Luxemburg's major economic work).[4] But the notion of anti-critique can be seen as having its

1 Darwin 1964, p. 62. On Haeckel's use of ecology, see Foster 2000, p. 195.

2 Marx and Engels 1988a, pp. 54–66.

3 Burkett 2014; Foster 2000.

4 Luxemburg 1972; Bukharin 1972.

roots even earlier in Engels's *Herr Eugen Dühring's Revolution in Science* (1878), better known as *Anti-Dühring*.[5]

Anti-Dühring was one of the formative works of Marxism. It was frequently seen in the early years of the socialist movement as the single most important work after Marx's *Capital* itself. Marx and Engels's writings prior to the publication of *Anti-Dühring* were seen as primarily economic and political. The connection of their analysis to the broader areas of philosophy and science was largely unknown even to their earliest, closest followers.[6] It was this seeming gap in historical-materialist analysis that allowed Eugen Dühring's development of a larger socialist and naturalistic philosophy to attain considerable influence, due to his crossing of all boundaries of thought. Dühring created a theoretical system (one that has long ceased to draw any interest) that sought simultaneously to explain such disparate realms as philosophy, economics, history, and the philosophy of science, while directly challenging the preeminence of Marx and Engels as socialist thinkers. Consequently, Marx and Engels concluded that there was no choice but to confront Dühring's analysis directly: a task that fell to Engels.

As Engels observed in his second preface to *Anti-Dühring*, he was 'compelled to follow' Dühring 'wherever he went and to oppose my conceptions to his. In the process of carrying this out my negative criticism became positive; it was transformed into a more or less connected exposition of the dialectical method and of the communist world outlook represented by Marx and myself'.[7] In his original preface, Engels noted that he was forced to engage in controversies in areas where his own ideas and knowledge remained undeveloped.[8] It was this traversing of the forest of modern thought (even though the trees were often obscured) that gave *Anti-Dühring* its great overriding importance for Marxists in the early socialist movement, and which turned a negative anti-critique into a positive one. *Anti-Dühring* with all of its faults – of which Engels was all too aware – became the widest-ranging presentation of Marx and Engels's overall historical materialism.

Luxemburg's *Anti-Critique* was a reply to socialist critics of her *Accumulation of Capital*, in which, building upon Marx's analyses of capitalist reproduction and accumulation, she had sought to develop the connections between imper-

5 Marx and Engels 1975a, Vol. 25, pp. xi–309.

6 This was partly due to the fact that such works as Marx's *Economic and Philosophical Manuscripts* and the *Grundrisse*, as well as Marx and Engels's *The German Ideology*, were at that time unpublished and unknown.

7 Marx and Engels 1975a, Vol. 25, pp. 8–9.

8 Marx and Engels 1975a, Vol. 25, p. 7.

ialism and economic crises. She wrote her *Anti-Critique* while in prison over her opposition to the First World War. It was only published posthumously, in 1921, two years after her brutal murder by reactionary forces. The nature of the debate over economic crises and imperialism with which her *Anti-Critique* was associated made it emblematic of the deep divisions within German Social Democracy at the time – in the context of the First World War and the Russian Revolution.

It is this sense of anti-critique, associated with Engels and Luxemburg – but not meant in any way to rival their great achievements – that we have tried to retain in the present book. The ecological problem is the great problem of the twenty-first century. We are in a period of planetary crisis and struggle that is unprecedented. Completely new challenges have arisen. How we choose to move forward on ecological questions is fundamental to the future of socialism and humanity. In our view, the underlying bases for a revolutionary materialist-dialectical critique adequate to the ecological challenges of our time are to be found in the classical Marxian tradition. This is because of the depth and range of its critique of capital, which opposes to the present system of economic commodity exchange an alternative conception of society based on sustainable human development.

Taking our cue from Engels and Luxemburg, we have thus gradually come to see our own efforts to define a historical-materialist ecology, in opposition to those ecosocialists who want to dump the greater part of the classical Marxist legacy, as taking on the overall character of an anti-critique. Moreover, what started out by necessity as a negative response to charges directed at Marx, ended up uncovering untold depths in the classical Marxian ecological critique of capitalism, arising out of the materialist and dialectical method of Marxism itself, and therefore taking on a positive character. In this view, the philosophy of praxis associated with Marxism in its most revolutionary sense offers important new weapons in the defining struggles of our time, pointing to the need for a society of sustainable human development, i.e. socialism.

In preparing this book we have drawn considerably on previously published writings. Chapter 1 is adapted from an article by the same title in *Organization and Environment* 13, no. 4 (December 2000): 403–25. Chapters 2 and 3 each draw on parts of three previous published articles: (1) 'Ecological Economics and Classical Marxism: The "Podolinsky Business" Reconsidered,"' *Organization and Environment* 17, no. 1 (March 2004): 32–60; (2) 'Metabolism, Energy, and Entropy in Marx's Critique of Political Economy: Beyond the Podolinsky Myth,' *Theory and Society* 35, no. 1 (February 2006): 109–56; and (3); and "The Podolinsky Myth: An Obituary: An Introduction to "Human Labour and Unity of Force" by Sergei Podolinsky," *Historical Materialism* 16 (2008): 115–61.

Chapter 4 is adapted from 'Classical Marxism and the Second Law of Thermodynamics: Marx/Engels, the Heat Death of the Universe Theory, and the Origins of Ecological Economics,' *Organization and Environment* 21, no. 1 (March 2008): 1–35. Chapter 5 is an edited version of 'Marx's Reproduction Schemes and the Environment,' *Ecological Economics* 49, no. 4 (2004): 457–67. The two appendixes, consisting of English-language translations of the Italian and German versions of Podolinsky's articles (which accompanied our analyses), first appeared, respectively, in *Organization and Environment* 17 (March 2004): 61–75 (translated by Angelo di Salvo and Mark Hudson), and *Historical Materialism* 16 (2008): 163–83 (translated by Peter Thomas).

Our intellectual debts in the present book are manifold. Our first and greatest debt is to Ryan Wishart, whose name is included on the title page. Ryan edited chapters 2–4 of this book, so as to create a coherent, sequential argument out of four separate articles that were necessarily repetitive when placed one against the other. All of this material needed to be massively edited and reorganised to form the backbone of the present book. He also helped with the editing of the book as a whole. It is literally true that without Ryan's editorial efforts, this work would not have come into being. We would also like to thank Jordan Fox Besek who contributed to the final editing in the preparation of the book.

Sebastian Budgen and Peter Thomas have been supportive of this project from the beginning; the latter translated the German version of Podolinsky's manuscript on human labour into English in association with *Historical Materialism*. Angelo Di Salvo translated the Italian version into English with the help of a rough translation from the French by Michael Hudson. Mikhail Balaev translated Podolinsky's letters to Lavrov of 24 March 1880 and 4 June 1880 from the Russian. Leontina Hormel translated the table of contents of the Russian version of Podolinsky's manuscript for us. We are immensely grateful for the help provided by all of these scholars.

We would also like to thank the many friends and colleagues who gave us encouragement in this project, including, most notably, Brett Clark, Hannah Holleman, John Jermier, R. Jamil Jonna, Fred Magdoff, John Mage, John J. Simon, and Richard York.

Most of all, we would like to thank our life partners, Carrie Ann Naumoff and Suzanne Carter, who are part of our circle of life and whose imprint is thus (indirectly) on every page.

12 December 2014
Eugene, Oregon
Terre Haute, Indiana

Introduction

> In order to use coal as a fuel, I must combine it with oxygen, and for this purpose transform it from the solid into the gaseous state (for carbon dioxide, the result of the combustion, is coal in this state: F.E.), i.e. effect a change in its physical form of existence or physical state.
>
> KARL MARX[1]

•••

> The working individual is not only a stabiliser of *present* but also, and to a far greater extent, a squanderer of *past*, solar heat. As to what we have done in the way of squandering our reserves of energy, our coal, ore, forests, etc. you are better informed than I am.
>
> FREDERICK ENGELS[2]

∴

Three Stages of Ecosocialist Analysis

Rosa Luxemburg once observed that Karl Marx's vast 'scientific achievements' with their 'immeasurable field of application' had so 'outstripped' the immediate concerns of the socialist movement of his day that it was almost inevitable that certain aspects of this critique would be neglected, to reemerge at later stages, as the historical contradictions of the capitalist system matured. 'Only in the proportion as our movement progresses, and demands the solution of new practical problems do we dip once more into the treasury of Marx's thought, in order to extract therefrom and to utilize new fragments of his doctrine'.[3]

The fate of Marx's ecological analysis in the century and a quarter following his death closely accorded to Luxemburg's assessment. Although Marx's trenchant critique of the degradation of nature strongly influenced some of

1 Marx 1978, p. 208.

2 Marx and Engels 1975a, Vol. 46, p. 411; quotation taken from a letter from Engels to Marx, 19 December 1882.

3 Luxemburg 1970, p. 111.

his early followers, knowledge of this part of his thought – although never disappearing entirely, particularly in the sciences – waned within the socialist movement over the course of the twentieth century, since it was not perceived as conforming to the immediate needs of the struggle.[4] This was especially the case in the years of the Second World War and the early Cold War, which led to the hegemony of technological modernism on both sides of the political divide. Marxist ecological thought only began to reemerge in a big way (by which time Marx's own distinctive contribution had been for the most part forgotten), as part of the practical struggle, with the development of the environmental movement, in the 1960s and '70s – itself mainly a response to the acceleration of planetary ecological contradictions.

The debate that was subsequently to emerge within the left over the significance of Marx's analysis to the contemporary ecological movement went through a number of stages. The first of these was a kind of prefigurative phase in the 1960s to early 1980s during the rise of the modern environmental movement, prior to the emergence of ecosocialism as a distinct form of inquiry. This was a time in which numerous socialist thinkers saw ecological concerns as blending naturally with the fundamental historical-materialist critique emanating from Marx. The convergence of Marxism and environmentalism was often viewed as an organic evolution, generating a kind of natural hybrid. This approach was evident in the work of such notable and varied thinkers as Scott Nearing, Barry Commoner, K. William Kapp, Shigeto Tsuru, István Mészáros, Herbert Marcuse, Paul Sweezy, the early Rudolf Bahro, Raymond Williams, Howard Parsons, Charles H. Anderson, Alan Schnaiberg, Richard Levins, and Richard Lewontin – all of whom drew heavily, and in a way that was seen as unproblematic, on Marx, Engels, and Marxism in order to promote the ecological critique of capitalism. It was mainly due to Schnaiberg's influence (drawing heavily on the earlier work of Anderson and on *Monthly Review*) that the new field of environmental sociology arising in the United States was to take on a neo-Marxian form.[5] The early work of Murray Bookchin, it should

4 On the early influence of Marx's ecological notions, see Foster 2000, pp. 236–51, and Foster 2009, pp. 153–60.

5 Nearing was an important ecological thinker as evidenced by numerous books and articles in which he dealt with ecological issues, and is most famous in the environmental movement as a leading exponent of self sufficiency: see Nearing and Nearing 1970. He authored a 'World Events' column in *Monthly Review* for two decades in the 1950s and '60s. An example of the ecological approach he fostered can be seen in his treatment of Rachel Carson's *Silent Spring* and its wider ecological implications at the time of its release (see Nearing 1962, pp. 389–94). See also Commoner 1976, pp. 236–8 and 243–4 (where Commoner drew

be noted, also drew on Marx, prior to Bookchin's development of his more distinctive anarchist 'social ecology'.[6]

A dramatic shift in the discussion occurred, however, with the explicit development of Green theory or 'ecologism' in the late 1970s and '80s.[7] The rise of deep ecology and related trends, along with the increasing incorporation of neo-Malthusian ideas into the environmental movement, led to a growing tendency to see Marxism and environmentalism as opposed to one another. This coincided with enhanced criticism of the environmental performance of Soviet-style societies. The result was the overly defensive and/or breakaway response of what we call 'first-stage ecosocialism', which sought to highlight the presumed ecological failings of Marx, and proceeded to graft Green theory onto Marxism (or in some cases to graft Marxism onto Green theory) as part of a process of *The Greening of Marxism*.[8] Starting in the late 1980s, there was an outpouring of such first-stage ecosocialist analyses – much of it extremely creative – in the work of such thinkers as Daniel Bensaïd, Ted Benton, John Clark, Jean-Paul Deléage, Robyn Eckersley, André Gorz, Enrique Leff, Alain Lipietz, the early Michael Löwy, Joan Martinez-Alier, Carolyn Merchant, the later Jason W. Moore, James O'Connor, Alan Rudy, Saral Sarkar, the early Ariel Salleh, Kate Soper, Victor Toledo, and Daniel Tanuro.[9] A key development was the founding in the late 1980s of the important journal *Capitalism Nature Socialism* under the leadership of James O'Connor.

However, in the late 1990s first-stage ecosocialism generated its own antithesis with the emergence of what could be called second-stage ecosocialism (also referred to as ecological Marxism). This encompassed such varied figures, in addition to ourselves, as Elmar Altvater, Brett Clark, Rebecca Clausen, Peter Dickens, Martin Empson, Hannah Holleman, Jonathan Hughes, Fred Magdoff, Andreas Malm, Philip McMichael, the early Jason W. Moore, the later Ariel Salleh, Kohei Saito, Mindi Schneider, Walt Sheasby, Del Weston, Ryan Wishart,

---

heavily on Marx); Tsuru 1976, pp. 269–93; Mészáros 1995, pp. 170–86 and 872–97; Sweezy 1973, pp. 1–18; Anderson 1976; Marcuse 1972; Parsons 1977; Schnaiberg 1980; Levins and Lewontin 1985.

6 Herber 1965 (Bookchin published this work under the pseudonym Lewis Herber).

7 On 'ecologism' or what is here more often called Green theory, see Dobson 1990; Smith 1998; Naess 1973, pp. 95–100; Rolston III 1988.

8 On first-stage versus second-stage socialism, see John Bellamy Foster's 'Foreword' in Burkett 2014, pp. vii–xiii; and Burkett 2006. For examples of the grafting of green theory onto Marxism, see Benton 1996.

9 Examples of first-stage ecosocialism include Gorz 1994; Benton 1989; Deléage 1994; Leff 1993; Löwy 1997; Salleh 1997; O'Connor 1998; Sarkar 1999; Lipietz 2000; Kovel 2002; Tanuro 2013.

and Richard York.[10] Since first-stage ecosocialism was founded in many ways on criticisms of Marx and Engels (as well as Marxism in general) for having neglected and even violated a Green worldview, it encouraged deeper explorations into these foundational questions by various thinkers connected with the broader ecosocialist (or red and green) project. These second-stage investigations led to the rediscovery of the ecological depths of classical Marxist thought and to the rejection within ecological Marxism of many of the presumptions of first-stage ecosocialism itself.

These explorations demonstrated that Marx and Frederick Engels, along with other early Marxian thinkers, had conceived of historical materialism in terms that were by any meaningful definition deeply ecological. Rather than falling prey to an anti-environmental 'Prometheanism' (or an uncritical promotion of hyper-industrialism), as was frequently charged by first-stage ecosocialists, Marx and Engels, it was discovered, had developed a dialectical theory of socio-ecological conditions and crises unequalled in their time, and arguably – where the social sciences are concerned – in ours as well. This extended to radical conceptions of sustainability, and to the definition of socialism/communism in these terms. Criticisms of Marx's value analysis for failing to take into account ecological variables were similarly discredited – with the results pointing rather to a powerful, ecologically nuanced value analysis underlying his theory of commodity values under capitalism. These discoveries nevertheless required a new outlook on the foundations of historical materialism, embracing elements which had hitherto not been fully understood by, or integrated into, the praxis of Marxism, with fateful consequences.

Although the debate on socialism and ecology necessarily focused on Marx and Engels's original contributions, the real question was not so much the ecological status of Marx's writings themselves, but rather the overall *method* governing a historical-materialist ecology, capable of influencing our praxis in the present. From a Marxian paradigmatic perspective, the idea of simply grafting Green conceptions – themselves an eclectic mix of idealistic, dualistic, and formalistic postulates – onto an unexamined, unreconstructed historical

10 On the recovery over the last decade and a half of Marx's ecological views and the relation of this to the development of ecosocialist analysis, see especially Altvater 1993; Burkett 2009; Burkett 2014; Dickens 2004; Empson 2009; Foster 1999; Foster 2000; Foster et al. 2010; Foster and Holleman 2014; Magdoff and Foster 2011; Moore 2011; Saito 2014; Salleh 2010; Weston 2014; Williams 2010. For references to other thinkers mentioned here, see Wishart et al. 2013.

materialism was at best a kind of ad hoc, patchwork, and not infrequently quasi-Malthusian solution; one that would inevitably fall short of a dialectical critique of the system as a whole, and that would leave fundamental questions about historical materialism itself unanswered.[11] Such partial, eclectic solutions were likely to be distinguished as much by what they discarded from the materialist conception of history, and by the resulting torn, patched-up fabric in what had once been cut out of whole cloth, as by what they added to the overall critique of the system.

A case in point was James O'Connor's brilliant conception of the second contradiction of capitalism. O'Connor argued that 'the first', or economic, contradiction of capitalism had mainly to do with the overaccumulation of capital in relation to demand. In contrast, what he called 'the second contradiction of capitalism' (presumed missing in Marx's own analysis) had to do with the undermining of the 'conditions of production' (human labour power, external nature, and the built environment) thereby increasing the costs of production, and generating supply-side economic crisis tendencies. The presumption was that capitalism, faced with growing environmentally induced economic costs and crises, would increasingly focus on supply-side contradictions induced by environmental degradation. Attempts by capital and the state to cope with rising environmental costs would generate various fault lines, creating an opening for new social movements to push the system in a green direction. This would, in turn, generate further impasses and even more costs due to the imposition of various environmental regulations, generating additional struggles and ultimately creating the potential for structural and revolutionary reforms. New social movements then could join in and get at the head of the parade of environmental redress. The rôle of the environmental movement and other new social movements was seen in this context as primarily one of the radical defence of the conditions of production.

Yet, as powerful as this analysis was in certain respects, its relation to the historical materialist tradition as a whole was questionable. It tended to subsume environmental contradictions within economic crisis, while failing to see *ecological crises* as serious problems in their own right. In reality, although ecological crises were ultimately generated by the system of capital, they superseded the question of purely *economic crises*. The deficiency of such a functionalist analysis was manifest in the fact that there was no feedback mechanism that would serve to translate ecological degradation, even on a planetary level, into eco-

11 On the dualistic, formalistic and idealist fallacies of the mainstream liberal tradition, see Mésázros 2010; Lukács 1980.

nomic crises, demanding an immediate response on the part of capital. In other words, a major ecological crisis – even to the point of threatening the earth as a place of human habitation – did not necessarily feed into the logic of economic crisis and recovery that characterised capital accumulation. Capitalism could advance and even prosper indefinitely while promoting what amounted to (from the standpoint of humanity as a whole) the irreversible degradation of the earth.[12]

All of this suggested that ecological disruption constituted a contradiction of capitalism in a deeper, more absolute, and more complex sense than was initially suggested by O'Connor's 'second contradiction' theory, which was seen as an extension (or a new form) of economic crisis theory. Ecological contradictions did not just influence economic crises; ecological crises (for example, the destruction of whole ecosystems and the accelerated extinction of species) represented rifts in the condition of human civilisation and life itself in a much broader sense. It was then possible to speak of an 'absolute general law of environmental degradation under capitalism' as a dialectical counterpart within the *ecological* realm of Marx's 'absolute general law of capital accumulation', which was conceived mainly in *economic* terms.[13] (None of this denied, of course, that economic problems could themselves in some cases have ecological causes or vice versa).[14] If such a critique may have seemed on the surface 'dualistic' – since economic and ecological crises were not reducible to a single logic within the system – this was not due in this case to an error in the analysis, but rather to the alienated nature of capitalism itself.

The intensifying ecological problem of capitalist society could be traced therefore mainly to the rift in the metabolism between human beings and nature (that is, the alienation of nature) that formed the very basis of capitalism's existence as a system, made worse by accumulation, i.e. capitalism's own expansion. It is precisely this approach, linked to nineteenth-century thermodynamic conceptions, that emerged in Marx's work itself. One of the characteristics of Marx's overall critique was that it saw ecological crises as caused by capitalism but not simply reducible to the internal logic of capital accumulation and crises. Capitalism exploited workers, but it also robbed the earth. Implicitly, Marx adopted a framework of 'social costs' in which the effects of the system were externalised onto society and nature, with the result that ecological con-

12 For more detailed versions of this criticism, see Foster 2009, pp. 201–12; Burkett 2014, pp. xviii–xxi.

13 Foster et al. 2010, pp. 207–11.

14 Burkett 2014, pp. xviii–xxi.

traditions could grow cumulatively, while remaining outside the economic-growth accounting of the system and thus rendered socially invisible.[15]

As a result of second-stage ecosocialist research, the extraordinary power of Marx and Engels's work in this area (which because of its complexity and its integration with the entire critique of capitalist society is still being slowly absorbed) is nowadays widely accepted. Thus few knowledgeable observers are willing to question the overall ecological contributions of classical historical materialism, or the fact that these contributions provide us with important bases on which to move forward to the critical analysis of the ecological present.[16] But since one of the key motivations of first-generation ecosocialism was always to distinguish itself from earlier socialism/Marxism – even presenting itself in some cases as a newer, better form of socialism, and as historical materialism's heir apparent – attempts to find irredeemable ecological faults in Marx and to connect this to the subsequent failures of previously existing socialist societies have continued. Thus much of the impetus for first-stage ecosocialism as a breakaway movement from the classical historical-materialist paradigm remains intact, even if its initial premises were proven unfounded. Moreover, such criticisms have now taken on new, often more sophisticated forms, acknowledging the ecological insights in Marx's work, while at the same time questioning the dialectical integrity of his analysis, and claiming to detect deeper, hidden flaws in this respect.

A number of these newer, more complex, more dialectically oriented criticisms of classical historical materialism and the environment form the focus of this book: (1) the claim made by John Clark and others that in describing nature as 'the inorganic body of man', Marx was being anti-ecological and demonstrating the instrumentalism and the anthropocentrism of his analysis; (2) the notion presented most notably by Martinez-Alier that in supposedly rejecting Sergei Podolinsky's attempts to link the labour theory of value to energetics, Marx and Engels fundamentally repudiated an ecological worldview; (3) the closely related charge propounded by Martinez-Alier, Bensaïd, and others

15 Kapp 1950, pp. 35–6.

16 Naturally, some of the old fallacies persist. Thus some still promote the Podolinsky myth refuted in this book. See González de Molina and Toledo 2014, pp. 48–9. Sarkar criticises Marx and Engels for being 'growth optimists' in the context of the nineteenth century (ignoring their concept of economic crisis and their critique of capitalist production), and claims that this negated their many ecological concerns. On this basis (and what he considers their mistaken rejection of Malthusianism), he proposes abandoning Marxism altogether for a theoretically amorphous ecosocialism, which he says can do just as well 'with or without Marx' (Sarkar 2012).

that Engels in particular rejected the second law of thermodynamics; (4) the criticism advanced by Herman Daly and some ecological economists that Marx's reproduction schemes excluded material flows; (5) Tanuro's recent argument that Marx's analysis suffered from the fatal flaw of failing to distinguish between fossil fuels and other, renewable forms of energy; and (6) the claim of Kovel and environmental historian Donald Worster that Marx's critique of capital slighted the 'intrinsic value' and the holism of nature.

Marx and the Earth grew primarily out of our inquiries into Marx's analysis, in response in large part to the first four criticisms mentioned here, all of which are dealt with in the following chapters. The last two criticisms, which are of more recent origin, are dealt with more briefly in this introduction. In each case we sought to determine whether the specific criticisms of Marx's ecology were correct or incorrect (and if the latter then to what extent), and to go on to what this tells us about the methodological foundations of historical materialism.

The process of confronting such challenges to Marx's ecology head on has led to what we have called in our subtitle (after Luxemburg) *An Anti-Critique*.[17] The systematic nature of this anti-critique will serve, we hope, to bring out both the enormous dialectical power of Marx's theory and its historically specific character. In each case we found that not only were the criticisms wrong or seriously misleading, but also, and much more importantly, that determining why took us closer to the foundations of Marx's thought. This shed new light on Marx's materialist analysis, and allowed us to perceive, in more fundamental ways, its dialectical structure as a whole.

That this should be the result is not surprising. As Sartre observed in *The Search for a Method*, Marx's revolutionary materialist critique of bourgeois society was so thorough in its ruthless critique of everything existing that it is impossible to surpass it without surpassing bourgeois society itself. Consequently, 'an "anti-Marxist" argument is only the apparent rejuvenation of a pre-Marxist idea'.[18] Again and again we find this to be the case. Nor should it surprise us, given Marx's materialist conception of history and its relation to the materialist conception of nature, that this should be apparent in the ecological realm most of all. The morphology of Marx's materialism is ultimately more ecological than economic, since its aim is to transcend the economic, as that has come to be defined by capitalist class society.

What makes Marx's analysis so valuable and worth exploring to the fullest is its absolutely uncompromising revolutionary materialist character. This is

17 Luxemburg 1972.

18 Sartre 1963, p. 7.

just as true with respect to the ecological contradictions of our present society as in other areas. For example, the rejuvenation of the pre-Marxian 'return to nature' consciousness – so fundamental to the development of the Green theory – has never quite caught up with Marx's revolutionary-critical approach to sustainability in its historical breadth and planetary scale.[19] As Marx wrote in *Capital*, Volume III:

> From the standpoint of a higher socio-economic formation, the private property of particular individuals in the earth will appear just as absurd as the private property of one man in other men. Even an entire society, a nation, or all simultaneously exiting societies taken together, are not the owners of the earth. They are simply its possessors, its beneficiaries, and have to bequeath it in an improved state to succeeding generations, as *boni patres familias* [good heads of the household].[20]

Throughout our analysis we take it as axiomatic that authenticity in historical materialism, as Georg Lukács wrote in *History and Class Consciousness*, 'refers exclusively to *method*', which is at the same time the *dialectical method*. It is not this or that particular thesis that defines Marxism, but rather the materialist dialectic that constitutes the basis of its interpretation of history.[21] While Marx and Engels pioneered the development of the materialist conception of history, they saw this as inextricably connected to the materialist conception of nature – just as human history is connected to natural history, and the human sciences to the natural sciences. Orthodoxy in the application of the dialectical method to historical materialism thus cannot be separated from questions of natural evolution, and the human relation to ecological systems. In this respect, Marx referred both to 'the universal metabolism of nature' and the more specific 'metabolism between man and nature'.[22] To follow this complex, variegated line of thought, we believe, is to capture much more fully the *Marxian method* in its classical form; and at the same time it means retrieving those lost parts of his 'scientific achievements' (as Luxemburg put it) that are related to 'new practical problems'.[23]

---

19 For a description of early romantic, Rousseauian 'return to nature' conceptions, see Mumford 1926, pp. 20–39.

20 Marx 1981, p. 911.

21 Lukács 1971, p. 1.

22 Marx and Engels 1975a, Vol. 30, pp. 40, 62–3.

23 Luxemburg 1970, p. 111.

Marx and Engels constitute the dual founders of historical materialism. There is no doubt that Marx, as Engels himself readily acknowledged, was the more powerful thinker. But Engels's contributions were also of extraordinary brilliance, even if frequently overshadowed by those of Marx. Although the two thinkers were not identical, and must be distinguished from each other, attempts to separate them entirely, which have become common in some circles of Western Marxism in recent years, are, in our view, self-defeating and misguided. In relation to the ecological critique of capitalism, both were major contributors.

The development of the complex materialist ecology at the root of classical Marxism therefore requires a reconsideration of the foundations of Marx and Engels's materialism, and especially of what Luxemburg called 'the Marxist method of research', incorporating that which in his overall social science has remained 'unused because, while ... inapplicable to bourgeois class culture, it greatly' transcended 'the needs of the working class in the matter of weapons for the daily struggle' – pointing rather to a much broader scientific critique of bourgeois civilisation.[24]

In this view it is crucial to rediscover – even to excavate in an almost archaeological fashion – certain neglected methodological foundations of classical historical materialism, which were set aside, ignored, or even in some cases practically unknown to the movement. The objective here is not primarily a scholastic one, but one of developing an ecological materialism organically connected to historical materialism itself. The goal is to bring this to bear on revolutionary praxis. The second stage of ecosocialist research, which required a return to foundations, and a reconstruction of Marx and Engels's materialist dialectic on those terms, reincorporating the ecological aspects of their thought, is therefore only meaningful to the extent that it can help us in the development of an ecological-materialist praxis – addressing the ecological challenges and burdens of our historical time.[25]

What can be called third-stage ecosocialist research thus seeks to utilise the richer, more ecologically nuanced understanding of classical historical materialism as uncovered by second-stage ecosocialist analysis, and the needed synthesis to which it points in our times, in order to explore our accelerating planetary environmental crisis, encompassing such issues as: climate change; species extinction; ecosystem destruction; destabilisation of the nitrogen cycle; fresh water loss; unsustainable energy use; ecological waste; urban decay; envir-

24 Luxemburg 1970, p. 111.

25 Mészáros 2008.

onmental injustice; species injustice – indeed the whole question of 'the alienation of nature' under the modern system of capital accumulation.[26] Such a theory of the alienated social metabolism between humanity and nature becomes a basis for understanding the vast changes that must take place to create an ecologically sustainable and socially just society. Moreover, a developed third-stage ecosocialism, in which the Marxian tradition has recovered its deeper materialist-humanistic roots, must also confront the dominant theoretical traditions – even those of the environmental movement itself – in an attempt to create a broader, more effective basis for the necessary epochal change.

Various radical ecological critics have begun to use Marx's classical ecological critique of capitalism to confront the major environmental questions of today, and as a spur for the burgeoning ecosocialist movement within twenty-first century socialism. This application of Marx and Engels's ecological dialectic can be seen, for example, in works engaged in the analysis of: capitalism and the carbon metabolism (Naomi Klein; Brett Clark and Richard York); ecological civilisation (Fred Magdoff); ecofeminism/environmental justice (Ariel Salleh); agro-fuels (Philip McMichael); marine ecology (Rebecca Clausen and Stefano Longo); nitrogen fertiliser dependency (Philip Mancus); solid waste management (Matthew Clement); fire management in forestry (Mark Hudson); land cover change (Ricardo Dobrovolski); the political economy of coal (Ryan Wishart); livestock agribusiness (Ryan Gunderson; Mindi Schneider); food systems (Michael Carolan); urban agriculture (Nathan McClintock); and La Via Campesina (Hannah Wittman). In our own work, we have drawn on Marx's critique to construct a Marxian approach to contemporary ecological economics (Burkett), and to develop a theory of ecological imperialism and of unequal ecological exchange (Foster writing with Brett Clark and Hannah Holleman).[27] A variety of thinkers (including Ian Angus, Patrick Bond, Simon Butler, Naomi Klein, Annie Leonard, Fred Magdoff, Ariel Salleh, Paul Street, Victor Wallis, Del Weston, and Chris Williams) have put this general analysis – based directly or indirectly on the classical Marxist contributions to ecology – to work, in promoting an emerging ecosocialist (ecological materialist) movement.[28] The result of all of this is the development of a powerful and rapidly growing Marxian ecological theory/praxis.

26 On the concept of 'the alienation of nature', see Mészáros 1970, p. 104 and p. 111.

27 See, for example: Burkett 2009; Clausen 2007; Clausen and Clark 2005; Dobrovolski 2012; Foster et al. 2010, pp. 121–50; Gunderson 2011; Klein 2011; Klein 2014, p. 177; Longo 2012; Mancus 2007; Wishart 2012; Wittman 2009. For further citations, see Wishart et al. 2014.

28 Willliams 2010; Angus and Butler 2011; Wall 2010; Wallis 2014; Weston 2014.

The Debate on Marx and Ecology a Decade and a Half Later

In 1999–2000, we each individually, but in close correspondence, launched major arguments on Marx and nature (Burkett, *Marx and Nature*; Foster, *Marx's Ecology*), seeking to overturn the then dominant claims with regards to the anti-environmental character of Marx and Engels's thought. At the time, a host of charges were levelled against Marx for being: (1) 'Promethean'; (2) rejecting or downplaying natural limits to capital accumulation; (3) neglecting eco-regulatory forms of production; (4) failing to incorporate ecological factors into his value analysis; (5) adopting a narrow anthropocentrism and instrumentalism that was unable to comprehend the need for ecological sustainability; (6) denigrating rural life and agriculture; and (7) incorporating ecological values only in his early works or in the margins of his writings. The critical response that we carried out along with others demonstrated that such allegations were entirely without foundation, and indeed were contradicted by mountains of evidence as well as by the logic of Marx's system. Meanwhile, so important has been the discovery of Marx's own metabolic analysis that these original erroneous criticisms are for the most part seldom heard today, a decade and a half later.

Yet, as noted above, new declarations of fundamental ecological flaws in Marx and Engels's analysis have partly taken the place of these earlier ones. These new objections or attempted critiques are addressed in this book. The persistence of such criticisms, even after the general ecological tenor of Marx and Engels's thought has generally been conceded, has to do, we believe, with the fact that those adopting the viewpoint of first-stage ecosocialism often seek to separate their analysis in some crucial way from the classical Marxian critique. This is accompanied by attempts to stipulate fundamental flaws in the latter – in order to justify the promotion of ecosocialism as a new paradigm, superior to classical socialism or Marxism, and as its heir apparent. Hence, after a few requisite acknowledgments are made to Marx's ecology, such first-stage ecosocialists often simply graft on to socialism/Marxism, in a purely ad hoc manner, many of the same basic assumptions that have come to characterise now fairly conventional Green perspectives, derived as offshoots from mainstream liberal thought. In the process, the strengths of a more revolutionary materialist ecological critique, building on the actual foundations of historical materialism, are frequently lost altogether.[29]

---

29 For an interesting piece querying how the term 'ecosocialism' is to be viewed within the Marxist tradition, calling to attention differing points of view, see Baker 2011. Some more

Indeed, the general stance of some of these first-stage ecosocialist analyses seems to be that classical Marxism, or socialism 'pure and simple' as Tanuro puts it, has only limited power today as a critical perspective, and is in the process of being supplanted by a more advanced ecosocialist view.[30] But in order to make such a case, it is necessary to demonstrate that classical Marxism suffers from irredeemable errors in this respect. Ecosocialism of this specific kind is seen not so much as a part of socialism, but as its intended replacement; and this often goes along with the notion that elements of the materialist conception of history and nature can be accepted or rejected at will.

For example, Sarkar says that, viewed from an ecosocialist standpoint, many of Marx's 'basic positions have become indefensible'. Marx and Engels are to be criticised for seeing the expansion of production and productive capabilities over earlier, seventeenth- and eighteenth-century levels as indications of historical progress – even if they did not advocate production for production's sake (or accumulation for accumulation's sake). Faced with such 'indefensible' positions, emanating from classical historical materialism itself, 'ecosocialists', Sarkar declares, should 'strive to create a socialist society, with or without Marx' – indeed the implication is 'without'. He goes so far as to contend that a 'great flaw in the thoughts of Marx, Engels, Lenin and their followers has been that they totally rejected the views of Malthus'. Malthus is thus interpreted, in accord with the general presumptions of mainstream Green theory, as an ecological thinker concerned with the effects of overpopulation on the environment – however much this is in conflict with the facts.[31]

To question such accusations is, of course, not to deny that first-epoch socialism (or twentieth-century post-revolutionary society) was often caught up in forms of ecological degradation that effectively mimicked capitalism. Although ecological destruction has certainly been carried out at times in the name of socialism (as well as in the name of capitalism), this is not, we argue, due to the inherently anti-ecological character of classical historical materialism. Rather these failures can be attributed to the specific historical limitations governing the early twentieth-century socialist revolutions. These took place under conditions of underdevelopment and external imperialist pressures, and

---

academic critics of Marx's ecology within the ecological economics or industrial metabolism traditions seem to be concerned less with defending a different kind of socialism than with promoting the now fashionable concept of 'social metabolism', which originated with Marx but has to be shorn of his influence. In this regard, see de Molina and Toledo 2014; also see the work by Martinez-Alier discussed in chapters 2 and 3 below.

30 Tanuro 2011.

31 Sarkar 2012.

were burdened by the accompanying maldevelopments of state and ideology.[32] Nevertheless, there is reason to hope that the uncompromising revolutionary nature of historical materialism in its classical conception, when put together with the emergent needs and possibilities of today, creates the *potential* for a radical reconstruction of the socialist project for the new millennium. Today, more than ever, it can be said that there is no socialism without ecology, and that ecological Marxism is not so much the child of socialism as its still beating heart. What is required, then, is the retrieval of what Williams called the 'ecologically conscious socialism' exemplified by figures like Marx, Engels, Luxemburg, and William Morris.[33]

But here we run into another common objection. It is sometimes said that Marx's nineteenth-century ecological critique is dated in the sense that he could not possibly have envisioned global climate change, nuclear power, dioxin, or sea-level rise.[34] There is no doubt that historical materialists must explore all of these ecological contradictions of our time and more. But to do so, it is necessary to employ the dialectical and materialist method – since it is this that gives us the critical basis for understanding capitalism's cumulative alienation of the earth, and its relation to the alienation of labour (society). Here Marx's own penetrating ecological dialectic, tied as it was to the critique of capitalism, remains the essential starting point.

The present book, as we have already noted, is mainly devoted to addressing the first four ecological criticisms of Marx and Engels stipulated above, pertaining to the organic/inorganic distinction; energetics; the entropy law; and Marx's reproduction schemes. In the various chapters below, we demonstrate that not only are these criticisms wrong, but that a close examination of them only serves to highlight the extraordinarily powerful ecological methodology embedded within classical Marxism – occupying a unique place in social science even today.

Yet, discoveries of new fundamental ecological fissures in Marx and Engels's analysis continue to be announced by first-stage ecosocialists – even while they concede the primary ground on the existence of Marx's ecology. The latest, most fashionable of these new fault lines are: (1) Marx and Engels's alleged failure to see the disjuncture in energy use associated with the turn to non-sustainable fossil fuels; and (2) the purported lack in their work of a concept of the intrinsic value of nature.

32 On the ecological degradation in the Soviet Union, see Foster 1994, pp. 96–101.

33 Williams 1989, p. 225.

34 See, for example, de Kadt and Engel-Di Mauro 2001.

In the remainder of this introduction, therefore, we shall briefly address these two latest criticisms, not covered in the main part of this book. This will be accompanied by a short, related discussion of a dimension in Marx's ecological analysis that has not yet received attention: namely, the materialist ecology embedded in his sensuous aesthetics. In the end, as Marx with his background in Epicurus was wont to emphasise, life is seen as active, sensuous existence, connecting *human species-being* to the metabolism of the earth itself.

Marx's 'Major Ecological Flaw': The Tanuro Thesis

One of the most influential and audacious recent contributors to first-stage ecosocialist analysis is the Belgian ecosocialist and agricultural engineer, Daniel Tanuro, author of *Green Capitalism: Why It Can't Work*, and numerous other writings. Tanuro argues that 'the only possible socialism is ecosocialism' and that 'the real challenge is not to integrate ecology into socialism, but rather socialism into ecology'. Marxism, he tells us at one point, cannot simply be 'ecologised' or greened (a view he mistakenly attributes to the present authors). Rather socialism should be made to 'converge' with ecologism or Green theory. 'Pure and simple socialism' must be replaced with a new ecosocialism, representing a later, more developed variety. Twenty-first century socialists 'must be "ecosocialists"'. Marxists may disclaim the environmental depredations of twentieth-century post-revolutionary societies, but this 'will be convincing only when they show their rupture with productivism, by raising the flag of ... "ecosocialism"'. Moreover, they must recognise that ecosocialism cannot be a product of the socialist worker's movement; it 'must be brought to the working class from outside'.[35]

The case that Tanuro makes for ecosocialism as the new form *sui generis* that socialism must take in the twenty-first century logically depends on purported deficiencies of the traditional Marxist perspective. This is, of course, necessary if ecosocialism is to constitute a fundamental departure from previous Marxist thought, so that the reference to socialism by itself (without the 'eco' prefix) is no longer adequate. In this way, ecosocialism is seen as an advance not simply on first-epoch socialism, but also on the theoretical framework inherited from Marx and Engels themselves. Thus while Tanuro praises the power of Marx's ecological vision, and particularly his analysis of the metabolism of nature and humanity, he simultaneously argues that Marx's ecological vision was contra-

35 Tanuro 2013, pp. 136–43; Tanuro 2009, p. 282; Tanuro 2011; Tanuro 2010; Tanuro 2007.

dictory at its core, and that references to Marx's ecology are an exaggerated 'reconstruction'; 'a brilliant reconstruction by John B. Foster, but as an implicit reconstruction, it ignores the tensions, unresolved issues, or flaws in the thinking of Marx'.[36]

In claiming one cannot speak without substantial reservations of 'Marx's ecology', Tanuro insists that Marx fell prey to what he variously calls: 'a major ecological flaw'; 'a serious error'; 'a defect'; 'a failure'; 'a blind spot'; 'a shadow zone'; 'a lack of understanding'; 'an ambiguity'; 'a confusion'; 'an inconsistency'; a 'contradiction'; 'an unacceptable flaw'; an inner 'antagonism'; a 'slippage'; and 'a Trojan Horse'. All of this in fact is meant to describe a *single contradiction* that Tanuro purports to have discovered: namely, Marx and Engels's alleged failure to distinguish between renewable and non-renewable (fossil fuel) energy. More specifically, Tanuro claims that Marx and Engels fell prey to the fallacy of energy neutrality – i.e. 'the implicit conclusion that energy sources are neutral' – ignoring altogether the *forms* of energy.[37]

In Tanuro's interpretation, Marx saw renewable and non-renewable energy as a single 'amalgam'. It supposedly never occurred to Marx – despite his argument for the rational regulation by the associated producers of the metabolic relation between nature and society – that such rational regulation was ultimately undermined by the fact that production had already in Marx's day come to rely on coal rather than a renewable energy source such as wood. Fossil fuel use, Tanuro argues, goes against rational regulation of energy forms, and instead requires the shift to renewables or solar energy – even presumably from the standpoint of Marx's day. Moreover, this stands as a major flaw in the whole conception of production and its rational regulation in Marx. By allegedly adopting the postulate of energy neutrality, Marx therefore destroyed the ecological coherence of his system and its capacity to enable us to approach today's ecological problems. Here it is useful to quote Tanuro at length:

> It is striking that, in their analysis of the Industrial Revolution, Marx and Engels simply did not grasp the enormous ecological and economic implications of the passage from a renewable fuel, produced through the photosynthetic conversion of the solar flux – wood – to an historically non-renewable fuel – coal – as a result of the fossilization of the solar flux …

36 Tanuro 2012.

37 Tanuro 2013, pp. 136–40; Tanuro 2010, pp. 94–7; Tanuro 2012.

> It would be overdoing it to criticize Marx and Engels for failing to foresee climate change. However, it is unfortunate that they did not extend their thinking about the limits of soil availability to equally systematic thinking about the limits to coal stocks. This inconsistency affects their 'ecology': the failure to grasp that the qualitative leap from wood to coal prevented them from seeing that the necessary 'rational management of material exchanges' offers a perspective of sustainable management if, and only if, one resorts to renewable energy sources ...
>
> The failure to take into account the difference between renewable and non-renewable energies leads more or less spontaneously to the implicit conclusion that energy sources are neutral ... That is why it can be said that the energy question represents a Trojan horse in 'Marx's ecology' and Marxism in general, irrespective of tendency.[38]

Drawing on a metaphor taken from a quote by Leon Trotsky, Tanuro contends that the 'major ecological flaw' in Marx's analysis was left largely unattended in subsequent Marxist thought, and thus went 'from a scratch to the danger of gangrene'. The infection that set in was of such a 'systematic nature', so we are led to believe, that the body of Marxian thought became gangrenous, seriously undermining its organic integrity and even longevity as a system of thought. 'Utilitarian', 'linear', and 'productivist' notions contaminated Marxian thought as a whole. Today, Marxism, even in its classical form, must be 'reconstructed' – but *not internally*, we are told, but rather primarily *from the outside*, drawing its essentials in this respect from the alternative paradigm of Green theory or ecologism.

Tanuro thus invites us 'to revisit the work of Marx, [and] to "green" his conclusions'.[39] Marx's theory of metabolic rift, as Tanuro acknowledges, was an extraordinary contribution to ecological understanding, providing a 'methodology [that] bears comparison with the best contemporary concepts of global environmental problems'. But by failing to extend his metabolic analysis to fossil fuels, Marx undid this great achievement, compromising his system and his ecological legacy.[40]

Classical Marxism is thus presented by Tanuro as in effect a 'degenerative' methodological research programme in Imre Lakatos's sense, rather than a 'progressive' research programme, in that it is seen as having a shrinking rather

38 Tanuro 2010, pp. 91–5.

39 Tanuro 2010, pp. 93–9; Tanuro 2013, pp. 136–40; Tanuro 2012; Tanuro 2009, pp. 278–80.

40 Tanuro 2013, pp. 137–40; Tanuro 2012.

than a widening empirical content, and is no longer able to generate novel facts. Classical historical materialism is said to be unable to maintain its theoretical 'hard core', even with respect to its central notion of the *rational regulation of production by the associated producers*. Faced with the ecological challenge, it is unable to carry out what Lakatos calls a progressive problemshift, by introducing 'auxiliary hypotheses' emanating from the logic of its scientific model, so as to increase its overall content (and create a 'protective belt' around its hard core). Instead it is compelled, like those research programmes caught in a degenerative problemshift, to resort to '*ad hoc* hypotheses', borrowing from outside.[41] This seemingly takes the form of ecosocialism, conceived as an analytical departure in which the *methodological borrowing* from the outside takes on a dominant role.

What is astonishing in all of this is that Tanuro advances these charges in a major article in *Capitalism Nature Socialism* criticising Marx for his ecological failings, and continues them in his book *Green Capitalism*, without providing a single item of proof, or even so much as a single reference to anything Marx wrote concretely on energy. His criticisms thus seem to be less a product of a genuine investigation into Marx's work, and rather to derive from the necessity of his own argument with respect to ecosocialism as the heir apparent to classical Marxism. To be sure, it would be hard to provide evidence of what is non-existent, and Tanuro tells us that any discussion of the specificity of non-renewable fossil fuels, and of their relation to other renewable forms of energy, is entirely missing from Marx's analysis, and, what is more, has no methodological place in that analysis. Yet, all that this tells us is that he has not sufficiently studied Marx and Engels's work in this area (including *Capital*) in which there are numerous references to the properties of coal vis-à-vis other alternative fuels, and that he is also unaware of much of the secondary work already written on Marx and Engels's treatment of thermodynamics.[42]

Indeed, we were surprised to note that while Tanuro cites our 2004 essay 'Ecological Economics and Classical Marxism', he seems oddly unaware of letters from Engels to Marx that we quoted in that article, in which Engels refers explicitly both to the process of stabilising present 'solar heat', and the using up of 'past solar heat' through the 'squandering of our reserves of energy, our coal'.[43] Surely, such statements go directly against the view that Marx and Engels promoted a view of energy neutrality and were unconcerned with the

41 Lakatos 1970, pp. 116–18, pp. 133–8.

42 For example, Wendling 2009.

43 Foster and Burkett 2004, pp. 50–4; Tanuro 2013, p. 163.

nonrenewable nature and scarcity of coal – or indeed with the question of solar budget. Chapters 2–4 of the present book are largely concerned with Marx and Engels's conceptions of energy, entropy, and thermodynamics. Indeed, it is not by mere accident that Marx and Engels's analysis played a foundational role in the origins of ecological economics, with nearly all the major early figures in this area being heavily influenced by their pioneering efforts.[44]

In our earlier published research on Marx's (and Engels's) treatment of energy, which we have brought together and consolidated in this book (see Chapter 3), we explored Marx's argument on how factory production in the Industrial Revolution initially had its energetic basis in water power (a renewable source associated with solar energy) and then shifted to coal – not because this was more efficient, economically or ecologically, but because it allowed the factories to be built near population areas. This was crucial to Marx and Engels's whole understanding of the Industrial Revolution. Tanuro himself refers to this argument, but only in relation to Lenin's *The Development of Capitalism in Russia* – not recognising that Lenin took this in part from Marx.[45] Andreas Malm's 2013 article, 'The Origins of Fossil Capital: From Water to Steam in the British Cotton Industry', not only provides a detailed historical analysis of these developments, but also refers to Marx's own argument on the development of steam power (based on coal) in opposition to water power. Rather than seeing energy as neutral in its forms, Marx examined the historical conditions governing each form, their natural properties, and the influence of geological location.[46]

But can't Marx and Engels still be criticised for their 'failure', in Tanuro's words, 'to grasp the qualitative leap from wood to coal'? Here Tanuro seems to adhere to a fairly standard neo-Malthusian argument on the role of energy in the Industrial Revolution. As R.G. Wilkinson famously presented this neo-Malthusian view in his 1973 book *Poverty and Progress: An Ecological Model of Economic Development*:

> The ecological roots of the English industrial revolution are not difficult to find. The initial stimulus to change came directly from resource shortages and other ecological effects of an economic system expanding to meet the needs of a population growing within a limited area. As the traditional resources became scarce new ones were substituted ... As landbased resources became scarce it became increasingly urgent to find substitutes

44 On this point, see Martinez-Alier 1987.

45 Tanuro 2010.

46 Malm 2013, p. 57; Marx 1981, pp. 784–5.

> for them. The substitution of coal for wood is the most important case ... The 'timber famine' [of the sixteenth and seventeenth centuries] was a result of purely ecological forces. Population growth and the consequent extension of the economic system led to the conversion of woodland into arable land and a simultaneous expansion in the demand for wood ... As the demand for coal increased [due to wood shortages] and production expanded to meet it, open-cast coal deposits were soon used up and mines had to be sunk deeper and deeper. The invention of the steam engine was a direct result of the new technical problems posed by deep mines.[47]

It is this simple story of a substitution of coal for wood (rooted in demographic pressures) that Tanuro thinks is entirely missing in Marx and Engels's analysis. Yet, the reason such an analysis is not evident in Marx and Engels's work is that the simple neo-Malthusian story that goes from overpopulation to the overconsumption of timber supplies to increased demand for coal to the steam engine – on which Tanuro implicitly relies – is extremely inaccurate as a historical explanation of the energetic basis of the Industrial Revolution. It de-emphasises crucial technological changes and shifts in forms of energy in order to fit these within the simple neo-Malthusian, population-based frame.

The timber famine of the sixteenth, seventeenth and early eighteenth centuries in England, which resulted in the increased use of coal, was not the direct result of overpopulation and consequent use of wood for heating or engines. Rather it was a product of charcoal smelting, which was at the time the only way of smelting iron – prior to the development in the eighteenth century of the coking process, which allowed coal (coke) to be used in smelting. Hence, technological and economic issues, in addition to questions of population and energy sources, played key roles in a combined ecological and economic transition – a historical complexity largely eluded in the crude neo-Malthusian analysis. As Vaclav Smil has written,

> During the early eighteenth century a single English blast furnace, working from October to May produced 300t[ons] of pig iron. With as little as 8 kg of charcoal per kilogram of iron and 5 kg wood per kilogram of charcoal, it needed some 12,000 t[ons] of wood ... In 1720, 60 British furnaces

47 Wilkinson 1973, pp. 112–18; Malm 2013, pp. 20–5; Wrigley 2010, pp. 26–52; Siferle 2001, pp. 112–24. It should be noted that Wilkinson's argument does address the revolution in iron production associated with the substitution of coke to charcoal, but attributes to it only a secondary importance in his causal explanation.

> produced about 17,000 t[ons] of pig iron, requiring 680,000 t[ons] of trees. Forging added another 150,000 t[ons], for a total of some 830,000 t[ons] of charcoaling wood ... Already in 1548 anguished inhabitants of Sussex wondered how many towns would decay if the iron mills and furnaces were allowed to continue (people would have no wood to build houses, watermills, wheels, barrels, and hundreds of other necessities) and they asked the king to close down many of the mills ... Widespread European deforestation was to a large degree a matter of horseshoes, nails, axes (and mail shirts and guns).[48]

What largely deforested England, then, was not so much population growth, as an early stage of proto-industrialisation relying on iron via charcoal smelting. It was this that led Max Weber to declare that the discovery of the coking process for smelting iron with coal had saved the German forests.[49] Marx and Engels were, of course, well aware of this particular qualitative leap from wood-based (charcoal) iron smelting to coal-based (coke) iron smelting, and the relationship between energy, technology, and industrialisation. So serious, Engels explained, was the shortage of wood for charcoal in the eighteenth century, until the means of smelting iron with coal became widespread, that the English were forced, when the environmental crisis peaked, 'to obtain all their wrought iron from abroad', particularly from Sweden. Engels further observed that the shift from wood-based charcoal to coal-based coke smelting of iron in 1788 led to a sixfold increase in iron production in six years.[50]

From the standpoint of the history of industrialisation, coal was at first all about iron. As Engels explained, the rapid introduction of improved forms of iron smelting from coke in the form of puddling ('withdrawing the carbon which had mixed with the iron during the process of smelting') in late eighteenth-century England gave an enormous impetus to the production of iron. The sheer size of smelting furnaces grew fifty times in a matter of decades. The Industrial Revolution was in many ways symbolised, as he noted, by Thomas Paine's designing of the first iron bridge in Yorkshire.[51] In 1869, at the very height of the Industrial Revolution, more coal in England was used in the iron and steel industry than for firing all the steam engines of general manufactures and trains combined.[52]

---

48 Smil 2008, p. 191.

49 Weber 2003, p. 304; Foster and Holleman 2012, pp. 1644–50.

50 Marx and Engels 1975a, Vol. 3, pp. 483–4.

51 Marx and Engels 1975a, Vol. 4, p. 317.

52 Jevons 1906, pp. 138–9; Hobsbawm 1969, pp. 70–1.

Marx and Engels, as already indicated (see also Chapter 3 below) analysed water and steam as contemporary power technologies, with waterpower forming the initial basis for industrialisation and the principal motive force for industry throughout Britain well into the nineteenth century. The original studies in thermodynamics were focused on waterpower rather than steam. Precisely because Marx and Engels recognised that industrial capitalism originated with waterpower, which they took careful note of and compared to steam power, it follows that they did not, like Tanuro himself, equate either industrial capitalism or industrialism generally with steam power from coal alone. Nor, by that token, did they ignore varying forms of energy or conceive of all forms of energy as a single, indistinguishable 'amalgam', as Tanuro claims.

By the end of the nineteenth century, it was clear that any country attempting to industrialise in the internationally competitive environment of the day would necessarily require substantial coal stocks. Capitalism had by this time become increasingly dependent on fossil fuels for its motive power. For this reason, Engels, far from neglecting the significance of coal, declared: 'The conditions of modern industry, steam-power and machinery can be established wherever there is fuel, especially coals. And other countries besides England – France, Belgium, Germany, America, even Russia – have coals'.[53]

Why then did Marx and Engels not go on to develop, as Tanuro tells us they should have, a systematic treatment of the natural and historical limits of nonrenewable fossil fuels? They were certainly aware of the problem and, as we have seen, referred to the limitations on coal supply. However, no one in the mid-nineteenth century – or indeed much later than that – had a clear notion of the extent of coal reserves (and other potential fossil fuels) that existed in Britain or in Europe as a whole, much less globally. Geological knowledge in this area was still rapidly expanding and had a long way to go. William Stanley Jevons, the one major economist in their day who, in *The Coal Question*, was to essay a guess on the speed with which coal supplies would be depleted in relation to Britain alone, generated estimates that were stunningly incorrect. Jevons believed that British coal production would soon peak and decline. Yet British coal production doubled in the first thirty years after the publication of his 1865 book; while US coal production increased by ten times over the same period. Indeed, coal production continues massively today, a century and a half later. Jevons's mistake was to fail to conceive of possible substitutes for coal, such as petroleum, natural gas, and hydroelectric power – as well as failing to understand the extent to which new technology would allow for the exploita-

53 Engels 1892, p. 32.

tion of coal deposits previously inaccessible. Hence, as John Maynard Keynes wrote two-thirds of a century later in the early 1930s, Jevons's projections of declining coal production were 'over-strained and exaggerated'.[54]

Marx and Engels were considerably more cautious than Jevons in their scientific assessments in this area. Although recognising that coal supplies were limited, they did not assume, short of real evidence, that they would run out in the immediate future. Marx carefully studied the geological manual of Joseph Beete Jukes, noting its discussions of coal strata on which Jukes was an authority. Such geological work made it clear that the actual limits of coal supplies were in no way ascertainable at the time. The fact that they were limited, however, was clearly recognised in Marx's analysis. Such scientific caution, given the lack of empirical knowledge in this area in his day, is hardly evidence of a major weakness. Indeed, it was axiomatic in Marx's critique of political economy that capitalism would seek to expand beyond all natural limits – including those posed by critical nonrenewable energy sources. To say, as Tanuro does, that Marx 'overlooked the ecological gold watch' of nonrenewable energy and its distinctiveness in comparison to renewable energy, or that 'he does not notice the incompatibility between … accumulation and the energy base on which it develops', is to advance contentions for which he can find no actual basis; which are contradicted by Marx's own analyses in these areas; and which are ahistorical in character.[55]

Indeed, in pointing out the obvious, that Marx and Engels could not be blamed 'for failing to foresee climate change', Tanuro almost seems to suggest ahistorically that they *might* be faulted for exactly that – particularly as his whole criticism of them for allegedly mistaking the importance of fossil fuels is presented not in terms of their own historical time, *but in the context of today's global warming problem*. Indeed, he goes so far as to indicate in a footnote – inserted at the very point that he says that 'it would be overdoing it to criticize them' for failing to anticipate global warming – that John Tyndall had already presented his experiments demonstrating the role of carbon dioxide in generating a greenhouse effect in London 'in 1861' (the actual date was 1859), with the implication that Marx and Engels somehow missed this, or failed to recognise the implications of energies changing forms.

As recently as 2006, Tanuro stated – no doubt with a degree of hyperbole – that given the power of Marx and Engels's ecological argument, 'it is amazing they didn't' anticipate today's 'global environmental crisis'. In subsequent

54 See Keynes 1951, p. 128; Foster et al. 2010, pp. 169–81.

55 Tanuro 2010, p. 93.

versions of his argument, this metamorphoses into an actual failure on their part – with their inability to go further in this area, despite the historical and scientific limitations of their time, attributed to an actual defect in their (mid-nineteenth-century) understanding of the importance of fossil fuels.[56]

Tanuro contends that 'Marx, *because he didn't know the global ecological crisis* ... could not draw "ecologically correct" conclusions from his genius anticipations'.[57] Yet this is to suggest that the planetary ecological crisis of our time (particularly climate change) is the whole of the environmental problem – and that there were not ecological crises in Marx's own day that allowed him to perceive a metabolic rift between capitalism and the environment. Surely, the fact that Marx, in the process of addressing the ecological crises and contradictions specific to his own epoch, unearthed the fundamental contradiction between the social metabolism of capital and the universal metabolism of nature sufficiently refutes such ahistorical interpretations. Moreover, Marx's analysis here stands out as a revolutionary ecological discovery in itself. From this ecological dialectic we can derive a methodological approach applicable to the quite different (but not unrelated) environmental problems of today.

Ironically, Tanuro, despite his focus on Marx and energy issues, seems to be unaware of just how intimately Marx and Engels were actually involved in the exploration of these problems in their day. Marx not infrequently attended Tyndall's lectures in London at the time of the latter's greenhouse effect discoveries, and is known to have been fascinated with the latter's experiments regarding solar radiation.[58] It is therefore conceivable that Marx was present in the hall at the Royal Institution when the role of carbon dioxide in the greenhouse effect was first explained. Whether or not Marx was actually present on that occasion is a mere historical curiosity. But the *very possibility* of Marx's presence is important in that it serves to highlight the fact that Marx and Engels carefully followed the major scientific experiments in relation to energy, with a focus on the forms of energy. They took careful note of carbon dioxide emissions from coal, and explained it as a change in the form of energy arising from combustion – though they could hardly have dreamed how significant this would be to us today.[59] They paid close attention to all the new applications of energy technology. In the last year of his life, Marx, with his usual forward-

56 Tanuro 2010, p. 94; Tanuro 2006.

57 Tanuro 2012, emphasis added.

58 See Hulme 2009; Yergin 2011, pp. 425–8; Lessner 1957, p. 161; Uranovsky 1935, p. 140; Weart 2003, pp. 3–4; Henderson 1976, p. 262.

59 Marx 1978, p. 208.

looking perspective, was particularly fascinated by, and took careful note of, the experiments of the French electrical engineer Marcel Deprez, who in 1882 demonstrated a long-distance electricity distribution system.[60]

Not content to criticise Marx for his supposed 'major ecological flaw', Tanuro also raises the issue of other 'tensions' that he says beset Marx's analysis, especially with regard to agriculture.[61] We are told that despite Marx's important critique (based on Liebig) of the metabolic rift – in which soil nutrients are removed from the soil in the form of food and fibre and exported hundreds, sometimes thousands, of miles to the towns, where the nutrients end up as pollution, failing to return to the soil – Marx did not go on to conclude 'that agricultural production should be relocated' in order to transcend this contradiction (though Tanuro assumes that this would logically follow).[62] However, he fails here to recognise that Marx and Engels, from *The Communist Manifesto* on, argued that population should be dispersed and the enormous urban concentration characteristic of capitalism broken down. The answer was not to be found in simply relocating agriculture, but in transcending the division of town and country through a 'more equable distribution of population'.[63] This necessarily arises in Marx's analysis from the fact that 'private ownership of land, and thus the expropriation from the land of the direct producers – private ownership of some, involving non-ownership of the land for others – is *the basis of the capitalist mode of production*' (emphasis added). Conversely, a society of associated producers demands the 'conscious and rational treatment of the land as permanent communal property, as the inalienable condition for the existence and reproduction of the chain of human generations'.[64] This can only be the case if society is de-proletarianised, that is, if humanity is reconnected to the earth, and the extreme division of town and country is dissolved.

Tanuro proceeds – in what constitutes a much more important criticism along this line – to fault Marx for having a negative attitude toward peasant production. In this respect, he acknowledges that in criticising large-scale capitalist farming, Marx had declared that agricultural production needed to be carried out instead through 'either small farmers working for themselves or the control of the associated producers'.[65]

60 Draper 1986, p. 55; Engels 1983, p. 39.

61 Tanuro 2012.

62 Tanuro 2010, p. 91.

63 Marx and Engels 1964, p. 40. See Foster 2009, pp. 221–2.

64 Marx 1981, pp. 948–9.

65 Marx 1981, p. 216. See also Magdoff et al. 2000, p. 21.

Nevertheless, we are informed that Marx contradicted himself on this elsewhere in *Capital*, when he pointed out that 'the capitalist mode of production' involved the 'conscious scientific application of agronomy', rather than 'a merely empirical set of procedures, mechanically handed down and practiced by the most undeveloped portion of society'.[66] For Tanuro, this reads as a denigration of indigenous, peasant agriculture, in which Marx 'tends to deny the peasant's knowledge'.[67]

Such a reading, however, negates the often-complex relation between science and traditional knowledge, in which each has its part to play. Marx believed that science could add immeasurably to all forms of agricultural production. But the fact that the development of the science of agronomy had emerged under capitalism did not in itself make capitalist agricultural production inherently superior to that of the small producer – precisely because capitalist agronomy was a response in many ways to the rift generated by capitalist production itself. Presumably, whatever forms of agriculture persisted under socialism (and Marx insisted that small producers working for themselves or the associated producers working collectively were both viable) would necessarily incorporate scientific developments into their production. In fact, Marx specified that the main limitation of small-scale agriculture in the capitalism of his day was 'a lack of the resources and science needed to apply the social productive powers of labour' – limitations that could be overcome at the level of small-scale agricultural production, in a higher form of society.[68]

Still, Tanuro argues on this exceedingly thin basis that Marx exhibited 'a kind of "contempt for the peasantry"'.[69] This is perhaps related to the common criticism of Marx and Engels for referring, in *The Communist Manifesto*, to 'the idiocy of rural life'. This is frequently interpreted as suggesting a negative attitude toward the peasantry. Such a reading, however, has been shown to be wrong on two counts. First, as classical scholars well know, the term 'idiocy' here is used in the sense of the original Greek meaning, whereby it referred to 'isolation' from the cultural life of the polis. Hence, later English-language translations have corrected this by substituting the term 'isolated'.[70] It thus becomes clear that Marx and Engels were not simply denigrating rural life, but were addressing the isolation of rural workers from the wider cultural-

66 Tanuro 2012; Marx 1981, p. 216, p. 754.

67 Tanuro 2012.

68 Marx 1981, p. 949.

69 Tanuro 2012.

70 Two translations of *The Communist Manifesto* have corrected this mistake: Draper 1998; Marx and Engels 2005. For a discussion of this error, see Foster 2009, pp. 219–23.

intellectual life of urban communities. (By the same token, urban workers were divorced from the means of physical wellbeing associated with the country – clean air, clean water, ample living space, etc.). All of this, then, is part of the critique of the division between town and country.

Second, the criticism of rural life in both *The Communist Manifesto* and in the passages from *Capital* that Tanuro cites were not in any way denigrating the peasantry, since the enclosures had already expropriated the land from the peasantry in the most developed capitalism of the day, the Britain on which Marx and Engels's analysis was based. In nineteenth-century England, constituting the primary focus of Marx's analysis of capitalist agriculture, a peasantry no longer existed in any real sense, since they had been replaced by rural wage workers – also providing proletarianised labour for the industrial towns. In the very section of *Capital* from which Tanuro quotes, Marx states that his analysis there 'presupposes the expropriation of the rural workers from the soil', and that therefore peasants no longer existed – replaced by capitalist tenant farmers and rural wage workers. In criticising the scattered, non-scientific rural agriculture, where scientific agronomy has not been applied, Marx had in mind, then, mainly the practices of capitalist tenant farmers in rural areas where science has not yet been incorporated.[71]

Tanuro specifically faults Marx for being 'very ironical against [Léonce de] Lavegne, an author who, in Marx's words, "believes in legends" such as soil enrichment by plants drawing from the atmosphere elements necessary to fertility'. Here Tanuro has in mind the phenomenon of gaseous nitrogen (N_2) fixation by bacteria inhabiting the roots of legumes. This, he admits, 'had not been scientifically established before the death of Marx'. But Tanuro claims that it 'had been discovered "empirically" as early as the fifteenth century (in Flanders): the first agrarian revolution of modern times – the abandonment of fallow, allowing a significant increase in productivity'.[72] Hence, this is seen as a 'tension' or 'contradiction' in Marx's ecological understanding.

Here Tanuro has a point – if an exaggerated, inaccurate, and altogether misleading one. Léonce de Lavergne was a French agronomist who, in 1854, wrote *The Rural Economy of England, Scotland, and Ireland*, a work that Marx studied carefully and critically in its English edition referred to in all three volumes of *Capital*.[73] Marx's great antipathy toward Lavergne – aside from

71 Marx 1981, pp. 751–2.

72 Tanuro 2012.

73 De Lavergne 1855. Although this book first appeared in French, Marx relied on the English edition.

the latter's failure to recognise the full depth of the nutrient-depletion problem in agriculture – had to do with the latter's strong defence of English high farming as the most advanced agriculture in the world, attributable to its greater emphasis on meat and its relative de-emphasis on cereal (corn or grain) production. At the time, meat was predominantly fed to the upper classes and cereals predominantly fed the working class. Meat production and agricultural production as a whole, Lavergne argued, was expanded in English agriculture by the introduction of the famous Norfolk rotation and the replacement of the traditional fallow field with crops devoted to certain legumes such as clover as well as ryegrass, on which sheep and cattle could graze.[74] All of this went hand in hand with the English enclosures, which, beginning in the late fifteenth century, replaced peasants with sheep – a process of primitive accumulation of which Marx was the greatest nineteenth-century critic.

In the passage in Volume III of *Capital* to which Tanuro refers, Marx disputed Lavergne's claims on the greater productivity of English meat-based over French grain-based agriculture, showing that this was simply due to price differentials between the two countries (Lavergne had presented his data in price terms). Marx went on to write:

> M. Lavergne shows on p. 48 [English edition] that he is not only acquainted with the economic successes of English agriculture, but also shares the prejudices of English farmers and landowners: 'One great drawback attends cereals generally ... they exhaust the soil which bears them'.
>
> M. Lavergne not only believes that other crops do not do this; he believes that fodder and root crops enrich the soil: 'Forage plants derive from the atmosphere the principal elements of their growth, while they give to the soil more than they take from it; thus both directly and by their conversion into animal manure contributing in two ways to repair the mischief done by cereals and exhausting crops generally; one principle, therefore, is that they should at least alternate with these crops; in this consists the Norfolk rotation'.
>
> pp. 50–51

No wonder then that M. Lavergne, believing these fairy stories about English rural conditions, should also believe that the wages of English rural

74 De Lavergne 1855, pp. 48–66.

labourers have lost their former abnormal character since the abolition of the Corn Laws.[75]

Here Marx's deep antipathy toward Lavergne's argument is based on its being used by English capitalist 'farmers and landowners' to justify the removal of an even greater number of rural wage workers from the land following the passage of the Corn Laws, and their replacement by pastures for cultivating sheep.[76] Tanuro mistakenly claims that Marx's argument is directed against the traditional knowledge of peasants, while Marx himself says that the argument is directed against the exploitative capitalist agricultural interests.

Most important here was Marx's dismissal as 'fairy stories' Lavergne's notion that forage plants could obtain from the atmosphere 'the principal elements of their growth', independent of the soil.[77] It is true that many people at the time when Lavergne and Marx were writing believed that all nutrients used by plants were obtained from the air. Yet it is now understood that for almost all plants, it is only carbon (as CO_2) that is derived from the atmosphere. Oxygen (O_2), of course, is used for respiration by all 'aerobic organisms', including all plants and animals, but atmospheric oxygen is used to get rid of waste electrons in the respiration process and does not become incorporated as part of plant constituents. Thus for almost all plants, the remaining sixteen essential chemical elements are *derived from the soil* and *not* the atmosphere. Of all the agricultural crops, only legumes such as clovers, peas, and beans can utilise nitrogen in the air (N_2) to make organic nitrogenous compounds (all the others need soil inorganic nitrogen such as nitrate (NO_3^-) or ammonium NH_4^+). As important as the nitrogen-fixing process is for maintaining soil fertility when legumes are used to support grain crops, it remains true that even legumes need to obtain fifteen essential elements from the soil.

None of this is to deny that legumes, especially those perennials grown for a number of years as pasture or hay crops, can improve soil for the crops that follow. This is mostly a beneficial effect on the structure of the soil, along with supplying nitrogen in a form that is usable by non-legumes (such as grains, most vegetables, root crops like turnips, etc.). Nevertheless, the important fact to recognise in the present context is that, despite the nitrogen-fixing ability of legumes (actually the bacteria living in legume roots), *all* crops, as

75 Marx 1981, pp. 768–9.

76 Marx 1981, pp. 768.

77 This paragraph and the following paragraphs on soil chemistry were both largely written by Fred Magdoff, emeritus professor of plant and soil science at the University of Vermont and a major contributor to Marxist ecological analysis.

Marx, following Liebig, argued, deplete the soil of essential nutrients (such as calcium, phosphorous, magnesium, potassium, iron, etc.).

Even though farmers had noted and utilised the soil improving ability of legumes, scientists did not demonstrate this experimentally until after Marx's death. Although Tanuro is right on the importance of indigenous knowledge, there are some areas in which such knowledge is clearly more useful than others. Small farmers generally know a lot about what works in their local environment and some have developed highly productive and ecologically sound farming systems. However, such indigenous knowledge is nowhere near the last word in the implementation of ecologically sound agricultural practices. Some practices by peasants (or small farmers) are not particularly good – such as using hillsides without proper terracing or other means to slow water flowing down the slope, leading to erosion. Most peasants (and many small farmers) are not aware of the tremendous usefulness of cover crops, or the importance of having plants whose main purpose is to attract beneficial insects. Thus although indigenous knowledge should be respected, it is critical to ferret out what should be kept and built on and what might be important to discard and to change. Marx's treatment of agriculture demonstrates a strong respect for science, together with an insistence on the relative sustainability of small farmers acting alone or organised as associated producers.

Marx's harsh criticisms of Lavergne's assessment of British industrial capitalist agriculture, as he indicated in *Capital*, Volume I, were connected to the conflict of Lavergne's views with the theory of metabolic rift. It was precisely Lavergne's failure to recognise that British high farming actually increased the rate of nutrient depletion of the soil, requiring its replenishment from outside in the form of various fertilisers, that was clearly, in Marx's view, his most serious weakness – coupled with the emphasis that Lavergne placed on meat over cereal production.[78]

Tanuro couples his criticisms of Marx's ecology with the additional charge, aimed at Marxism more generally – meant to reinforce the notion of a 'major ecological flaw' throughout Marxism – that socialists almost universally failed to meet the challenge of the environmental movement when it reemerged in the 1960s and '70s. Thus we are told that 'Marxists were caught unprepared when the ecological question appeared as a major issue in the 1960s'. Indeed, 'all currents of Marxist thought', he contends, 'missed the opportunity to engage with the ecological question during the 1970s'. Marx's theory of the metabolic

78 Marx 1976a, p. 636.

rift by this time had 'sunk into oblivion'.[79] On such grounds he suggests it is necessary for socialists to make good their failures by signing on to the ecosocialist cause as opposed to socialism proper.[80]

In contending that socialists/Marxists were slow to respond to ecological problems, Tanuro singles out Barry Commoner as a notable exception, indicating that Commoner had written in his 1971 book, *The Closing Circle*: 'Marx in *Das Kapital* does point out that agricultural exploitation in the capitalist system is, in part, based on its destructive effects on the cyclical ecological process that links man to the soil'.[81] Tanuro seems unaware, however, of the fact that Commoner was inspired here by ecologist and socialist K. William Kapp's 1950 treatment in *Social Costs of Private Enterprise* of the Liebig-Marx relation and Marx's theory of metabolic rift.[82]

Nor is there any mention of other early prefigurative socialist ecological contributions, preceding what we have called the three stages of ecosocialist research. A few notable examples should suffice: (1) Scott Nearing's numerous ecological interventions include, notably, his immediate praise for Rachel Carson's *Silent Spring* and its significance for the critique of capitalism, but also such works as his 1952 *Economics for the Power Age* in which he declared: 'The earth is a common heritage of various forms of life including human beings', and in which he argued for sustainable forms of production; (2) Marxist economist Shigeto Tsuru's articles on ecological issues beginning around 1970, and

79 Tanuro 2010.

80 Tanuro has, as he admits, gone back and forth on how he sees 'ecosocialism'. Originally, as he explained in 2006, he 'believed', in accordance with first-stage ecosocialism generally, that 'Marx and Engels had not given enough importance to the relationship between mankind and nature, had no global consciousness of natural limits ... and that at best they only had "brilliant intuitions"'. Nevertheless, by 2006, after he became acquainted with second-stage ecosocialist (or ecological Marxist) research, he reached a quite different conclusion, one that he said was 'broadly' in agreement with ours. Consequently, although originally ecosocialism had entailed for him a strong criticism of classical Marxism for its neglect of ecology, by 2006 he was using it simply as a way of challenging 'productivist' and mechanistic forms of socialism, and not Marx and Engels themselves, claiming that 'there is something like "an ecology of Marx"'. This position seemed to continue until 2008 (see Tanuro 2008). In his more recent work, since around 2009, he has adopted a third position, reverting to something closer to his original, first-stage ecosocialist stance; now insisting that classical Marxism contained a 'major ecological flaw', and that it is improper to refer to 'Marx's ecology'. Ecosocialism thus represents once again a decisive, qualitative break with classical Marxism (see Tanuro 2006).

81 Commoner 1971, p. 280.

82 Kapp 1950, pp. 31–6.

his role as principal organiser of the path-breaking 1972 Symposium on Environmental Disruption of the International Social Science Council; (3) István Mészáros's powerful ecological critique of unlimited economic growth under capital, in his 1971 Isaac Deutscher lecture, which built on Marx's concepts of alienation and social metabolism – appearing a year before the Club of Rome's *The Limits to Growth*; (4) *Monthly Review* editor Paul Sweezy's landmark article 'Cars and Cities' in *Monthly Review* in 1973, followed up in 1974 with the argument (coauthored with his coeditor Harry Magdoff) that 'instead of a universal panacea it turns out that growth is itself a cause of disease' – and later by his article 'Capitalism and the Environment', where he insisted that it was environmentally necessary to 'reverse' capitalist growth trends; (5) Marxist socialist Charles H. Anderson's book *The Sociology of Survival*, appearing in 1976, which provided perhaps the first full-scale socialist attempt at an ecological critique of capitalism as a global system, arguing for a stationary state (degrowth) and introducing the concept of 'ecological debt' – a year before Herman Daly wrote his seminal *Steady-State Economics*; (6) Howard Parson's 1977 *Marx and Engels on Ecology*, in which the systematic, overarching nature of the classical historical-materialism ecological view was presented; (7) Alan Schnaiberg's 1980 *The Environment*, which was influenced by all of these arguments, and introduced the critical framework of 'the treadmill of production', which was to play a large role in ensuring that the new Environmental Sociology (later Environment and Technology) section of the American Sociological Association was to be heavily – even predominantly – neo-Marxist in its orientation down to the present day; and (8) Raymond Williams's nuanced 1982 discussion of the various aspects of 'Socialism and Ecology'.[83]

The examples could be multiplied many times over. Although it took another generation before second-stage ecosocialist research began to excavate the foundations of Marx's own ecology (which, however, had never been fully forgotten), socialist ecological critiques developed throughout the 1960s and '70s and '80s. If they were sometimes overlooked, as in the case of Anderson's work, this had to do primarily with the fact that they were not only ahead of the working-class movement (and most socialist and Marxist parties), but also well ahead of the environmental movement itself – which is only

83 Nearing 1962, Nearing 1952, p. 125; Tsuru 1994, pp. 233–309; Mészáros 1995, pp. 872–97; Sweezy 1973, Sweezy 1989, p. 4; Magdoff and Sweezy 1974, pp. 9–10; Anderson 1976; Schnaiberg 1980. On Schnaiberg, see Foster et al. 2010, pp. 193–206; on Williams, see Williams 1989, pp. 210–26.

now beginning to catch on with what some socialist environmentalists were already arguing in the 1960s and '70s.

If Marxism as a whole was nonetheless slow to take up the environment as an issue or to give it the attention it deserved (a problem not confined to the left, since ecology was still very much a minority movement across the board), this does not alter the fact that many of the most important analyses of the contemporary global 'environmental disruption', as Tsuru called it, were socialists or individuals strongly influenced by Marxian thought, who made major contributions long before the term 'ecosocialism' came into widespread usage.[84] Indeed, socialists had played a large role in the original development of ecological theory itself, though this influence was subsequently undermined by the weakening of Marxist ecological contributions in both the West and the East beginning in the 1930s.[85]

The ecological critique of economic growth developed by Marxian/neo-Marxian thinkers such as Anderson and Schnaiberg in the 1970s was much more thoroughgoing in its systematic challenge to the system than most European degrowth analyses today. Unfortunately, the main environmental movement, with its neo-Malthusian emphasis, was not willing at the time to embrace so thoroughgoing an ecological critique of the prevailing social system, and hence these more radical social-ecological critiques were largely ignored or marginalised. The Cold War setting that still dominated up through the 1980s was scarcely conducive to systematic anti-capitalist environmental critiques in the West. This goes a long way toward explaining the neglect of analyses like those of Anderson and Schnaiberg.

---

84 On the role of Marx's ideas in the genesis of modern ecological thought see Foster 2009, pp. 153–60.

85 Tanuro writes with respect to one of us 'that J.B. Foster is wrong to attribute the loss of continuity with "Marx's ecology" to "Western Marxism" alone' (Tanuro 2010, p. 98). However, far from neglecting the 'loss of continuity' in the East, Foster emphasised that many of the Soviet Union's most important Marxist ecological thinkers, such as Nikolai Bukharin, Nikolai Vavilov, Boris Hessen, and Y.M. Uranovsky, were purged. 'Bukharin was executed. Vavilov died of malnutrition in a prison cell in 1943' (Foster 2010, p. 113; Foster 2010, pp. 243–4). That the literal killing of its ecological thinkers was bound to create a discontinuity in ecological analysis in the USSR should hardly need to be pointed out.

Marx and the Foreshortening of Intrinsic Value: The Kovel Thesis

If Tanuro's criticisms of Marx's ecology, due to their concrete character, are fairly easily dealt with, those of Kovel raise somewhat more challenging questions, since they are more abstract and philosophically based. Kovel provides perhaps the most ambitious recent attempt to present ecosocialism as 'the logical successor to the socialism that agitated the last century and a half before sputtering to an ignominious end'. For Kovel, 'total revolution, which I would call *ecosocialist*' is 'related to but distinct from the socialisms of the past century'. Here the specific 'name given ... to the notion of a necessary and sufficient transformation of capitalist society for the overcoming of the ecological crisis is *ecosocialism*'. Conceived in this way, he tells us, 'ecosocialism is a transformation of the original socialist project'. It is 'this-epoch socialism' – which remains socialist mainly in the sense of still holding on in some sense to the idea of freely associated labour. 'If socialism of the "first-epoch" was not able to encompass the ecological crisis', Kovel remarks, 'then there needs to be a "next-epoch" socialism that does. For this notion we reserve the word, *ecosocialism*, to signify "where we want to be going". Ecosocialism is socialism made ecologically rational'. Ultimately, he concludes, 'the test of a post-capitalist society is whether it can move from the generalized production of commodities to the production of flourishing, integral ecosystems. In doing so, socialism will become *ecosocialism*'.[86]

However, in order for ecosocialism to be crowned as the successor to first-epoch socialism and in order to prevent challenges to its new dynasty, a kind of severing of the Marxian genealogical tree is required. It is thus necessary to transcend classical Marxism by demonstrating a major theoretical gap. Kovel praises Marx's ecological insights, but argues that they were lacking where it counted most – in the identification with external nature. Marx, he contends, was more advanced ecologically in many ways in his early writings than his later writings – despite the theory of ecological crisis and the argument for sustainability that he developed in more mature works. Marx can therefore be seen as slipping over his lifetime in this respect. Kovel favourably cites Martinez-Alier's contention (calling this a 'good point') that Marx cannot be seen as 'a realized ecologist' because he failed to incorporate the ecological implications of the second law of thermodynamics (a charge refuted in Chapters 2–4 of the present work).

86 Kovel 2002, pp. viii–ix, 7, 151; Kovel 2011a; Kovel 2011b; Kovel 2005, p. 2.

Nevertheless, in Kovel's view the crucial question with respect to ecology is less a question of materialism or science than one of ethics. Hence, he seeks to downplay the whole significance of Marx's now well-known development of the concept of socio-ecological metabolism. What Marx called the 'irreparable rift in the interdependent process of social metabolism' between humanity and nature is reduced to a mere side issue, a matter of quantitative, mechanical analysis. As Kovel writes:

> The term, 'metabolism', (*Stoffwechsel*) appears frequently in Marx, and is frequently cited in works by Foster and others, such as Alfred Schmidt (1971), to show that Marx was at home with concepts of contemporary science, and as indications of the analogy between labor and transformations in nature. To this is frequently added the phrase, 'metabolic rift', as descriptive of aberrations in our relationship to nature. These terms may be used for descriptive purposes, but they belong to the dimensions of physiology and chemistry and are bound to the notion of material exchange, that is, they tend to reduce our vision to the quantitative movement of matter and energy through nature and between society and nature, rather than helping us understand the essentially structural and formal questions posed by ecosystems. Life is best defined as self-replicating form, and while metabolic processes are necessary for comprehending life, they are not sufficient. Terms like metabolism are not more than analogical metaphors, in my view, for the Heraclitean belief that change and transformation is the most fundamental feature of reality, whether in nature or society. Marx saw things this way, as should we all, but his theory of alienation went further, to demonstrate which kinds of transformation conduce to the flourishing evolution of society and nature, and which spell doom. Mere recitation of 'metabolism', or 'metabolic rift', to indicate the presence of ecological damage finesses the key questions. It indicates, to my view, the limitations of Marx within the framework of 19th Century Science.[87]

However, such an assessment misunderstands the whole development of ecological science. Marx's use of metabolism was not 'analogical', but was meant to promote the basis for a materialist and dialectical understanding of the human productive relation to nature. Just as we commonly see a bird's nest as part of the metabolism of the bird, so human production is to be understood in

87 Kovel 2011c.

such systemic-organic terms.[88] The concept of metabolism, which by Marx's day was integrated with thermodynamics, was crucial to the development of ecosystem analysis. Here it is worth noting that it was Arthur Tansley – a materialist and a socialist (albeit of a Fabian variety) and a student of Marx's friend, the great zoologist E. Ray Lankester – who was to introduce the ecosystem concept.[89] The notion of metabolism became the foundation for twentieth-century ecosystem ecology. As indicated in *Marx's Ecology*: 'The concept of metabolism is used to refer to the specific *regulatory processes* that govern this complex interchange between organisms and their environment. Eugene Odum and other leading system ecologists now employ the concept of "metabolism" to refer to all biological levels, starting with the single cell and ending with the ecosystem'.[90] Today, NASA, hardly an organisation of nineteenth-century science, measures the 'earth's metabolism' in the context of the carbon cycle.[91]

The concept of 'social metabolism' derived from Marx has become a key category throughout socio-ecological analysis and of ecological economics – as well as entering into Marxian philosophy primarily through the work of Lukács and István Mészáros.[92] This form of analysis has been widely used by Marxian ecological thinkers and environmental sociologists to explore a wide range of socio-ecological issues, such as the carbon metabolism, the ocean metabolism, land cover loss, soil degradation, fertiliser use, and industrial meat production.

Indeed, it is difficult to exaggerate the critical-theoretical importance of the metabolism category in Marx's thought. It was on this basis that Marx constructed his notion of sustainability as the rational regulation by the associated producers of the metabolic relation between human beings and nature – his most complete definition of socialism/communism.[93] Metabolism also played a critical role, as Lukács emphasised, in both Marx's ontology of the labour process and in his specific approach to the question of the dialectic of nature.[94]

Rather than use Marx's analysis of the metabolic rift to understand ecological crisis, Kovel relies on O'Connor's first-stage ecosocialist notion of the second contradiction of capitalism. The concern here, in contradistinction to

88 Fischer-Kowalski 1997, pp. 121,131.

89 Foster et al. 2010, pp. 324–34.

90 Foster 2000, p. 160.

91 NASA/Goddard Space Flight Center 2003.

92 See Lukács 2000, pp. 97–108; Mészáros 1995, pp. 39–71. See also Schmidt 1971.

93 Marx 1981, p. 959.

94 Lukács 2000, pp. 97–108.

Marx's theory, is never environmental crisis directly, but only the way it generates in some cases supply-side economic crises. The failure to utilise Marx's own analysis is justified in Kovel's case, much like in Tanuro's, by arguing that there was a specific 'history to which he [Marx] had not been exposed, namely, of the ecological crisis'.[95] Yet, to focus simply on *the* ecological crisis and see it as a purely contemporary phenomenon (to be subsumed within a notion of economic crisis) is to belie the fact that Marx and Engels were sufficiently aware of the genuinely ecological crises of their day to address them in their diverse manifestations (as well as within a larger metabolic perspective), encompassing: the degradation of the soil, deforestation, regional climate change, loss of biological diversity, natural resource shortages, etc.[96] Engels wrote of the 'total pollution of the air' from coal-fired factories in the environs of working-class housing in Bolton, less than a dozen miles north-east of Manchester.[97] It is true – as we noted with respect to Tanuro – that Marx and Engels could not perceive today's planetary ecological crisis, involving such disruptions as climate change, ocean acidification, and fracking. But what Marx's ecology provided was exactly what environmental thought even today most critically lacks, namely a historical-theoretical critique of capital as an alienated form of social-metabolic reproduction, and of what Marx called the 'irreparable' effect that this is bound to have on the earth as a place of human habitation if allowed to continue unhindered.[98]

To suggest further, as Kovel does, that Marx's theory of metabolism is somehow unrelated to his theory of alienation is to make a major category mistake. For Marx (and more recently Mészáros), alienation is the estrangement of the necessary organic relation between human beings and nature. In this view, the alienation of nature (ecological metabolism) and the alienation of labour (social metabolism) are simply two sides of the same coin.[99] As Marcuse forcefully put it: 'Marxist theory has the least justification to ignore the metabolism between the human being and nature'.[100]

For Kovel, in contrast, 'Marx became the prisoner of a "scientistic" methodology and lost the fluidity of the dialectic', which prevented him from developing a wider ecosocialist view. Hence, 'at this one point, his genius abandoned

95 See Kovel 2002, pp. 40–1, 211.

96 See Foster 2011, pp. 1–17.

97 Marx and Engels 1975a, Vol. 4, p. 346.

98 Marx 1981, p. 949.

99 Mészáros 1970, pp. 104, 110–11; Mészáros 1995, pp. 137–9; Foster 2014, pp. 11–12, 16; Burkett 2014, pp. 60–2.

100 Marcuse 1978, p. 16.

him'.[101] It would therefore seem from this that Kovel's disinterest in Marx's discussion of the metabolism of nature and society is not altogether arbitrary. Rather the core issue for Kovel is to be found not in science or material relationships, but in the ethics of our relation to nature. It is on this score that Kovel sees Marx as most vulnerable:

> Here it needs to be observed that, however Marx may not have been Promethean, there remains in his work a foreshortening of the intrinsic value of nature. Yes, humanity is part of nature for Marx. But it is the active part, the part that makes things happen, while nature becomes that which is acted upon. Except for a few entrancing anticipations, chiefly in the *Manuscripts* of 1844, nature to Marx appears directly as use-value, not as what use-value leaves behind, namely, recognition of nature in and for itself.
>
> In Marx, nature, is so to speak, subject to labour from the start. This side of things may be inferred from his conception of labour, which involves an entirely *active* relationship to what has become a kind of natural substratum.[102]

We can understand this more fully by looking in some detail at how Kovel treats Marx's famous opening description of the labour process in *Capital*, Volume I, Part 3. There, Marx had written:

> Labour is, first of all, a process between man and nature, a process by which man through his own actions, mediates, regulates and controls the metabolism between himself and nature. He confronts the materials of nature as a force of nature. He sets in motion the natural forces which belong to his own body, his arms, legs, head and hands, in order to appropriate the materials of nature in a form adapted to his own needs. Through this movement he acts upon external nature and changes it, and in this way he simultaneously changes his own nature. He develops the potentialities slumbering within nature, and subjects the play of its forces to his own sovereign power ...

101 Kovel 2011d, p. 13.

102 Kovel 2002, p. 211.

> An instrument of labour is a thing, or a complex of things, which the worker imposes between himself and the object of labour, and which serves as a conductor, directing his activity onto that object. He makes use of the mechanical, physical and chemical properties of some substances in order to set them to work on other substances as instruments of his power, and in accordance with his purpose ... Thus nature becomes one of the organs of his activity, which he annexes to his own bodily organs, adding stature to himself in spite of the Bible. As the earth is his original larder, so too it is his original tool house. It supplies him for instance, with stones for throwing, grinding, pressing, cutting etc. The earth itself is an instrument of labour, but its use in this way, in agriculture, presupposes a whole series of other instruments and a comparatively high stage of development of labour-power.[103]

Kovel's comment on this passage, which he quotes from extensively, treats it entirely from the standpoint of what it supposedly excludes, i.e. a view of nature's own activity and power. Marx is seen as one-sided in this respect, lacking a dialectical perspective:

> In one of the most definitive statements of his life's work, then, we see that though nature indeed plays a role for Marx, it is a highly asymmetrical and unequal one, and radically passive. Marx sees nature as an organ subordinated to the master's mind and an instrument of labor: indeed, the whole earth is seen as such an instrument and even a kind of slave. Though the worker is a force of nature, he is a force *opposed* to nature, and this opposition is chosen of his own accord, hence not just opposed to, but outside of, nature. Man, *Homo faber*, is purely active for Marx here, as nature is passive – indeed, it is hard to see how Man can be a force of nature, if in the labor process he acts of his own accord on a passive nature. Nature is not just passive, but dumb, inertly waiting for Man to be fashioned into objects of use to him.

---

103 Marx 1976a, pp. 283–5. It should be noted that Kovel, in quoting extensively from this passage, used the earlier Moore-Aveling translation and excluded some sentences included here. We have chosen to utilise the Penguin version here, which explicitly captures Marx's references to metabolism [*Stoffwechsel*] and to also include some sentences excluded from Kovel's quotation, since it is essential to show that Marx moves from the discussion of labour (or the labour process in general) to the more specific issue of 'the instruments of labour'.

> It is a striking indication of how much work needs to be done in rethinking the ecological dimension of Marxism that this famous passage has drawn so little critical attention, despite its logical and ontological incoherence.[104]

Nevertheless, one may legitimately wonder how Kovel is able to support, based on these selected passages from *Capital*, his contention on their 'logical and ontological incoherence'. Nature, in Marx's conception, he tells us, is entirely 'passive' and 'dumb'. But if nature is passive, for Marx – a proposition that could scarcely be defended even on the basis of the passages from Marx that Kovel highlights – then what are we to make of Marx's references elsewhere in his work to 'nature's metabolism' as a universal, active force of which humanity is a mere part?[105]

Marx's whole conceptual framework emphasised the power and activity of nature in relation to (and as exhibited by) human production, and the dialectical interdependence between the two. For this reason he distinguished between production as a whole and the labour process, or the specifically human contribution to production. For Marx, production depends on eco-regulatory processes that supersede specifically human production, and on which the latter is dependent. The conditions of all production, he insists, are 'furnished by Nature without human intervention'. This means that 'when man engages in production, he can only proceed as nature does itself, i.e. he can only change the form of the materials'. In terms of 'the universal laws of physics', nature is seen as the active power: human beings do not in this sense actually create in their production, as, for example, plants do through photosynthesis, but simply alter the material forms of what nature provides.[106]

There are other issues here, related to Marx's larger ontological conceptions. It is true that Marx discusses the human activation of 'the potentialities slumbering within nature' and the bringing of these within the distinctly human domain. But are we to condemn this absolutely? Would it be better that modern medicine had not developed nature's potential in this area (for example, modern vaccines), which nature obviously does not provide already ready-made? Would it be better that today's science did not exist? Is it not possible, as Marx clearly believed, to develop human and natural potentials and then

---

104 Kovel 2011d, pp. 10–11.

105 Marx and Engels 1975a, Vol. 30, p. 78.

106 Marx 1976a, pp. 133–4. On eco-regulation in Marx's theory, see Burkett 1998; Burkett 2001a; Burkett 2014.

place them in the service of a sustainable and just society? Is this supposition 'logically and ontologically incoherent', as Kovel says of Marx's argument quoted above?[107]

Marx does not ignore the concept of intrinsic value, or 'intrinsick vertue', which he connected to use value, and mentioned in a discussion of the work of the political economist Nicholas Barbon – on the very first page of *Capital*, Volume I. But he incorporated this, in the context of his critique of political economy, within the category of 'use value', which stood for the entire realm of production in general and human needs in general, independent of capitalist production and bourgeois conceptions of utility.[108]

Here the key points from Marx's perspective are: (1) Intrinsic values are part and parcel of use value in the broader sense of the conditions of sustainable human development. (2) Not all use values are created by labour. (3) Intrinsic valuation of nature is not given the same importance in the labour and production process at all points in history because there are systems like capitalism where such intrinsic values play no essential role in production decisions. (4) Nonetheless, Marx's labour process conception does leave room for intrinsic valuation since it is part of use value broadly defined, and can be treated as a form of primary appropriative labour, including that of a mental and even spiritual type. Still, the concept of intrinsic valuation, for Marx, is uncharacteristic of capitalism, where all economic valuation comes to be formed on a commodity basis, while it is characteristic of a higher social formation, i.e. socialism/communism.

In nonetheless criticising Marx for his supposed 'foreshortening' of the intrinsic value of nature, Kovel compares him unfavourably to Rosa Luxemburg (who could hardly have been expected to agree with Kovel's criticisms of Marx in this regard). Luxemburg wrote a couple of ecologically nuanced letters in prison, in which, looking out from her prison walls at life outside, she commented on the loss of habitat of songbirds and the abusive treatment of a buffalo.[109] In alluding to her letter on the buffalo, Kovel writes: 'When Rosa Luxemburg felt for the buffalo she was being receptive to its anguish. There was recognition there, which meant a taking in of the buffalo's being, and its re-awakening inside her'. We are thus told that Luxemburg was one of the few socialists who had what might be called an 'ecocentric way of being'.[110] He qualifies this, however, by saying that this was 'existentially' the case, but not critically and

107 Marx 1976a, p. 283.

108 Marx 1976a, p. 125.

109 Luxemburg 2013, pp. 457–8.

110 Kovel 2002, pp. 209–12.

intellectually so, since Luxemburg clearly did not explicitly incorporate such ecocentric views into her scientific writings in political economy or into her overall conception of socialism.

Kovel introduces these comments on Luxemburg into the very same pages in which he is questioning Marx's 'ecological *bona fides*'. We are told that Luxemburg's 'capacity to express a *fellow-feeling* for non-human creatures ... is quite exceptional in the Marxist tradition'. The implication, since the focus here is directly on Marx, is that this failure of the Marxist tradition applied to Marx himself. Yet Kovel gives us no reason to think that Marx lacked such 'fellow feeling' for non-human life. Perhaps he thinks we need no such proof. If it is true that there is no contrary evidence to be found, then his point might be considered valid. But what is the nature of the evidence in question here? The letters of Luxemburg were personal letters, written to a friend. Isn't it likely that if we were to look thoroughly at Marx's life and letters (if not his work itself), we would find similar examples of receptivity toward nature and '*fellow-feeling* for non-human creatures'?

In fact, Marx's concern for external nature is evident from his earliest writings. 'Antiquity', he observed in his youth, 'was rooted in nature, in materiality. Its degradation and profanation means in the main the defeat of materiality, of solid life'.[111] Marx celebrated the German peasant revolutionary Thomas Müntzer's criticism of the commodification of non-human nature, writing: 'The view of nature which has grown up under the regime of property and of money is an actual contempt for and practical degradation of nature ... In this sense Thomas Müntzer declares it intolerable that "all creatures have been made into property, the fish in the water, the birds in the air, the plants on the earth – all living things must be free"'.[112] He emphasised that human beings relate to nature sensually and that this sensual (also rational) relationship has its highest, fullest expression in human love, which is extended to other spheres. It is 'love', Marx wrote, 'which first really teaches man to believe in the objective world outside himself'.[113]

Marx conceived of nature not only in terms of instrumental production, but aesthetically, in notions of beauty: 'Man therefore also forms objects in accordance with the laws of beauty'.[114] Marx's youthful poetry exudes a sense of the beauty of nature, often focusing directly on the natural world, while harmon-

111 Marx and Engels 1975a, Vol. 1, p. 423; Lifshitz 1938, p. 16; Marx 1974, p. 239.

112 Marx 1974, p. 239; Müntzer 1988, p. 335.

113 Marx and Engels 1975a, Vol. 3, pp. 295–6, 300, 304; Marx and Engels 1975a, Vol. 4, p. 21.

114 Marx and Engels 1975a, Vol. 3, p. 277.

ising it with human existence.[115] In his correspondence over the years, and particularly some of his later letters, Marx occasionally paused to evoke nature and its beauty, and evinced concern for environmental sustainability. While convalescing in Monte Carlo, he wrote to Engels: 'You will know *everything* about the charm exerted by the beauties of nature here ... Many of its features vividly recall those of Africa'.[116]

In one of his trips to Karlsbad, now in the Czech Republic, Marx was disturbed to find that there were no birds, and surmised that it might have to do with the vapours from the ten mineral springs in the area.[117] Human beings, '*like animals*', Marx insisted in his *Notes on Adolph Wagner*, also learn 'to distinguish "theoretically" from all other things the external things which serve for the satisfaction of their needs' (emphasis added). In attributing to non-human animals not only the ability to think, but also the ability to think 'theoretically', and by referring to their 'needs', Marx was deliberately attacking any sharp separation of human and animal life even in relation to thought. This suggests that he was not being merely flippant – but expressing a real existential contradiction, even a wider conception of suffering embedded in the relation of human beings and other animal species – when he noted rather grimly only a paragraph later that 'it would scarcely appear to a sheep as one of its "useful" properties that it is edible by man'.[118] Viewed in the context of Marx's strong affirmation of high levels of animal cognition in which he saw these as related to human cognition, one might perhaps recognise a kind of 'fellow-feeling'.

Marx was intrigued early on by Herman Samuel Reimarus's treatment of animal psychology and particularly his analysis of the *Drives of Animals* and incorporated this kind of thinking into his own work, allowing for a more subtle understanding of the relation between human and animal work.[119] We know that in the notes that he took on Lavergne in his excerpt notebooks (MEGA IV/18), Marx indicated the abhorrence with which he viewed the attempts of English farmers to find forms of artificial selection that would speed up the bodily growth rate of cattle, pigs, and sheep – in order to produce meat more rapidly.[120] In his personal life, Marx was devoted to his three small dogs (whose names included Whiskey and Toddy) with whom he walked daily and

115 See, for example, Marx and Engels 1975a, Vol. 1, pp. 535–6, 561, 580–1.

116 Marx and Engels 1975a, Vol. 46, pp. 253, 255.

117 See Sheasby 2001.

118 Marx 1975, pp. 190–1.

119 Foster et al. 2008, pp. 85–90.

120 Information provided by Kohei Saito (17 August 2014) in relation to Marx's notes on Lavergne in forthcoming MRGA volume IV/18.

to which he was much attached; he was known to indicate that they displayed an intelligence akin to that of humans.[121]

The same insistence on the human connection to non-human species is to be found in Engels, who wrote to Marx that 'comparative physiology gives one a withering contempt for the idealistic exaltation of man over the other animals. At every step one is forced to recognize the most complete uniformity of structure with the rest of mammals, and in its main features this uniformity extends to all vertebrates and even – in a less distinct way – to insects, crustaceans, tapeworms, etc.'.[122] In the *Dialectics of Nature*, Engels wrote: 'Animals ... change the environment by their activities in the same way, even if not to the same extent, as man does, and these changes ... in turn react upon and change those who made them. In nature nothing takes place in isolation ... It goes without saying that it would not occur to us to dispute the ability of animals to act in a planned, premeditated fashion'.[123]

Karl Kautsky, in 1906, integrated animals into his treatment of *Ethics and the Materialist Conception of History*, basing his treatment on Marx and Darwin. The higher animals, such as dogs and sheep, along with many other varieties of animals, he argued, share the qualities 'we should call moral in men': love, sympathy, conscience, duty, courage, and sociability. These then had a naturalistic basis and affect our own view of natural community.[124]

Indeed, the fact that Marx, Engels, and Kautsky, as well as Luxemburg, all gave evidence of an existential *receptivity* for external nature and a 'fellow-feeling' for non-human animals should hardly surprise us, since it is a general human trait. If such fellow-feeling did not stand out at an emotive level in Marx's scientific writings (any more than in the case of Luxemburg), this did not have to do primarily with some existential lack on his (or her) part, but with the specific political-economic focus of many of those writings and the nature of such scientific critiques – where such detachment is customary. Still, Marx's scientific research carried him to domains far from human production and human history, causing him to take detailed notes on the interrelation of climate change and species extinction (associated with shifts in the earth's isotherms) in earlier geological epochs, long before the appearance of human beings on the earth. This could only be interpreted as a concern with life in general.[125]

121 Comyn 1922.

122 Marx and Engels 1975b, pp. 101–2 (Engels to Marx, 14 July 1858).

123 Marx and Engels 1975a, Vol. 25, pp. 459–60.

124 Kautsky n.d., pp. 99–104.

125 Marx and Engels 2011, pp. 214–19. See also Beete Jukes 1872, pp. 476–512.

To be sure, in their criticisms of 'bourgeois socialism' in *The Communist Manifesto*, Marx and Engels pointed to the hypocrisy in promoting 'societies for the prevention of cruelty to animals', while allowing the most inhuman cruelty to be directed at human beings in their midst. But arguing in this way was not to deny animals their 'rights', but rather to emphasise the injustice to human beings themselves.[126]

The whole notion of 'intrinsic value', raised by Kovel, is of course philosophically complex and raises numerous conceptual issues. Idealist thinkers, from Plato to the present, and mainstream moral philosophers have often relied on ultimate, essentialist notions of (normative) value, in which things are 'good' or 'bad' in their own right, in conformity with a strong foundationalistic conception of ethics. In contrast, consistent materialist thinkers, such as Epicurus and Marx, have resisted such essentialisms in ethics, seeing values as socially and historically determined by human beings themselves within the context of changing material conditions and struggles.[127]

The 'intrinsic value of nature', as seen from an environmental perspective, however, is somewhat distinct from this more general issue of intrinsic value in moral philosophy as a whole – since the chief concern is no longer principally the question of the foundationalist conceptions of ethics, but rather the extension of human moral responsibility to external nature, including other species. This is often treated as a question of an 'anthropocentric' versus an 'ecocentric' relation to nature.

Here we agree broadly with Benton when he says: 'My own position is that the ecocentric/anthropocentric opposition can be overcome by acknowledging (*contra* some versions of ecocentric "intrinsic value" theory) that only creatures such as ourselves, capable of culture, can assign values. However, we can value (elements and relations in and with) non-human nature in virtue of their actual or potential use for us, or for what they are in themselves'.[128] The question of anthropocentrism versus ecocentrism thus becomes a question in reality of the extent to which humanity is able to move beyond an instrumentalist approach to nature – toward one that is inscribed within a broader sense of community with life as a whole (after providing for basic human needs). Human consciousness, human capacities, and human needs are irrevocably human-based, and in that sense inescapably 'anthropocentric'. But there is a great deal of difference between an anthropocentrism that promotes clear-cuts

126 Marx and Engels 1964, p. 52.

127 See Zimmerman 2010; West 1991.

128 Benton 2001, p. 316.

for purposes of unconstrained economic expansion, and one that attempts to sustain old-growth-forest ecosystems for the sake of the species within. 'All of history', Marx once stated, 'is nothing but a continuous transformation of human nature', and human nature can develop in this direction as well – or regain at a higher level what it has previously lost to an alienated world.[129]

According to the philosopher John O'Neill, the notion of intrinsic value – particularly with respect to the environment – can carry a number of different, logically separable, meanings: (1) non-instrumental value; i.e. the valuing of something in and of itself apart from human productive ends; (2) the recognition of certain intrinsic properties; and (3) the recognition of objective existence, independent of human perception.[130]

None of these distinct meanings pose any particular difficulty for a consistent historical materialist.[131] Marx saw the material world as existing prior to and logically independent of human beings. He pointed to a non-instrumental, although necessarily human-mediated, relation to nature as the characteristic of a non-alienated society. His work as a whole presents a strong critique of the alienation of nature and points to the 'universal metabolism of nature', and thus the necessity of ecological sustainability – or human coevolution with the earth.

All of this brings us to Kovel's own notion of intrinsic value, and what he perceives as Marx's specific 'foreshortening' in this respect. For Kovel, 'intrinsic value is distinct from use- and exchange-value in not being immediately tied to production at all. It may be likened to the attitude of wonder with which infants regard the world. As such, it is impossible for us to live by intrinsic value alone'.[132] He thus defines intrinsic value as something inborn in each human being, inherent in our own nature or our original psychological makeup from infancy. It therefore precedes our socialisation, and indeed (since we are speaking of 'infants') a clear consciousness of the objective world as something that exists apart from us. Elsewhere, Kovel writes: 'I would define I–V [intrinsic value] as an assertion that we should *value nature for itself, irrespective of what we do to it* – value it intrinsically and thereby as a function of its inherent right'.[133]

---

129 Marx 1963a, p. 147.

130 O'Neill 1993, pp. 8–10. For an exchange on Marx's ecology and intrinsic value, see Hughes 2000, pp. 16–35; Hughes 2001; Burkett 2001b; Burkett 2002.

131 See Bhaskar 1983, pp. 324–9.

132 Kovel 2011d.

133 Kovel 2014.

Kovel's treatment of the intrinsic value of nature here seems to owe much to contemporary deep ecology.[134] Arne Naess's application of the concept of intrinsic value to the human relation to the environment, in the process of developing the notion of 'deep ecology', was to play a formative role in contemporary Green thinking and mainstream environmental ethics.[135] Yet, although a non-instrumentalist approach to nature is essential to any ecological perspective, the abstract notions of intrinsic value prevalent within some versions of deep ecology have been known to feed into reactionary views, including mysticism, alienated forms of spiritualism, idealism, and anti-humanism. Even thinkers on the left can easily lose their way once they pass down this road. Thus the misanthropic notion that human beings are nothing but a cancer on the earth, with the implication that the human species should be eradicated, has at times infected some parts of the deep ecology movement.

Kovel himself draws in a number of places on the seventeenth-century mystic and vitalistic thinker Jakob Böhme, whom he connects, in a very dubious fashion, to Marx.[136] We are informed that Böhme was a prefigurative ecological theorist who combined spirit [*Geist*] with nature: 'Böhme was able to transcend the split between flesh and spirit that haunted Christianity ... Böhme's God does not create heaven and earth, It (though called "He") is itself created from non-being – the "Unground" – in a process that bears an uncanny resemblance to the Big Bang of current cosmological theory'. Based on an oblique reference to Böhme's play on words in the use of the word '*qual*' for both pain/torture and vital powers in Marx and Engels's *The Holy Family* (and in a short explanation of this later on in Engels's *Anti-Dühring*), Kovel tells us that the young Marx 'followed one of the most radically hermetic thinkers in the Western tradition' in adopting the vitalistic notion of 'the active internal potentials of nature'.[137]

134 On the wider basis of Kovel's application of the notion of the intrinsic value of nature, see Peterson 2010.

135 Naess 1973; Naess 2008, pp. 95, 140–1, 296, 300–1; Light 1997, p. 74; Rolston III 1988, pp. 112–25.

136 In two different sentences in two different works, Marx and Engels in describing the nature of matter (the material) had referred to the play on words in Böhme's use of the word 'qual' in its German and Latin meanings to indicate both torture and inherent forces of being. However, in these statements they do not indicate any additional interest in Böhme's thought beyond this useful dialectical play on words (Marx and Engels 1975a, Vol. 4, pp. 128, 691).

137 Kovel 2001, 79–81, 'Ecology'. For a criticism of Kovel's suggestion that there is a textual basis for claiming Marx imbibed Böhme's mysticism and vitalism, see Burkett 2001c; Marx and Engels 1975a, Vol. 4, p. 128.

It is true that Marx treated nature as alive and its own force, and thus had this much in common with vitalism (though he could hardly be seen as adhering to vitalism as a philosophical-scientific view in general). However, it is not clear, despite Kovel's best efforts, how recourse to a Christian mystic like Böhme can add anything truly meaningful to Marx's ecology. As G.W.F. Hegel observed of Böhme: 'Because no order or method is to be found in him, it is difficult to give an account of his philosophy'.[138]

Nevertheless, it is clear that for Böhme 'the abyss of Nature and creation, is God himself'. God as the Trinity (ternary) is revealed in all things:

> Ye blind Jews, Turks and Heathens, open wide the eyes of your mind: I will show you in your body, and in every natural thing, in men, beast, fowls, and worms, also in wood, stone, leaves, and grass, the likeness of the holy ternary in God [i.e. God, the Son, and the Holy Spirit] ... Now observe: in either wood, stone, or herbs, there are three things contained, neither can anything be generated or grow, if but one of the three should be left out ... Now if any of these three fail, the thing cannot subsist.[139]

Such views of the threefold essence of nature arising from God as the Holy Trinity doubtless played a key role historically in the break with Aristotelian scholasticism and medieval Christian theology. However, it is hard to see how ecological Marxism can be further developed on this basis. What are we to make of Kovel's odd claims that the 'genius' of the seventeenth-century mystic Böhme somehow allowed him to anticipate 'the big bang' theory of scientific cosmology, or that his work represented 'an intuitive and symbolic way of describing the awesomeness of nature that could stand in, so to speak, until the physics of general relativity and quantum mechanics could catch up with it'? Is theosophy then the principal road to ecology, science, and cosmology?[140]

A more important example of the dangers facing ecological analyses that uncritically adopt idealistic/teleological notions of nature/ecology can be seen in Donald Worster's *Nature's Economy*. Worster seeks to make an abstract adherence to holism the acid test for ecological thought. As Kovel points out, Worster contends that Marx lacked such a holistic view of nature's economy connecting all living organisms. 'You cannot ... find' in Marx and Engels, Worster exclaims, 'much concern about preserving any ancient feeling for nature

138 Hegel 1995, p. 189.

139 Böhme quoted in Hegel 1995, pp. 212–13.

140 Kovel 2001, p. 81.

or even any concern for environmental preservation'.[141] From the standpoint of nature's economy, theirs was, Worster believes, too mechanistic and technological a vision. Interestingly enough, Worster is critical on essentially this same basis of Tansley, the originator of the ecosystem concept, since the latter supposedly adopted a mechanistic-scientific approach to nature, going against a truly holistic approach to life.[142]

Worster, however, is on extremely shaky grounds here. He reserves some of his most extravagant praise in this respect for Tansley's major opponent, General Jan Smuts of South Africa (perhaps best known as the man who arrested Gandhi), due to Smuts's ecological 'holism' – a term he coined – and his appreciation of nature's intrinsic value. Yet Smut's reactionary idealist philosophy served in fact as a theoretical justification (as a mere reading of his work makes clear) for his rabid ecological-racism. This was implemented in murderous form through multiple mass attacks, including aerial bombings, on black populations in South Africa, killing hundreds – and in the process laying the political foundations for the apartheid system.

In contrast, Tansley introduced the concept of ecosystem (in line with Marxist ecological thinkers, such as Hyman Levy and Lancelot Hogben) as a direct materialist refutation of Smuts's proto-apartheid, idealist 'holism'. In Tansley's ecosystem analysis, Smuts's ecological-racist holism (along with the teleological approach to ecology in general) was dethroned, while at the same time the new concept of ecosystem was meant directly to question the human depredations of nature. All of this suggests the importance of a materialist approach to issues of use value/intrinsic value.[143]

Kovel offers by far his most pointed ecosocialist criticism of Marx for neglecting the intrinsic value of nature in a discussion of the etymology of the concept of use value in *Theories of Surplus Value*. For Marx, use value is related to production in general. Thus the concept of 'value' in its broadest, transhistorical conception meant use value (and also intrinsic value). '*Exchange-value*, as a result of the social development that created it, was later superimposed on the word value, which was originally synonymous with use-value'. Viewing

141 Kovel 2011c; Worster 1994, pp. 426–7. Kovel claims that Worster's view needs to be qualified, since, as Kovel indicates, Marx was capable of pointing to the intrinsic value of nature. He then turns around, however, and uses Worster's notion of ecology as reflecting life's holism as a way of indicating Marx's shortcoming as an ecologically oriented thinker – and goes on to criticise Marx for 'foreshortening' intrinsic value.

142 Worster 1994, pp. 239–42, 301–4.

143 Worster 1994, pp. 322–3; Kovel 2011c; Foster et al. 2010, pp. 289–343; Anker 2001; Carolyn Merchant, like Worster, lauded Smuts over Tansley. See Merchant 1980, pp. 292–3.

the word 'value' in terms of its etymological origins, and in its full, rich, non-capitalist form, Marx insists: 'The value of a thing is, in fact, its own *virtus* [virtue], while its exchange value is quite independent of its material properties'.[144]

Kovel remarks that this conception 'clearly reveals that for Marx use-value is embedded in natural ecologies, but at the same time that he sees no need to differentiate use-value from any notion of intrinsic value in nature'. This leads Kovel to draw the conclusion that in Marx's case 'a term [use value] belonging to economic discourse suffices to embrace the entirety of what nature means to humans'.[145]

Yet Marx's etymological discussion here explicitly states that the notion of 'value' in its broadest etymological and also natural-material sense (equivalent to use value or intrinsic value) originated *prior to* and engaged material issues *beyond* that of economic, i.e. exchange value, discourse. For Marx, all commodities have use values (as well as exchange values). But nature produces use values too, outside the commodity universe – the air we breathe, the earth we live on, and the stars above which guide us. These are not simply reducible to narrow economic rationality, but are associated with the fulfilment of human needs of the most varied kinds, including beauty, art, spirituality, and the love of nature. Without the alienation of nature brought to a head in capitalist society, value, intrinsic value, and use value *become one*: a process of social valuation, which supersedes mere instrumentalist ends, encompassing the rich world of human sensuous existence and sustainable human development.[146]

Marx, Aesthetics, and the Sensuous Value of Nature

As the foregoing suggests, there are numerous ways in which the direct, human valuation of nature enters into Marx's conception of revolutionary social transformation, which depends on overcoming the alienation of nature (together with alienation of labour). Revolutionary praxis in its widest sense encompasses the full range of human active, sensuous experience. In this book, the emphasis is on how *science* requires the transcendence of the alienation of nature, i.e. of the estrangement of human beings from the full diversity of life. But another vital realm in which the alienation of nature must be transcended, in Marx's view, is *art*, or what Kant called the 'dialectic of aesthetic judgment'.[147]

144 Marx 1971, pp. 296–7.

145 Kovel 2002, p. 221.

146 Burkett 2005b, pp. 34–62.

147 Kant 1952, pp. 204–27.

Indeed, it is in his 'lost aesthetics' that Marx's dialectical and emancipatory approach to nature-human relations is perhaps most powerfully expressed.[148]

The notion 'aesthetics' was first introduced by A.G. Baumgarten in his *Aesthetica* (1750–8) to refer to the 'study of sensory beauty', encompassing the beauty of nature as well as art. In *The Critique of Judgment*, Immanuel Kant saw aesthetics as dependent on the sensuous (or supersensible) [natural] substratum, and as encompassing the beauty of nature, as well as the beauty of art.[149] But this was to change radically in German Idealism after Kant, particularly with F.W.J. Schelling and Hegel. Hegel acknowledged that as a discipline (and in its etymological origins), '"Aesthetic" means more precisely the science of sensation or feeling'.[150] But his *Lectures on Aesthetics* were explicitly designed to contest this, and to remove the aesthetic from this sensory basis, separating it from external nature.

Central to Hegel's view is what Theodor Adorno called his 'theorem that art is inspired by negativity, specifically by the deficiency of natural beauty'.[151] Hegel's *Lectures on Aesthetics* thus began by asserting the need to 'exclude the *beauty of Nature*' from aesthetics, as not conforming to the self-conscious spirit, in which beauty is the product of human consciousness and action. For Hegel, 'artistic beauty starts *higher* than nature'. Indeed, 'everything spiritual is better than anything natural'. Natural beauty and sensuousness were to be condemned for their immediacy and the fact that they were 'too destitute of *criterion*'. Nature, which was 'contaminated and infected by the immediate sensuous environment', was in fact impervious to any meaningful aesthetics: 'The hard rind of nature and the common world gives the mind more trouble in breaking through to the idea than do the products of art'. Not surprisingly, Hegel viewed the 'province' of aesthetics as first and foremost 'fine art', while displacing the supposedly lower or decorative arts along with natural beauty.[152]

148 Marx wrote a number of works on aesthetics in his early formative period, around the time of the *Economic and Philosophic Manuscripts*, which have been lost, hence the reference to his 'lost aesthetics'. Scholars have subsequently reconstructed his aesthetics by examining his various early writings, notably the *Economic and Philosophical Manuscripts*, where his critical relation to Hegel's aesthetics is evident, as well his scattered aesthetic discussions in his later work. Margaret A. Rose, in a remarkable work, was able to contextualise and enhance the appreciation of Marx's aesthetics, through an examination of left-Hegelian and utopian-socialist aesthetic theory, which formed the immediate ground for Marx's own historical-materialist approach to aesthetics (see Rose 1984).

149 Inwood 1992, pp. 40–3; Kant 1952, pp. 35, 72–5, 206–13; Adorno 1997, p. 61.

150 Hegel 1993, p. 3.

151 Adorno 1997, p. 66.

152 Hegel 1993, pp. 3–5, 11, 34.

The anti-sensuousness of Hegel's idealist philosophy of art was its most prominent characteristic. Art was seen as taming the senses and creating in that way reconciliation between immediate sensuous nature and the spirit in history. 'Art', he wrote, 'by means of its representations, while remaining in the sensuous sphere, delivers man at the same time from the power of sensuousness'. It does this by 'mitigating the fierceness of the desires'. It thus achieves 'the *purification* of the passions'. Humanity is relieved of the 'mere sunkenness of nature'.[153]

The zenith of painting, consistent with his notion of 'the end of art' in modern bourgeois society, was represented, in Hegel's aesthetics, by Raphael's modernised, desensualised madonnas, for example the *Sistine Madonna* (1513–14) with its deeply pensive figures. These were characterised by carefully controlled, desensualised, and artificially 'spiritualised' passions.

Marx was to emphasise that Raphael's art, like all art, was social in character. He clearly saw it as an outgrowth of early modern bourgeois society, in relation to Hegelian aesthetic ideas. In Marx's immediate circle, represented in this respect by the poet and critic Heinrich Heine, Raphael was seen as the idol of spiritualism in history, promoted by the Prussian court. This stood opposed to the sensual beauty favoured by revolutionary social movements, as depicted, for example, by Eugène Delacroix's famous painting of *Liberty Guiding the People* (1831) – in which a woman with exposed breasts carried the tricolour flag at the head of the victorious French revolutionary forces. In Heine's language, the conflict within aesthetics was one between the spiritualist 'Nazarenes', represented above all by Raphael, versus the sensuous 'Hellenes', represented by Delacroix.[154]

'In Hegel's transition from nature to art', Adorno observed, 'the much touted polysignifcance of *Aufhebung* [simultaneously conveying transcendence, suppression, preserving, overcoming, superseding] is nowhere to be found'.[155] Hegel's aesthetics lack a clear dialectical transcendence in which nature is partially overcome, and yet preserved, to be carried forward in a new form. Rather art, as the life of the spirit, is seen as superseding nature and sensuousness in a more absolute way. For Hegel, 'the work of art, although it has sensuous existence ... does not require concrete sensuous existence and natural life; indeed, it even *ought* not to remain on such a level, seeing that it has to satisfy only the

153 Hegel 1993, pp. 53–5.

154 Rose 1984, pp. 15–16, 34–7: Hegel 1975c, pp. 814, 881–5; Marx and Engels 1975a, Vol. 5, pp. 391–3.

155 Adorno 1997, p. 76.

interest of mind, and is bound to exclude from itself all desire'.[156] Written in 1817, a year before *The Philosophy of Right*, Hegel's *Introductory Lectures on Aesthetics* represent his later, more conservative phase, when he sought to reconcile with the Prussian state and social reality.

Marx in his materialist, sensuous aesthetics radically rejected Hegel's idealistic, desensualised aesthetics. As in much of the socialist movement – including Saint-Simon and Heine, as well as left-Heglians like Bruno Bauer (who sought to reinterpret Hegel's aesthetics in sensuous terms) – Marx saw the overthrow of alienated art as a necessary counterpart of the overthrow of alienated nature and alienated labour, and the return to the wealth of human-natural interconnections. This is most evident in the *Economic and Philosophic Manuscripts* and has been captured in the scholarly analysis over many decades into Marx's 'lost aesthetic'.[157]

The ruling-class aesthetics of bourgeois society were, for Marx, characterised by the irretrievable loss (outside of revolutionary transformation) of genuine, non-alienated sensuous existence:

> In the place of *all* physical and mental senses there has therefore come the sheer estrangement of *all* these senses, [in] the sense of *having* ... The abolition of private property is therefore the complete *emancipation* of all human senses and qualities ... Need or enjoyment has consequently lost its *egotistical* nature, and nature has lost its mere *utility* by use becoming *human* use ... Only through the objectively unfolded richness of man's essential being is the richness of subjective *human* sensibility (a musical ear, an eye for beauty of form – in short, *senses* affirming themselves as essential powers of *man*) either cultivated or brought into being.[158]

As Mikhail Lifshitz wrote in explanation of Marx's aesthetics: 'Artistic modification of the world of things is therefore one of the ways of assimilating nature ... An aesthetic relation to reality is one of inner organic unity with the object, equally as remote from abstract, contemplative harmony with it as from arbitrary distortion of its own dialectic ... Whereas Feuerbach, whenever he dealt with the subject of art, always started with contemplation, Marx invariably stressed the significance of the productive factor, which determines aesthetic needs, and evolves them through practice' (praxis).[159]

---

156 Hegel 1993, p. 40.

157 Rose 1984, pp. 12–23, 37–64.

158 Marx and Engels 1975a, Vol. 3, pp. 300–1.

159 Lifshitz 1938, p. 65.

In Marx's materialist philosophy, and notably in his aesthetics, human beings were conceived of as both *human* and *natural* beings – insofar as they were not alienated social beings. He called for not only 'the emancipation of all human senses and qualities', but also at the same time their active cultivation, through the education of the senses, which was nothing other than the freeing up of human creative powers in history. In sharp contrast to Hegel, Marx explicitly declared: 'That abstract thought is nothing in itself; that the absolute idea is nothing for itself; that only *nature* is something'. A human being is a directly sensuous being: 'a human and natural subject endowed with eyes, ears, etc., and living in society, in the world, and in nature'.[160]

The specifically human-material *relation* exists only through the senses, and as a condition of life itself. It is expressed intellectually both in human science (understanding) and human art (sensuous imagination); but more importantly it is expressed in non-alienated social production, i.e. in the human creative process. 'The first object of man – man – is nature, sensuousness; and the particular human sensuous essential powers can only find their self-understanding in the science of the natural world in general, just as they can find their objective realization only in *natural* objects. The element of thought itself – the element of thought's living expression – *language* –is of a sensuous nature. The *social* reality of nature, and *human* natural science or the *natural science of man*, are identical terms'.[161]

The valuing of nature in this perspective was not that of an abstract, distant contemplation, reflecting the reality of alienation; but rather something that was to be realised in a real, sensuous way, in the form of an active, material interdependence, within a higher form of society. As Marcuse was to emphasise, art, as the realm of subjective imagination, is capable of providing a 'feast of sensuousness', pointing to another 'reality principle' beyond the received reality.[162]

For Marx, sensuousness is the insurmountable condition of human objectivity, which means having 'sensuous objects outside oneself – objects of one's sensuousness'. Hence, 'to be sensuous is to *suffer*. Man as an objective, sensuous being is therefore a *suffering* being – and because he feels that he suffers, a *passionate* being. Passion is the essential power of man energetically bent on its object. But man is not merely a natural being: he is a *human* natural being', and a thinking being.[163] As such, human species-being constantly generates new needs and values.

160 Marx and Engels 1975c, pp. 300–3, 343–4; Hegel 1975b, pp. 33–4.

161 Marx and Engels 1975c, p. 304.

162 Marcuse 1978, p. 14.

163 Marx and Engels 1975c, p. 337.

Mészáros argues, in *Marx's Theory of Alienation*, that Marx develops his theory of need and his concept of sensuous value most clearly in his aesthetics. Values, for Marx, Mészáros writes, 'have their *ultimate* foundation and *natural* basis in human *needs*. There can be no values without corresponding needs. Even an alienated value must be based on a – correspondingly alienated – need ... Art, too, represents value only insofar as there is a human need that finds fulfillment in the creation and enjoyment of works of art'.[164] And such needs, of course, develop historically, through struggle, with the development of human nature (the many-sided human ontology) itself.

But what material human need, then, corresponds to intrinsic value, i.e. the realm of use values freed from commodity exchange? It can only be the fundamental human need for a free sensuous existence, which is predicated on love for objects or beings outside of ourselves, and which necessarily requires their continued existence outside of ourselves. It is reflected at its most impassioned level in the aesthetic dimension that connects us sensuously with each other, both in an immediate relation to nature, and through the development of our senses within society, expanding our connectedness and the depth of our perceptions. This suggests a community with nature (which is both a natural relation and a human-social relation) in which our human values become organic values, no longer those of mere exchange value or utilitarian calculus. In this sense, there is no contradiction between Marx's critique of the alienation of nature and Aldo Leopold's land ethic based on extending our community (as much as is *humanly* possible) to nature.[165]

Ironically, the very notion of use value/intrinsic value – standing opposed to the law of value of the cash nexus – is itself the specific product of an alienated society. In a higher form of society, it would be replaced with far richer representations of value and need, encompassing a widening community of life and a world of sustainable human development. Under present conditions, however, it is essential to engage in a ruthless critique of capital's rift in 'nature's metabolism', which is predicated on its rift in the 'social metabolism', and vice versa. Capital as a system is guilty not only of the exploitation of the workers, but also 'of the *robbing of the soil*: the acme of the capitalist mode of production is the undermining of the *sources of all wealth*: the soil and the worker'.[166] To solve either of these great problems requires solving the other, since they are dialectically related. The way out lies in the transformation of production,

164 Mészáros 1970, pp. 191–2.

165 See Foster 2002, pp. 83–90.

166 Engels 1937, p. 95.

including the conditions of science and art. This is the inescapable conclusion of Marx's entire critique: the need for the revolutionary creation of community on the earth.

In the words of Christopher Caudwell, capturing this classical Marxian sense of the human relation to the earth:

> But men cannot change Nature without changing themselves. The full understanding of this mutual interpenetration or reflexive movement of men and Nature, mediated by the necessary and developing relations known as society, is the *recognition* of necessity, not only in Nature but in ourselves and therefore also in society. Viewed objectively this active subject-object relation is science, viewed subjectively it is art; but as consciousness emerging in active union with practice it is simply concrete living – the whole process of working, feeling, thinking and behaving like a human individual in one world of individuals and Nature.[167]

167 Caudwell 1937, p. 279.

CHAPTER 1

The Dialectic of Organic and Inorganic Relations

Cartesian dualism gave to Western thought an enduring split between science and philosophy, between the physical-mechanical realm of science on the one hand, and the metaphysical realm of pure reason on the other. Ecological analysis, like most other modes of thought, still in many ways reflects this split today. Thus, it can be seen as being divided into (a) a *science of ecology*, which deals primarily with the relationship between organisms and their environments (including other organisms as well as 'inorganic nature'), and (b) a *philosophy or metaphysics of ecology*, which attempts to draw on the notions of interdependence and holism and to apply them to all of existence, while also attributing to them an ethical content.[1] The philosophical approaches to ecology most commonly propounded today, such as deep ecology and environmental ethics, naturally claim to be inspired by recent developments in science that point to the need for greater holism. But by the same token, philosophical ecologists often see the mainstream of Western science, with its mechanistic, reductionistic, and deterministic orientation, as the ultimate source of the ecological problem. For many ecological scientists, in contrast, the leading philosophical approaches to ecology appear to be metaphysical and spiritualistic in nature, having little to do with ecological science as such.

Various philosophical attempts have been made over the centuries to heal the deep division that originated with Cartesian dualism. The most ambitious attempts to bridge or transcend the gap arose within classical German philosophy. Kant responded to seventeenth-century Continental metaphysics and seventeenth- and eighteenth-century British empiricism and scepticism by creating a more transcendent dualism, which was seen as a necessary condition of critical reasoning itself. Hegel sought to overcome the divide within thought through a dialectical method that privileged an idealist ontology. Marx offered a dialectical method that privileged a materialist ontology and praxis.

In evaluating Marx's contribution to ecological thought, a tendency has emerged within metaphysical ecology to see him as a thinker who embraced certain aspects of a holistic, organic perspective, while ultimately giving in to a mechanistic and deterministic understanding of human-natural relations.

1 Brennan 1988; Hughes 2000.

Thus, in *The Turning Point*, Fritjof Capra refers to an 'ecological Marx', pointing out that Marx had 'profound insights into the interrelatedness of all phenomena', but goes on to contend that Marx ultimately succumbed to the deterministic perspective characteristic of mechanistic science.[2]

Indeed, Marx is frequently criticised for having sinned against ecology as early as 1844 in his *Economic and Philosophical Manuscripts* by referring to nature as 'man's inorganic body'.[3] This is often interpreted as evidence of a mechanistic view that set human beings against the rest of nature, and that justified the domination of nature. The fact that Marx was developing a dialectical view which, although inspired by Hegel, took into account the alienation of nature from a materialist perspective and thus linked up with developments within nineteenth-century science (while explicitly rejecting mechanistic materialism) is simply missed in these criticisms.

A close study of the development of Marx's thought in this area will therefore serve to highlight the dualistic mode of thinking that characterises much of contemporary ecology. At the same time, it will demonstrate the power of Marx's own ecological method and how it might serve as a guiding thread for a more revolutionary ecological praxis.

The Critique of 'Marx's Inorganic Body'

The term *organic* more than any other serves to denote the aspirations of philosophical ecology. Within contemporary Green theory, organic is often seen as a virtuous notion that reflects the essence of a deep-ecological perspective. Organic connotes naturalness, connectedness, respect for living processes, a noninstrumental approach to nature, and so forth. In contrast, *inorganic* suggests something that is nonliving, unconnected, and maybe even unnatural. The whole notion of 'organic farming' – that is, farming without pesticides and other harmful synthetic chemicals – further reinforces this conception of the organic as somehow representing the natural as opposed to the synthetic. This fits with the generally romantic, vitalistic, and spiritual character of much of today's Green theory, which seeks to impart a dualistic environmental ethics in which there is a sharp divergence between ecocentrism and anthropocentrism. Ecocentrism in this view is always on the side of the organic. Anthropocentrism, however, partakes of the inorganic; it relies

2 Capra 1982, pp. 208–9.

3 Marx 1974.

instrumentally on dead nature (say, petroleum-driven machinery) to manipulate living nature and living species.

Such distinctions may strike one as crude, even meaningless. It is difficult, if not impossible, to maintain a hard and fast distinction between the organic and the inorganic within ecological analysis, nor does one represent nature and the other not represent it. But given the overriding importance assumed by such distinctions in alternative ecological approaches today, the problem of the 'organic' as somehow the object of ecological thought and practice is not so easily dismissed.

Out of this general outlook, which privileges the organic, has arisen one of the most ambitious criticisms of Marx's ecological thought – one that raises issues that go back to the roots of ecological understandings within antiquity and that extend forward into the very heart of contemporary debates over ecology and the alienation of nature. Contemporary ecological critics commonly claim that in referring to man's inorganic body, Marx created a dualistic conception of the human-nature relationship in which human beings and nature exist in perpetual antagonism. Marx, it is suggested, is a thinker who is anthropocentric in the extreme sense of insisting on human exemptionalism – that is, the notion that human beings are not really part of nature, but are somehow above it, able to dominate it and escape its laws, which do not pose limits to humanity. At the same time, Marx is often criticised for being instrumentalist or 'Promethean' in his view of nature, believing in the almost infinite capacity to manipulate nature for human ends through the development of technology – even siding with the machine and productivism against nature.

Thus, in an influential 1989 article titled 'Marx's Inorganic Body', social ecologist John Clark argued that Marx's *Economic and Philosophic Manuscripts of 1844* had employed a 'dualistic view of humanity and nature and an instrumentalist view of the latter'.[4] These early manuscripts of Marx, in which he developed his conception of alienation (including the alienation of nature), have often been characterised as deeply ecological. For Clark, however, Marx's embrace of the concept of nature as man's inorganic body represented the beginnings of an antiecological perspective that was to pervade all his work.

In the *Economic and Philosophical Manuscripts*, Marx argued that

> the universality of man manifests itself in practice in that universality which makes the whole of nature as his *inorganic* body, (1) as a direct means of life and (2) as the matter, the object and tool of his activity.

4 Clark 1989, p. 251.

> Nature is man's *inorganic body*, that is to say, nature in so far as it is not the human body. Man *lives* from nature, i.e. nature is his *body*, and he must maintain a continuing dialogue with it if he is not to die. To say that man's physical and mental life is linked to nature simply means that nature is linked to itself, for man is a part of nature.[5]

For Clark, Marx's statement here – despite its emphasis on the interdependent and continuous exchange between human beings and nature – offers definitive proof of the antiecological character of Marx's thought, simply because nature (outside of the human body) is characterised as man's *inorganic* body. As Clark himself puts it,

> Marx distinguishes between nature as 'organic body', that is, as human body, and nature as 'inorganic body', that is, the rest of nature. While a mere distinction between two such realms within material nature is not, obviously, in itself an ontological false step, the valuation underlying the distinction is another question. The 'inorganic' quality of 'external' nature signifies its instrumental character in relation to an abstracted humanity, which is taken to be the source of all value.[6]

Although Marx explicitly recognises an estrangement (or alienation) between human beings and nature, Clark still asserts that the reference to 'inorganic nature' as 'man's' extended body is symptomatic of a kind of ecological imperialism. Hence, according to Clark,

> Estrangement from nature [for Marx] is in no way taken to mean nonrecognition of intrinsic value throughout nature or of the interrelatedness between human values and the larger unfolding of value over the course

---

5 Marx 1974, p. 328.

6 Clark 1989, p. 251; Marx's notion that labour is the sole source of value under capitalism (a view that he shared with the other classical economists) is sometimes taken as an indication of the antiecological nature of his thought – a view that Clark (1989) exploits here. Yet, the significance of Marx's argument in this respect is frequently misunderstood, because Marx also repeatedly insisted that labour is not the sole source of use value or wealth – nature being just as important (or more important) in that respect. Indeed, in Marx's view, the fact that capitalistic values disregard nature's contribution to wealth only points to the one-sided, alienated reality of capitalist society and its law of value, which needs to be transcended in postcapitalist society. For further discussion, see Foster 2000, pp. 167–8, and Burkett 2014, chapters 6–8.

> of natural history; rather it means the failure of 'man' to utilize nature self-consciously and collectively in productivity, that is, in 'the objectification of man's species life'.[7]

Clark is fully aware that defenders of Marx, such as Donald Lee and Howard Parsons, have emphasised the organic connection that Marx was trying to express through his reference to external nature as man's inorganic body.[8] Clark rebuts,

> Presumably, we are to focus all our attention on the 'organic' term *body*. Yet this still leaves us with the rather perplexing and embarrassing fact that the evidence for a reality being 'organic' is that it is described as being 'inorganic'![9]

Marx had written in the *Grundrisse*,

> Nature builds no machines, no locomotives, railways, electric telegraphs, self-acting mules, etc. These are products of human industry; natural material transformed into organs of the human will over nature, or of human participation in nature. They are *organs of the human brain, created by the human hand*; the power of knowledge, objectified.[10]

Clark refers to this statement, in which Marx builds on his notion of nature as the inorganic body of humanity, as exhibiting 'at best … a highly distorted body consciousness'. It merely suggests, Clark claims, that 'mechanized nature becomes more "organic" to "man" than the living whole of nature can ever be'.[11]

Other critics of Marx have followed suit. Robyn Eckersley refers to Marx's treatment of nature as man's inorganic body in order to promote the view that Marx relied heavily on the notion of an 'antagonistic dialectic' inherent to the human-nature relationship:

> In the so-called Paris Manuscripts (i.e. *The Economic and Philosophic Manuscripts of 1844*), Marx referred to the labor process as effecting the progressive 'humanization' of nature and 'naturalization' of humanity.

7 Clark 1989, p. 251.
8 Lee 1980; Parsons 1977.
9 Clark 1989, p. 244.
10 Marx 1973, p. 706.
11 Clark 1989, pp. 243, 254.

> Nature was described as 'the inorganic body' of humanity that had been increasingly assimilated, through work, into an 'organic' part of humanity ... Marx's treatment of humans as *homo faber* is a central feature of the antagonistic dialectic between humanity and nature set out in these early writings ... Although more and more areas of nature would come under human control through technological development, the *antagonistic* dialectic between humanity and nature would never be entirely resolved.[12]

In Eckersley's interpretation, Marx developed a one-sided concept of freedom in which nature is merely an instrument for the extension of the human body.[13] Human society has as its goal 'the further subjugation of the nonhuman world' as the route to human freedom.[14] Similarly, Kate Soper associates Marx's notion of nature as inorganic body with the 'Promethean' goal of the human, mechanistic domination of nature that supposedly characterised his later thought.[15]

Val Routley, writing like Clark for *Environmental Ethics*, objects to the mere fact that Marx referred to extrahuman nature as 'man's body'.[16] This is interpreted as an extreme Enlightenment humanism that transfers God's traditionally conceived determining role in nature to humanity itself. The treatment of nature as the 'body' of man can, in Routley's words,

> usefully be seen as the product of Marx's well-known transposition of God's features and role in the Hegelian system of thought onto man ... Thus Marx's theory represents an extreme form of the placing of man in the role previously attributed to God, a transposition so characteristic of Enlightenment thought.[17]

In Routley's view, Marx's claim that nature is

> man's body seems to carry also the unattractive implication that nature is man's *property* – one's body is, after all, one's own, and usually considered

12 Eckersley 1992, pp. 78–9.

13 Ibid.

14 Eckersley 1992, p. 90.

15 Soper 1996, p. 91. The charge that Marx adopted a 'Promethean' (mechanistic, productivist) view of nature has come under heavy criticism in recent years. See Sheasby 1999, Foster 2000, pp. 126–36, and Burkett 2014, Chapter 11.

16 Routley 1981; Clark 1989.

17 Routley 1981, pp. 239–40.

> to be entirely at one's disposal, subject to only very minor qualifications. The analogy thus reinforces damaging 'human property' views of the natural world.[18]

Ariel Salleh has argued that the extreme anthropocentric view that 'plants and animals are supplied by evolution as a means of human subsistence' can be found in Marx's notion that nature is man's inorganic body – and that this ontological embrace of human domination over nature goes hand-in-hand with an uncritical acceptance of the domination of men over women.[19] Marx's analysis is thus 'riddled with ontological notions derived from the [medieval, scholastic] Great Chain of Being. This ancient theological rationale established a value structure based on God's domination over Man, and men's domination over women, the darker races, children, animals and wilderness'.[20]

A somewhat different criticism of Marx's organic/inorganic distinction was forwarded by John O'Neill in an article in *Radical Philosophy*. O'Neill claims that the treatment of nature as 'our "inorganic body"' is the part of Marx's thinking 'most compatible with recent green thought'.[21] Nevertheless, he argues that it has to be rejected as untenable on both scientific and ethical grounds. O'Neill writes:

> Nothing in the science of ecology entails that there is no significant division between an individual organism and its environment. Ecology studies the relationships between different populations that are made of just such individual organisms. It entails no radically holistic ontology. Hence it does not entail that 'I and nature are one' or that the 'the world is my body'.[22]

---

18 Routley 1981, p. 243.

19 Salleh 1997, pp. 71–4.

20 Ibid. The assertion that inscribed within Marx's thought was the medieval notion of the Great Chain of Being, traceable ultimately back to Aristotle, is a peculiar one, given that Marx was part of the materialist revolt against such teleological conceptions; so much so in fact that as early as his doctoral dissertation, he had turned to Epicurus, the great enemy of Aristotelian and Christian teleology. On this, see Foster 2000, pp. 21–65. As Clement of Alexandria had pointed out, Epicurus was the great enemy of all those who argued from the standpoint of providence (Marx and Engels 1975a, Vol. 1, p. 37).

21 O'Neill 1994, p. 26.

22 Ibid.

However, in making this argument, O'Neill implicitly uses what scientists Richard Levins and Richard Lewontin have called a 'reductionist' approach, according to which ecology must emphasise either holism or differentiation, as opposed to a more 'dialectical' approach that encompasses both (by treating nature as a complex *unity-in-difference*).[23] Along similar lines, O'Neill suggests that it is 'ethically untenable' to describe nature as the 'inorganic body' of humanity because to do so is to treat 'distinct natures' as mere 'extensions' of human beings, thereby downgrading their distinct intrinsic values.[24] Here again, O'Neill's argument presumes that Marx's approach was not a dialectical one but a purely instrumentalist-anthropocentric view running counter to any ecological ethics. Such are the grounds on which O'Neill suggests that 'Marx's view of nature as our "inorganic body", together with those "holistic" components of recent green thought to which it is similar, should be rejected'.[25]

So pervasive have been environmentalist criticisms of Marx for employing the concept of nature as inorganic body that even a strong defender of Marx such as Donald Lee, whose ecohumanist Marxist essay for *Environmental Ethics* helped set off the debate in this area, chided Marx for this formulation. Although noting Marx's recognition that humanity is 'intrinsically tied to nature', Lee laments 'Marx's homocentrism ... [that is] so pronounced here when he speaks of the nonhuman animal and vegetable kingdom as man's "inorganic" body'.[26]

Yet, the problem with all such criticisms of Marx's 'organic/inorganic' outlook lies in their refusal to engage fully with the complex issues that it raises. For Clark, as we shall see, a body of thought can only be considered ecological to the extent that it is 'teleological' and thus distances itself from a consistent materialism – and a similar one-sidedness afflicts the other ecological critics.[27] By contrast, Marx treats the human relation to nature as both materialist and dialectical, so that the crucial question becomes the nature of the interaction that takes place (in Marx's later vocabulary, the 'metabolic' relation between society and nature). The question of nature as man's inorganic body is not to be dismissed as a mere instrumentalist, anthropocentric view. To do so is to succumb to a static foundationalist approach to environmental ethics, divorced from history and evolution. Furthermore, the organic/inorganic question should not be rejected on the grounds that it contradicts a scientific perspective on ecology,

23 Levins and Lewontin 1985.
24 O'Neill 1994, p. 26.
25 Ibid.
26 Lee 1980, pp. 14–15.
27 Clark 1989.

as O'Neill claims.[28] Rather, the issue is the coevolution of human society and nature – that is, the very possibility of an ecological understanding of human history.

The Organic/Inorganic Distinction and Hegel's Philosophy of Nature

It is somewhat odd, given the foregoing charges with regard to Marx's concept of inorganic nature, that none of the critics have made any inquiry into the history of the concepts of organic and inorganic, their relation to Hegel's texts, or even the systematic way in which the organic/inorganic distinction is variously applied in Marx's writings themselves. Yet, to proceed in this way, examining the history and use of this conceptual distinction, is to sharpen our understanding of the origins of ecological knowledge and of Marx's own contributions.

In ancient Greek usage, the word *organ* (*organon*) also meant tool, and organs were initially viewed as 'grown-on tools' of animals – whereas tools were regarded as the artificial organs of human beings.[29] Characteristic of the natural-dialectical worldview of the ancient Greeks was the recognition of a close relationship between tools as extensions of human beings and the organs of animals, because they were both part of the general process of species adaptation to natural conditions. Indeed, the connection between what Marx in *Capital* was to call 'natural technology' (that is, physical organs) and 'artificial technology' (the tools created by human beings) was to play a part in the earliest accounts of evolutionary adaptation in the writings of the ancient Greek and Roman materialists: Empedocles, Democritus, Epicurus, and Lucretius. Marx, who wrote his doctoral thesis on Epicurus, was well aware of these ancient contributions to evolutionary thinking.[30]

---

28 O'Neill 1994.

29 *Oxford English Dictionary* (OED) 1971, p. 2007; Pannekoek 1912, p. 50.

30 See Foster 2000. In addition to the ancient materialists, notably Epicurus, Marx was also influenced in complex ways in the development of his materialist philosophy by Aristotle, despite the latter's general teleological perspective. Not only was Aristotle a dialectical thinker, but it was owing to his work that some of the earliest materialist ideas were known, and Aristotle's corpus (particularly his *Physics, Metaphysics* and *History of Animals*) can be seen as encompassing materialist conceptions at points – so much so that Marx himself did not hesitate to refer to Aristotle as 'a materialist' in some qualified sense (Marx, 1934, p. 80; see also Farrington 1944, pp. 114–15, 120). Nevertheless, Aristotle's strong adherence to teleological views, most glaringly apparent in his *On the Parts of Animals*

Indeed, Marx and Engels extended this line of argument, developing it into a full-fledged theory of human evolution following Darwin's great contribution (see the next section).

By early modern times, as Carolyn Merchant notes in *The Death of Nature*, 'the term *organic* usually referred to bodily organs, structures, and organization of living beings'.[31] Within physiology in the nineteenth century, *organic* meant having to do with bodily organs or an organised physical structure, especially with regard to plants and animals. In his *Philosophy of Health*, T.S. Smith wrote that 'the organic actions consist of the processes by which the existence of the living being is maintained'.[32] By the same token, the word *organically* was usually used to refer to 'bodily organs or their functions; in the manner of an organized or living being'.[33]

This use of the term *organic* naturally had its counterpart in the use of *inorganic*. According to the OED, the principal meaning of inorganic in the seventeenth, eighteenth, and nineteenth centuries was that of being 'not characterized by having organs or members fitted for special functions; not formed with the organs or instruments of life'.[34] In his *Essay Concerning Human Understanding*, Locke had referred to 'the lowest and most inorganical parts of matter' as lying at the bottom of the great chain of being.[35] Similarly, *inorganically*, in the early modern period and up through the nineteenth century, generally meant 'without organs'.[36]

Where Marx was concerned, the use of the terms *organic* and *inorganic* did not derive simply from ancient philosophy and from contemporary scientific usage, but rather was directly affected by his confrontation with Hegel's *Philosophy of Nature*.[37] Marx had taken notes on the section of Hegel's *Encyclopedia* dealing with the philosophy of nature in 1839, while working on his doctoral thesis.[38] The philosophy of nature is the most problematic point in the Hegelian system because here the Idea is most alienated from itself. Indeed, nature is

(Aristotle 1882), in which crude teleological explanations for the development of animal organs were used at every point, made his work a fountainhead for later antimaterialist views, which were to dominate medieval Christian (Scholastic) thought.

31 Merchant 1980, pp. xix–xx.

32 Smith 1835, as cited in OED 1971, p. 2008.

33 OED 1971, p. 2008.

34 OED 1971, p. 1443.

35 Locke 1959 [1690], Vol. 2, p. 68.

36 OED 1971, p. 1443.

37 Hegel 1970 [1830].

38 Hegel 1959.

viewed as the alienation of the Idea, which imposes its own conscious structure on nature (in conformity with nature's mechanics) and thus returns to itself. Hegel thus sought to demonstrate that nature is in its essence 'self-alienated Spirit'.[39] Needless to say, it is here that the conflict with materialism is most severe. As Auguste Cornu observed, although

> it might be relatively easy to establish a radical concatenation and dialectical order among concepts; it is already harder to do so in history, where the contingent and the accidental play a greater part; and by the time we come to the realm of nature, the assimilation of the real to the rational can be carried out only by extremely arbitrary procedures.[40]

Nevertheless, although rejecting Hegel's philosophy of nature, Marx drew critically on its more dialectical insights. This was particularly the case where Hegel's central dialectic of the organic and inorganic was concerned. The attempt to develop an understanding of the organic unfolding out of the inorganic pervades Hegel's entire philosophy of nature. It is in its 'Organics', as Marx observed, where one finds in Hegel's philosophy of nature 'the determination of subjectivity, in which the real distinctions of form are ... brought back to ideal unity, which is self-found and for itself'.[41] It is thus in the realm of organics that the estrangement of the spirit, which has gone over into the exteriority of nature, is overcome as it returns into its own higher unity (of consciousness). The organism (particularly the animal organism), in other words, comes to stand for subjectivity and self-dependence – that is, for rational life connected to the life of the spirit within nature. Here, animate species are the means by which the spirit discovers itself in nature and overcomes its estrangement.

Hegel argues that the organic is connected to the inorganic in three ways. First, the organic and the inorganic are one (unity) because each organism has its inorganic within itself as a part of itself. Second, the organic and inorganic are in opposition (difference) because the organic lives only by feeding off the inorganic as its condition of existence. Third, the organic and inorganic are combined as a unity-in-difference (the divisions between them dynamically transformed and to some extent resolved) in reproduction, development, and death.[42] 'In its inner formation', Hegel writes, 'the animal is an unmediated

---

39 Hegel 1970 [1830], Vol. 1, p. 206.

40 Cornu 1957, pp. 37–44.

41 Marx and Engels 1975a, Vol. 1, p. 510.

42 See Taylor 1975, p. 359.

self-production, but in its outwardly oriented articulation, it is a production mediated by its inorganic nature'.[43] Nature in its subjectively determinate form exists, for Hegel, only when it

> individualizes inorganic things, or relates itself to those already individualized, and assimilates them by consuming them and destroying their characteristic qualities, i.e. through *air* entering into the process of respiration and of the skin, *water* into the process of thirst, and the particular formations of individualized *earth* into the process of hunger. Life, which is the subject of these moments of the totality, constitutes a state of tension between itself as Notion and the external reality of these moments, and maintains the perpetual conflict in which it [consciously] overcomes this externality.[44]

It is in this sense, Hegel suggests, that 'organic being, which is an individuality existing for itself and developing itself into its differences within itself, constitutes totality as found in nature' – a totality that forms the basis for the spirit's transcendence of nature's exteriority.[45]

As Hegel explained in his smaller *Logic*, part of the *Encyclopedia of the Philosophical Sciences*, the living thing has as its 'presupposition ... an inorganic nature confronting it. As this negative of the animate is no less a function in the notion of the animate itself, it exists consequently in the latter ... in the shape of a defect or want'.[46] Hence, the 'self-assured living thing' maintains and develops and objectifies itself only 'against an inorganic nature', which it 'assimilates to itself'.[47] This inorganic nature is subdued and 'suffers this fate, because it is *virtually* the same as what life is *actually*. Thus in the other the living being only coalesces with itself'.[48]

Hegel's understanding of organic/inorganic relations is set out most concretely in his treatments of plant and animal existence. The unity of the 'vegetative' realm (captured in Goethe's *Metamorphosis of Plants*) is, according to Hegel, evident in 'the processes of its quantitative growth', whereas 'its qualitative metabolism of elements are at the same time the processes of its

43 Hegel 1970 [1830], Vol. 3, p. 185.
44 Hegel 1970 [1830], pp. 147–8; see also Hegel 1977 [1807], pp. 154–80.
45 Hegel 1970 [1830], Vol. 1, p. 219.
46 Hegel 1975a [1830], Vol. 3, p. 281.
47 Ibid.
48 Ibid.

decay; cells, fibers, and the like multiply until they smother the plant in dead wood'.[49] Furthermore,

> the plant is interwoven with its environment, whereas the animal breaks this immediate context. It is alive for itself. The animal soul is the inner unity of the whole animal, wholly present in all of its functions. Corresponding to this concentration in itself, the environment becomes for the animal an outer world to which it has to adapt itself. Whereas in the plants the elementary life of nature in earth, water, air, and light is directly absorbed; the animal, on the contrary, transforms the elementary life of organic and inorganic nature into stimuli to which it responds in many ingenious ways.[50]

What Marx took from Hegel (and also from Feuerbach) in this regard was the dialectical perception that human beings, as objective, organic creatures, are also dependent on inorganic nature as part of their own species-being. Marx wrote:

> *Hunger* is a natural *need*; it therefore requires a *nature* and an *object* outside itself in order to satisfy and still itself ... The sun is an *object* for the plant, an indispensable object which confirms its life, just as the plant is an object for the sun, an *expression* of its life-awakening power and its *objective* essential power. A being which does not have its nature outside itself is not a natural being and plays no part in the system of nature.[51]

For Marx, human beings are active, living, transformative creatures in charge of their own bodies and drives; at the same time, they are 'natural, corporeal, sensuous, objective' beings who suffer, whose natural objects, the conditions of their existence, the inorganic body of nature by which they seek to extend themselves, are to be found outside of themselves.[52] In general, 'species-life, both for man and for animals', Marx argues, 'consists physically in the fact that man, like animals, lives from inorganic nature; and because man is more universal than animals, so too is the area of inorganic nature from which he lives more universal' – both physically and spiritually.[53] The relation between

---

49 Hegel 1959, p. 183.

50 Hegel 1959, pp. 185–6.

51 Marx 1974, pp. 389–90.

52 Ibid.

53 Marx 1974, pp. 327–8.

the organic body of a human being and the inorganic world is one that is conditioned by the subsistence needs of human beings and their capacity through social labour to transform the 'external' conditions of nature into means of satisfying these needs. Rather than postulating a sharp ontological break between human beings and nature (a break that, as we shall see, only arises through the alienation of nature), Marx thus attempted to describe the material interconnections and dialectical interchanges associated with the fact that human species-being, similar to species-being in general, finds its objective, natural basis outside of itself, in the conditioned, objective nature of its existence.

In Marx's dialectical understanding, in which he was heavily influenced by Hegel, all of reality consists of relations, and any given entity is therefore the product of the complex, ever-changing relations of which it is a part.[54] In this sense, the organic body of humanity (like all species) is incomprehensible apart from the inorganic conditions of its existence, which may at first appear one-sidedly (in a society characterised by the alienation of human beings and of nature) as mere 'external' things – opposing forces. It is this many-sided dialectical conception, then, that informs Marx's ecological thought. However, in Marx (in contrast to Hegel), the organic/inorganic dialectic was always influenced by the 'immanent dialectics' of materialism, going back as far as the ancient Greeks (particularly Epicurus). Hence, Marx's dialectical conception of nature never took the idealist form it assumed in Hegel in which the object of analysis was simply the estrangement of spirit, going out into nature and returning to itself. For Hegel, as Marx observed, nature is 'defective' insofar as it represents 'externality' or 'antithesis to thought'.[55] But for Marx himself, it was necessary to explain how human history and natural history were interconnected within sensuous existence.[56]

Marx's Dialectic of Organic/Inorganic: The Conditions of Human Existence

In developing his overall analysis of capitalism and communism, Marx employed the organic/inorganic distinction in three different but related senses, which can be designated as (a) *scientific*, (b) *dialectical*, and (c) *materialist*.

54 See Ollman 1976.

55 Marx 1974, p. 400.

56 Marx 1974.

First, he referred to nature (other than the human body) as the inorganic body of humanity in conformity with the scientific vocabulary of his day, wherein organic referred to bodily organs, whereas inorganic meant unrelated to bodily organs. From the *Economic and Philosophical Manuscripts* and on, for the rest of his life, Marx at various points treats nature, insofar as it enters directly into human history through production, as an extension of the human body (that is, the inorganic body of humanity).[57] Of course, this bodily relation to nature is mediated not only through human-social labour, but also by means of the tools – themselves products of the human transformation of nature through production – that allow humanity to appropriate and use nature in ever more universal ways (more will follow on this point). But the present, more basic point is that the human-nature relationship physically transcends, at the same time that it practically extends, the actual bodily organs of human beings – hence, the reference to nature as man's inorganic body. Here, inorganic simply means external to, yet in constant interchange with, the human body itself, in a basic material and biological sense. As Marx indicates, to say that humanity '*lives* from nature' is to say that nature is 'man's *inorganic body*' and that 'nature is linked to itself, for man is a part of nature'.[58] Marx's reference to nature as the inorganic body of man was meant, then, to convey that human beings and nature were connected to each other *bodily* (i.e. in the most intimate way possible), but that human beings through tool-making were able to extend their material capacities beyond their own bodily organs (i.e. 'inorganically' in this sense).

That nature is both external to and the material and biological substance of the human condition leads directly to a second way in which Marx employs the organic/inorganic distinction. This can be characterised as dialectical, emanating in particular from Hegel's *Philosophy of Nature*.[59] Here, inorganic is used to refer to the inherent 'exteriority' or 'objectivity' of nature as a condition of human subjective activity (labour) and the fulfilment of human bodily needs; hence, it appears as a condition of the development of humanity as a distinct species. Thus, in the *Grundrisse*, Marx refers to 'the *natural* conditions of labour and of reproduction' as 'the objective, nature-given inorganic body' of human subjectivity.[60] Further along in the same text, Marx explains that the 'first objective condition' of labour appears to the worker 'as nature, earth, as

57 Ibid.

58 Marx 1974, p. 328.

59 Hegel 1970.

60 Marx 1973, p. 473.

his inorganic body; he himself is not only the organic body, but also the subject of this inorganic nature', which represents an objective force external to the worker.[61] As society develops, the human producers in a given historical social formation come to identify 'a specific nature (say, here, still earth, land, soil)' as their own 'inorganic being, as a condition' of their own 'production and reproduction'.[62] As we have seen, Marx argued that a being that does not have its object outside of itself, in the objective, inorganic conditions of its existence, is not a natural, organic being.[63]

It is interesting to note that Marx's holistic perspective on the human-nature dialectic at times led him to reverse the ordering of organic/inorganic by applying the former term to extrahuman nature. These terminological reversals normally occurred when Marx was considering natural conditions as necessary and, at least partly, uncontrollable conditions of human production. When analysing material price fluctuations in Volume III of *Capital*, for example, Marx refers to raw materials derived from 'organic nature', whose production is in large part determined by 'uncontrollable natural conditions, the seasons of the year, etc.'.[64] Such 'raw materials supplied by organic nature' include 'plant and animal products, whose growth and production are subject to certain organic laws involving naturally determined periods of time'.[65] These naturecentric applications of the organic term have their counterpart in Marx's reference (in *Capital*'s discussion of ground rent) to the terrestrial body, comprising 'the earth's surface, the bowels of the earth, [and] the air' as the basis for 'the maintenance and development of life'.[66] The *Grundrisse* similarly refers to 'the earth' as 'the source of all production and of all being ... the seat, the *base* of the community' and describes 'the soil itself' as 'the direct well-spring of subsistence'.[67] Here, human beings are basically treated as (conscious, socially developed) extensions of nature's body.[68] In short, Marx's dialectical usage of the organic/inorganic distinction and his general analysis of natural conditions as conditions of human production and

61 Marx 1973, pp. 488–90.

62 Ibid.

63 Lenin, in his *Philosophical Notebooks*, carefully scrutinises Hegel's treatment of the inorganic conditions of human existence – further highlighting the Hegelian roots of Marx's dialectic here (Lenin 1961, p. 212).

64 Marx 1981, p. 213.

65 Marx 1981, pp. 213, 216.

66 Marx 1981, p. 909.

67 Marx 1973, pp. 106, 276.

68 Cf. Schmidt 1971, pp. 16, 42–3; Smith 1984, pp. 18–19.

human life hardly involve the kind of one-sided, antiecological anthropocentrism claimed by his critics.[69]

This brings us to the third sense in which Marx employs the organic/inorganic distinction. Ultimately, Marx's references to nature, external to the human body, as both the inorganic body of humanity and the precondition of human-social existence, are meant to get at the materialist foundation of human species-being. For Marx, the human-nature relation develops through tool-making (technology) – that is, the appropriation and use of inorganic nature to extend the organs of the human body in the production of the means of subsistence. As Marx put it in *Capital*,

> Leaving out of consideration such ready-made means of subsistence as fruits, in gathering which a man's bodily organs alone serve as the instruments of his labour, the object the worker directly takes possession of [within the labour process] is not the object of labour but its instrument. Thus nature becomes one of the organs of his activity, which he annexes to his own bodily organs, adding stature to himself in spite of the Bible.[70]

The role of human-manufactured tools, analogous to the productive function of human bodily organs – mediating between human beings and nature – is further highlighted elsewhere in *Capital* when Marx states: 'Just as a man requires lungs to breathe with, so he requires something that is the work of human hands in order to consume the forces of nature productively'.[71] The point is developed even more clearly in the 1861–3 draft of *Capital* in which Marx characterises the labour process as a 'process of appropriation' of nature 'as of the animated body, the organs of labour itself. Here the material appears as the inorganic nature of labour, and the means of labour as the organ of the appropriating activity'.[72] We have already referred at the beginning of this chapter to Marx's well-known statement in the *Grundrisse* in which he describes 'machines … locomotives, railways, electric telegraphs, self-acting

---

69 See Burkett 2014, Chapters 2–4.

70 Marx 1976a, p. 285. The concept of *extension* as a property of matter or the body has long been a crucial part of the materialist conception of nature. As Hobbes wrote in his *De Corpore*, 'a body is that, which having no dependence upon our thought, is coincident or coextended with some part of space' (1929, Vol. 1, p. 102).

71 Marx 1976a, p. 508.

72 Marx and Engels 1975a, Vol. 30, p. 58.

mules, etc.' as '*organs of the human brain, created by the human hand*; the power of knowledge, objectified'.[73] Despite Clark's previously cited claim that this constitutes a highly distorted body consciousness, Marx's point can now be seen as an essentially dialectical and materialist one, specifying the conditions underlying the labour process that alone allows human beings to exist and to develop in relation to nature.[74]

For Marx, this analysis of tools as 'organs' was central to a materialist account of human evolution. Accordingly, in *Capital*, Marx used Darwin's comparison of the development of specialised organs in plants and animals to that of tools to draw a distinction between natural technology and human technology.[75] Here, both Darwin and Marx were undoubtedly influenced by the original Greek notion of organ (*organon*), which also meant tool, making the organs of animals grown-on (adnated) tools.[76] For Marx, this approach offered clues to the development of human technology and the labour process. Whereas animals had for the most part evolved through the intergenerational development of their organic bodies directly, in the case of human beings, the capacity to make tools and thus to extend their bodies into inorganic nature had been of greater historical importance. This specifically human characteristic had allowed for more universal forms of development, which were obviously related to the social process of tool-making and the gradual development of the brain, language, and so forth.[77]

Engels was to expand this argument further in his important posthumously published essay, 'The Part Played by Labour in the Transition From the Ape to Man' (1896). According to Engels's analysis – which derived from his materialist philosophy but was also influenced by views introduced by Ernst Haeckel a few years before – when the primates who were to be the ancestors of human beings descended from the trees, erect posture developed first, prior to the evolution of the human brain, freeing the hands for tool-making.[78]

> *The hand became free* and could henceforth attain ever greater dexterity and skill, and the greater flexibility thus acquired was inherited and increased from generation to generation. Thus the hand is not only the

73 Marx 1973, p. 706.

74 Clark 1989.

75 Darwin 1968, in Chapter 5, p. 187, of *The Origins of Species*, on 'Laws of Variation'; Marx 1976a, pp. 461, 493–4.

76 Ibid.

77 Foster 2000, pp. 200–2.

78 Engels 1940.

> organ of labour, *it is also the product of labour*. Only by labour, by adaptation to ever new operations, by inheritance of the resulting special development of muscles, ligaments, and, over longer periods of time, bones as well, and by the ever renewed employment of these inherited improvements in new, more and more complicated operations, has the human hand attained the high degree of perfection that has enabled it to conjure into being the pictures of a Raphael, the statues of Thorwaldsen, the music of Paganini.[79]

As a result, early humans (hominins) were able to transform their relation to their local environment, radically improving their evolutionary adaptability. Those who were most ingenious in making and using tools were most likely to survive, which meant that the evolutionary process exerted selective pressure toward the enlargement of the brain and the development of speech (necessary for the social process of labour), leading eventually to the rise of modern humans. Hence, the human brain, in Engels's view, evolved through a complex, interactive set of relations now referred to by evolutionary biologists as *gene-culture coevolution*.[80] All scientific accounts of the evolution of the human brain, Stephen Jay Gould has explained, have been theories of gene-culture coevolution, and 'the best nineteenth century case for gene-culture coevolution was made by Friedrich Engels'.[81]

The contrast between this materialist explanation of human evolution and ancient teleological accounts could not be sharper. 'Man alone of all the animals', Aristotle had written,

> is erect, because his nature and his substance are divine. To think, to exercise intelligence, is the characteristic of that which is most divine ... Now Anaxagoras has said that it is the possession of hands that has made man the most intelligent of animals. The probability is that it was because he was the most intelligent that he got hands. For hands are a tool, and nature, like an intelligent man, always distributes tools to those that can use them. The proper thing is to give a genuine flute-player a flute rather than to give a man who happens to have a flute the skill to play; for that is to add the lesser to the greater and more august instead of adding the greater and more precious to the lesser. If, then, it is best that it should

79 Engels 1940, p. 281.

80 Engels 1940.

81 Gould 1987, pp. 111–12.

> be so, and if nature, out of what is possible, always does the best, it is not because he has hands that man is wise, but because he is the wisest of the animals he has hands.[82]

Such was the prejudice – what Engels referred to as the idealist emphasis on the seat of cognition in the understanding of human evolution – that the significance of the freeing of the hands for tool-making (hence, labour) was downplayed in nineteenth- and early twentieth-century science, and the belief of most evolutionary scientists continued to be that the brain had led the way in the evolution of the human species, so that our earliest ancestors would distinguish themselves first and foremost by their cerebral development.[83] The expectation was that the 'missing links' between primates and human beings, when they were discovered, would exhibit a brain at an intermediate level of development. These expectations collapsed with the discovery, beginning in the 1920s and more fully in the 1970s, of the genus *Australopithecus*, dating back as many as four million years. The brain of *Australopithecus* was enlarged only very slightly, and was generally of ape-like proportion in relation to the body. Nevertheless, the australopithecines were clearly hominin species, standing erect, exhibiting evolved hands (and feet), and already – many paleoanthropologists believe – making tools. As a result of these discoveries, much of modern anthropological theory has come around to the materialist-coevolutionary view pioneered by Engels in the nineteenth century, summed up by the phrase 'Tools Makyth the Man'.[84] It is labour – and the specific social relations in and through which it takes place – that constitutes the secret, from the very first, not only to the development of human society, but also to 'the transition of ape to man'. It is socially developed labour, moreover, that defines the distinctive ecological niche occupied by humanity. Marx and Engels thus see the human-social relation to the earth in coevolutionary terms – a perspective that is crucial to an ecological understanding because it allows us to recognise that human beings transform their environment not entirely in accordance with their choosing, but based on conditions provided by natural history (of which human-social history is a part).

If Marx and Engels emphasise tool-making as an evolutionary extension of human bodily organs, by which human society uses elements of inorganic nature, their understanding of this material-social relationship did not simply

82 Cited in Farrington 1944, pp. 128–9.

83 Engels 1940.

84 Washburn and Moore 1974, p. 186.

start or stop there. Equally important, because of its wider significance, was the whole human relation to the land, especially through agriculture and the connections between agriculture and other industries. 'Only cultivation of the soil', Marx wrote in the *Grundrisse*, 'posits the land as the individual's extended body'.[85] The existence of nature as a material precondition of human development is immediately apparent, according to Marx, in the case of the fertility of the soil. Agricultural production, the most basic form of production (because the physical subsistence of the labourers always depends on it), 'rests on qualities of its inorganic nature', that is, the chemistry of the soil and its nutrients.[86] It is the separation of human beings from the soil (and hence from the organic products of the soil) and their agglomeration into huge cities that constitutes, for Marx, the *differentia specifica* of capitalism.[87] This severing of prior social-material connections between people and the land (which Marx and other classical economists called *primary* or *primitive accumulation*) underpins not only the specific forms of class exploitation that characterise capitalism, but also this system's severe antagonism between town and country and its degradation of the soil. Indeed, for Marx, capitalism's alienation of labour was dependent on (and could only be developed in accordance with) the alienation of human beings from nature.[88]

In this connection, Marx insists that

> it is not the *unity* of living and active humanity with the natural, inorganic conditions of their metabolic exchange with nature, and hence their appropriation of nature, which requires explanation or is the result of a historic process, but rather the *separation* between these inorganic conditions of human existence and this active existence, a separation which is completely posited only in the relation of wage labor and capital.[89]

To understand capitalism, it is necessary to grasp its dual alienation of nature and labour, the extreme separation of the mass of the population from the natural, inorganic conditions of their being – a separation exhibited, according to Marx, in the antagonism of town and country. If human evolution has taken a form in which inorganic nature is appropriated through increasingly complex tools (extended organs) of human labour, it is also true that these conditions

85 Marx 1973, p. 493.
86 Marx and Engels 1975a, Vol. 34, p. 155.
87 Marx 1976a, pp. 769, 929.
88 Foster 2000, Chapter 5; Burkett 2014, Chapters 5–6.
89 Marx 1973, p. 489.

of production (inorganic nature and tools) have come under the control of a very few. In this way, the mass of the population has been deprived of any birthright connection to the earth and even to air, food, sunlight, health, and so forth, insofar as these connections contradict the profitable exploitation of wage labour in the production of privately vendible commodities.[90]

What arises from Marx's materialist dialectic of organic/inorganic relations, then, is an understanding of the ecological rift that forms the foundation of modern capitalist society. This rift can only be eliminated through a replacement of class-exploitative production, property, and market relations with a system of co-operative-democratic worker-community control over the conditions of production – a system that, predicated on the transcendence of the alienation of nature and labour, alters the human relation to the earth in ways that encourage sustainable forms of human development. 'From the standpoint of a higher socio-economic formation', Marx wrote,

> The private property of particular individuals in the earth will appear just as absurd as the private property of one man in other men. Even an entire society, a nation, or all simultaneously existing societies taken together, are not owners of the earth. They are simply its possessors, its beneficiaries, and have to bequeath it in an improved state to succeeding generations as *boni patres familias* [good heads of the household].[91]

The Ecological Transformation of Marx's Nature-Dialectic

The foregoing analysis has shown that Hegel's dialectic of organic/inorganic relations played a central role in the development of Marx's understanding of human-natural relations. Yet, in Marx, the idealistic cast of Hegel's philosophy of nature was rejected from the start in favour of a more materialist approach, reflecting Marx's systematic encounter with materialism via Epicurus and Feuerbach.[92] For Marx, the alienation of nature does not entail an

90 Marx 1974, pp. 359–60.

91 Marx 1981, p. 911. Marx's argument here – namely, that the earth/land should never be treated as individual or even communal property – contradicts completely Routley's (1981) contention (previously discussed) that Marx viewed nature as property in the standard sense of free disposability and exploitability (as opposed to communally limited user rights). See Burkett 2005b and 2014, Chapter 14, for further discussion in the context of a complete ecological evaluation of Marx's vision of communism.

92 In commenting critically on the materialist tradition in philosophy, which he associated

estrangement of the spirit from a nature that is exterior to it. Rather, it is the real historic process that creates a *social-material separation* between the inorganic conditions of human existence and the active existence of human beings, a separation that is fully realised only within bourgeois society. The philosophy of nature, which in its Hegelian form turns on the relation between organic and inorganic, was thus transformed by Marx into a question of human alienation and freedom as historical, material, and social products.

To understand how Marx's materialism affected his understanding of the dialectic of organic/inorganic relations, it is necessary to look more closely at Marx's materialism itself. Maurice Mandelbaum has usefully defined nineteenth-century materialism, of which Marx and Engels were among the greatest representatives, as follows:

> Materialists, like idealists, seek to state what constitutes the ultimate nature of reality, and are willing to distinguish between 'appearance' and that which is self-existent and underlies appearance. Taken in its broadest sense, materialism is only committed to holding that the nature of that which is self-existent is material in character, there being no entities which exist independently of matter. Thus, in this sense, we would class as materialist anyone who accepts all of the following propositions: that there is an independently existing world; that human beings, like all other objects, are material entities; that the human mind does not exist as an entity distinct from the human body; and that there is no God (nor any other non-human being) whose mode of existence is not that of material entities.[93]

There can be no doubt that Marx adopted such a broad 'materialist conception of nature' (as Engels called it), and that this constituted the basis on which he erected his materialist conception of history – that is, the notion of a practical materialism in which society was understood in terms of the development of human productive forces and relations rooted in human praxis.[94] Indeed,

in particular with Epicurus and his modern adherents, Hegel went so far as to concede that the aim of materialism was, in his terms, a dialectical one: 'We must recognize in materialism the enthusiastic effort to transcend the dualism which postulates two different worlds as equally substantial and true, to nullify this tearing asunder of what is originally One' (1971 [1830], p. 34).

93 Mandelbaum 1971, p. 22.

94 Engels 1941 [1888], p. 67; see also Bhaskar 1983; Foster 2000; Mandelbaum offers a further definition for those who he refers to as 'strict materialists' (1971). Strict materialists adhere

it was Marx's materialist conception of both nature and history that led him to develop his distinctive understanding of the dialectic of organic/inorganic relations as encompassing both physical phenomena and the historical development of human-social relations. It was this also that made Marx sensitive to developments in the natural sciences, with their increasingly materialist emphasis. Clearly, the issue was no longer one of the alienation of spirit from nature (as in the idealist philosophies of nature propounded by Schelling and Hegel), but the 'really earthly question' of human material existence.[95] Hence, in Marx's work, the dialectic of organic/inorganic relations represented by Hegel's philosophy of nature is gradually transformed into a materialist ecology concerned with the rifts in the metabolic relation between human beings and nature. For Marx, Hegel's idealism had taken the form of an attempt to restore the seventeenth-century metaphysics of Descartes, Spinoza, and Leibniz in opposition to eighteenth-century Enlightenment materialism, represented in ancient times by Epicurus and in more modern times by Bacon, Hobbes, Gassendi, Locke, Holbach, and Helvetius.[96] The answer to Hegel, for Marx, lay in the development of a materialism that was dialectical and emphasised historical praxis. In addition to its political significance, such an approach had the virtue of a close affinity to the main (materialist) currents in natural science.[97]

---

to the notion that there are, in addition, definite laws for all material phenomena. Thinkers such as Moleschott, Vogt, and Büchner were strict materialists in this sense. Engels also might be characterised as a 'strict materialist', according to Mandelbaum, in the sense that he believed in definite physical laws governing the material universe. But Engels's laws, in contrast to those of mechanistic materialists such as Moleschott, Vogt, and Büchner, were not fixed and unchanging – and therefore mechanistic in nature – but rather dialectical and thus consistent with an emergentist naturalism (Mandelbaum 1971, pp. 25–7).

95 Marx and Engels 1975a, Vol. 1, p. 225.

96 Hegel, as Marx was well aware, had counterpoised Epicurus to Descartes, Spinoza, and Leibniz in his *Encyclopedia of the Philosophical Sciences*, contrasting Epicurus, the materialist who gave the gods no role within the world of nature, to the seventeenth-century metaphysicians with their 'ontological proof' of God's existence (see Hegel 1971 [1830], p. 30). Locke's inclusion by Marx in the pantheon of materialists may surprise some, because it is more common nowadays to see Locke as an empiricist but not as a materialist. Marx, however, clearly associated Locke with Gassendi (the restorer of Epicurus) and the great opponent of Descartes. Gassendi rejected both Cartesian metaphysics and the mechanistic materialism of Descartes's physics. Instead, he emphasised a materialism that stressed the role of sensation. Modern scholarship has confirmed the close connection between Gassendi and Locke in this respect (see Ayers 1991, pp. 34–5, and 1999, p. 4).

97 Foster 2000, pp. 62–4; Marx and Engels 1975a, Vol. 4, pp. 128–9.

The great scientific revolutions of the nineteenth century in cell physiology, chemistry, the discovery of the conservation of energy, evolutionary theory, and paleontology all contributed to the further dissolution of the 'rigid system of an immutable, fixed organic nature', which had characterised medieval thought, opening the way to more dialectical conceptions.[98] The great chemical discoveries of Liebig and others tended to blur the former distinctions between animate and inanimate nature. At the same time, it became more and more clear that to study living things independent of the material environment in which they lived led to fallacious results. Central to the scientific progress of the period was the simultaneous discovery of the conservation of energy by Julius Robert Mayer, Hermann von Helmholtz, and James Prescott Joule. In this conception, the ancient materialist principles of Democritus and Epicurus – that nothing comes from nothing, and nothing being destroyed can be reduced to nothing – were given new meaning.[99] By placing emphasis on the transformation of energy, the idea of the conservation of energy freed physics of imponderables associated with underlying substance. As Ernst Cassirer put it, 'the permanence of relations replaced the permanence of matter'.[100] Closely interconnected with the discovery of the conservation of energy was the development of the concept of metabolism in the work of Liebig, Mayer, and others. It was quickly recognised that the fundamental biological processes of metabolism involved exchanges between organisms and their environments in conformity with the principle of the conservation of energy.[101] Out of this arose an early ecological understanding of the relation between organisms and their environmental conditions, exemplified by Liebig's research into the soil nutrient cycle. As one of Liebig's biographers was to write,

98 Engels 1940, p. 12.

99 In developing the notion of the conservation of energy and connecting this to metabolic relations within physiology, Mayer adopted as his foundation the principle of conservation as enunciated in ancient materialism, most clearly by Epicurus. Thus, in 1842, Mayer wrote (in terms Epicurus had made famous) that 'no given matter is ever reduced to nothing and none arises out of nothing' (as cited in Rosen 1959, p. 251).

100 As cited in Rosen 1959, p. 251.

101 Rosen 1959, pp. 253–61. The development of new knowledge concerning both the conservation of energy and metabolic processes, in a society that emphasised monetary exchange above all, was no mere accident. As George Herbert Mead pointed out, the discovery of the principle of the conservation of energy was rooted in concerns that arose out of the labour theory of value of classical economics (see Mead 1936, pp. 243–6).

> there is [expressed in Liebig's thought] a beautiful connection between the organic and the inorganic kingdoms of nature. It is inorganic matter mainly which affords food to plants, and they, on the other hand, yield the means of subsistence to animals.[102]

These scientific developments were to exert a profound influence on Marx who, in his later writings, tended to refer less frequently to the dialectic of organic/inorganic relations as such and emphasised rather the notion of the 'metabolic' relations between humanity and nature. Marx's analysis, under the influence of Liebig, of the metabolic rift in agriculture, resulting from the break in the soil nutrient cycle brought on by industrialised agriculture (and emanating from the whole antagonistic division between town and country under capitalism), led him to a much more directly ecological understanding of the relationship between human beings and their environment. By promoting an antagonism between town and country, capitalist production, Marx wrote,

> disturbs the metabolic interaction between man and the earth, i.e. it prevents the return to the soil of its constituent elements consumed by man in the form of food and clothing; hence it hinders the operation of the eternal natural condition for the lasting fertility of the soil. Thus it destroys at the same time the physical health of the urban worker and the intellectual life of the rural worker ... Capitalist production, therefore, only develops the techniques and the degree of combination of the social process of production by simultaneously undermining the original sources of all wealth – the soil and the worker.[103]

Here, the old fixed opposition between the organic and inorganic fully gives way to an understanding of ecological processes of exchange – raising the question of sustainability. If nature remained 'man's inorganic body', this human-natural dialectic was now, in new and more complex ways, conceived as arising out of a coevolutionary process.[104]

Labour became for Marx not simply the extension of human powers over inorganic nature, but rather a process of the transformation of energy in which human beings were dependent on larger material and/or ecological conditions. Foreshadowing ecological economics, *Capital*'s energy analysis proceeds from

---

102 Shenstone 1901, p. 84.

103 Marx 1976a, pp. 637–8.

104 See Foster 2000, pp. 141–77.

the *endosomatic* level of human labour power and its bodily organs to the *exosomatic* level of tools and machines as extended organs of human labour.[105] In considering machinery as a means of extracting more work from labour power, Marx was forced to confront the role of extra-human energy flows and energy conversions. This is evident from Marx's opening definition of machine: 'The machine, which is the starting-point of the industrial revolution, replaces the worker, who handles a single tool, by a mechanism operating with a number of similar tools and set in motion by a single motive power, whatever the form of that power'.[106]

Machines are thus means of converting both materials and (human and extra-human) energy into commodities bearing surplus value. This took his analysis even further away from purely instrumentalist perspectives in which nature's role was merely passive.[107] Rather, the issue became one of sustainability (and coevolution): the way in which agricultural improvement, for example, was tied to the necessity of sustaining 'the whole gamut of permanent conditions of life required by the chain of human generations'.[108] The materialist-dialectical approach to the philosophy of nature was thus gradually transformed in Marx's later work, as a result of ongoing developments in materialism and science, into a modern ecological vision.

Instrumentalism and Teleology: Contradictions in the Ecological Critique of Marx

Again and again, ecological critics of Marx have employed his reference to nature as the inorganic body of humanity to suggest that Marx adopted an

105 See Daly 1968, pp. 396–8 on the importance of the endosomatic/exosomatic distinction in ecological economics. Martínez-Alier asserts that 'Marx does not seem to have considered the metabolic energy flow, so he could not trace the distinction … between endosomatic use of energy in nutrition and the exosomatic use of energy by tools' (2005, p. 3). This has to rank as one of the most uninformed statements ever made by a scholar of Martínez-Alier's reputation. Not only did Marx make such a distinction (which went back to the ancient Greeks), but in his hands and those of Engels it became the basis for an original conception of human evolution in line with Darwin's analysis. See Foster 2000, pp. 196–207; Winder, McIntosh, and Jeffrey 2005, pp. 351, 354–5.

106 Marx 1976a, p. 497. This definition was adapted from the work of the English engineer and economist Charles Babbage (1791–1871).

107 See Burkett 1998, pp. 120–33, and 2014, Chapters 2–4.

108 Marx 1981, p. 754.

instrumentalist approach in which nature was no more than a mechanism to be appended to human productive needs. Related to this is the charge of dualism: that Marx saw the relation between nature and humanity as one of absolute opposition, rejecting any dialectical conception in this area. In the words of Clark, 'Marx's image of the relationship between humanity and nature remains the proprietary one bequeathed to us when the God of ancient Israel gave Adam dominion over the earth'.[109] Hence, for Clark, Marx simply replicates the Judeo-Christian ethic of the Bible in his attitude toward nature, seeing nature as an object to be exploited. At the same time, Marx is seen not as a 'prophet of resurrected nature, but rather of triumphant enlightenment' – that is, as a representative of the Enlightenment humanistic view in which nature was simply subordinated to human reason.[110] 'Nature, apart from "man", is therefore necessary', Clark writes in his interpretation of Marx, 'only as an instrument in this self-creation' of humanity.[111] Similar arguments have been advanced, as we have seen, by such thinkers as Eckersley, Salleh, Routley, and O'Neill.[112] Salleh argues that Marx drew on 'ontological assumptions derived from the Great Chain of Being' of medieval scholasticism, and even Aristotle, in order to develop an anthropocentric (and androcentric) account of nature as man's dominion.[113] Furthermore, in characterising nature as 'man's inorganic body', Marx purportedly treated nature as the '"instrument" of his needs'.[114]

Such conclusions, however, simply read into Marx an instrumentalism that is assumed to be there; they are not informed by a close and comprehensive examination of Marx's texts. Marx wrote in the *Economic and Philosophical Manuscripts* about the dialectic of organic/inorganic relations and of the alienation of human society from nature, drawing on both the Hegelian philosophy of nature and materialist philosophy extending back to Epicurus.[115] What emerges from such an analysis is a complex dialectical and coevolutionary view that focuses on ecological interdependencies. This analysis immediately transcends simple dualistic and instrumentalist accounts of the relationship between human beings and nature. Moreover, in Marx's case, the argument is

109 Clark 1989, p. 251.
110 Clark 1989, p. 252.
111 Ibid.
112 Eckersley 1992; Salleh 1997; Routley 1981; O'Neill 1994.
113 Salleh 1997, pp. 71, 74.
114 Ibid.
115 Marx 1974.

a materialist one and hence becomes integrated with a growing body of knowledge about ecological relations in the overlapping realms of human history and natural history.

In Marx's view, the growth of bourgeois society with its commodification of nature represents 'an actual contempt for and practical degradation of nature'.[116] This is exemplified by the fact that, as Thomas Müntzer declared, 'all creatures have been made into property, the fish in the water, the birds in the air, the plants on the earth'.[117] For Marx, a dialectic of organic/inorganic relations and of human-nature relations was assumed from the start. Furthermore, the dialectical unity of human beings with nature required no explanation. What needed to be explained was the severing of this unity – the alienation of human beings from nature or what Marx was later to call the 'rift' in the metabolism connecting human beings to nature. Marx's approach was dialectical, but it also presented the dialectic as a material problem arising from the alienated development of human society itself. Marx was thus driven by the very nature of his theoretical perspective to absorb the major ecological insights of his day – through the work of thinkers such as Liebig and Darwin.

The fact that Marx's approach to human-nature relations was dialectical does not, of course, completely elude critics such as Clark. Thus, Clark admits that 'on rare occasions' Marx's analysis moves 'in the direction of ... an ecological dialectic'.[118] He even refers to Marx's discussion of the degradation of the soil and the metabolic rift in the cycle of nutrients, arising from the antagonistic relation between town and country, as presented in Volume I of *Capital*.[119] Yet Clark claims that 'Marx himself fails to go very far in developing these rudiments of an ecological dialectic'.[120] Here, Clark is perhaps hampered by the fact that his references to Marx's writings in this area rely heavily on excerpts provided by Howard Parsons' book, *Marx and Engels on Ecology*.[121] He also utilises conventional assumptions, such as Marx's so-called 'Prometheanism', that have been refuted by more recent scholarship.[122] He fails to recognise, therefore, that Marx's analysis of the metabolic rift, which grew both out of his understanding of the work of Liebig and in response to the crisis of the soil in nineteenth-century agriculture, took the form of a complex, many-sided eco-

116 Marx 1974, p. 239.

117 As cited in Marx 1974, p. 239.

118 Clark 1989, p. 250.

119 Marx 1976a.

120 Clark 1989, p. 250.

121 Parsons 1977.

122 See, for example, Foster 2000; Burkett 2014.

logical dialectic in his later writings (particularly *Capital*), encompassing the concept of sustainability as well as the need for social-ecological transformation.[123]

Of greater importance in Clark's assessment, however, is his presumption that ecological thought is only dialectical to the extent that it is teleological. Clark suggests that Marx derived from his 'Aristotelian and Hegelian heritage' a teleological viewpoint in which 'a phenomenon ... must be comprehended as a being in process or movement in which its *ergon*, or peculiar behavior, is related to its *telos*, or completed form of development'.[124] Disregarding the fact that Marx, as a consistent materialist, followed Epicurus rather than Aristotle in this respect (and thus explicitly rejected all teleological analysis of nature), Clark argues that it was only to the extent that he incorporated such a 'teleological dialectic' that Marx developed an 'organicist dimension' in his thought and thus bordered on the ecological – although ultimately, according to Clark, Marx abandoned teleology and organicism.[125] In Clark's view, although Marx does exhibit 'a recognition of teleology in nature' at certain points, 'Marx does not develop this teleological conception'.[126] Hence, his thought remains antiecologically anthropocentric.

What such criticisms demonstrate is not Marx's inadequacy as an ecological thinker, but rather the criteria for ecological analysis embodied in much of contemporary philosophical ecology. In Clark's view, it is impossible to be a consistent ecological thinker from a materialist or realist standpoint, which rejects teleology; rather, ecological analysis is by definition teleological and essentialist.[127] For Clark, it is Marx's materialist conception of nature that is the enemy of ecology. We are thus led to believe that ecology must follow a mystical, spiritual direction, exemplified by Plato, Aristotle (in his more teleological analysis), and Hegel. Thinkers who developed more materialist perspectives, associated more with the development of modern science, such as Epicurus, Hobbes, Marx (insofar as he broke with Hegel), and Darwin are seen as antiecological in their thinking, despite the fact that scientific ecology has always been more closely connected with the latter than the former.

One is thus struck by the strange irony presented by thinkers who criticise Marx – along with Darwin, one of the greatest materialist thinkers of the nineteenth century – for supposedly taking his image of human-nature relations

123 Foster 1999, 2000, pp. 141–77; Burkett 2014, pp. 126–8.

124 Clark 1989, pp. 249–50.

125 Ibid.

126 Clark 1989, p. 255.

127 Ibid.

from 'the God of ancient Israel' and for his adherence to the 'great chain of being' of medieval Scholasticism. At the same time, it is claimed that Marx's primary failure was his abandonment of a teleological concept of nature. It is, of course, the latter criticism that should be taken most seriously. Marx is condemned here primarily for his materialism, which is assumed to be at variance with an ecological outlook.

This rejection of materialism accounts for much of the mysticism in contemporary Green theory, in which the issue is no longer material relations and sustainability, but rather an abstract, moral division between anthropocentric and ecocentric views. In the words of Murray Bookchin,

> Mystical ecologists who dualize the natural and the social by contrasting 'biocentricity' with 'anthropocentricity' have increasingly diminished the importance of social theory in shaping ecological thinking. Political action and education have given way to values of personal redemption, ritualistic behavior, the denigration of human will, and the virtues of human irrationality. At a time when the human ego, if not personality itself, is threatened by homogenization and authoritarian manipulation, mystical ecology has advanced a message of self-effacement, passivity, and obedience to the 'laws of nature' that are held to be supreme over the claims of human activity and praxis.[128]

If the main criteria for dealing with the present world ecological crisis is one of creating a more sustainable society, which means a more sustainable relation to nature, this cannot be achieved by means of a one-sided mystical, spiritual, romantic perspective and an emphasis on undifferentiated holism, abandoning all bases for meaningful praxis – any more than it can be achieved through a reliance on mechanism. What is needed, rather, is a nondeterministic materialism and ecological humanism that recognise the dialectical linkages between humanity and nature, between human consciousness and the natural world. What cannot be accepted is a 'passive' relation to nature, rooted in a perpetuation of dualistic conceptions. Unfortunately, much of philosophical ecology suggests precisely this kind of passive and dualistic standpoint.

In many ways, the flipside of Clark's argument is to be found in O'Neill's surprising contention that scientific ecology is inherently mechanistic and reductionist, opposed to approaches that are dialectical and holistic.[129] According to

128 Bookchin 1990, p. 47.

129 Clark 1989; O'Neill's 1994.

O'Neill's account, Marx's emphasis on the ultimate oneness of organic and inorganic nature contradicts science. Like many thinkers in the Hegelian-Marxist tradition, O'Neill essentially cedes nature (and the whole realm of physical and natural science) to positivism. The dialectic is confined to the realm of society and social science alone. Yet, for those among Marx's critics who insist, nonetheless, on dealing dialectically with nature, this has often led, as in the case of Clark, to the view that one can only be dialectical by being nonscientific, hence teleological, mystical, and so forth – a position that O'Neill himself rejects.[130]

Toward Ecological Materialism

At issue in the standard critique of Marx's organic/inorganic distinction then are two different and strongly opposed visions of ecological philosophy: one that is materialist, historical, and essentially scientific in character; the other that derives its emphasis from mystical distinctions between anthropocentric and ecocentric and from spiritualistic allusions to nature's teleology. From the latter standpoint, it is impossible to perceive the real class-exploitative alienation of nature. Hence, the social problem underlying ecological destruction disappears, giving way, as Bookchin aptly puts it, to a philosophy of 'personal redemption, ritualistic behavior', and the like.[131] For ecology to be related to social transformation, it must adopt a material-social standpoint that emphasises the reality, the this-sidedness of the degradation of nature – not as a mere ethical problem, but as a problem of real existence and human praxis. Marx's approach provides just such a standpoint. 'The conventional antinomies of nature/culture, environment/society, human/nonhuman, and subject/object', Timothy Luke has written, 'all implode in Marx's rendition of these links as one active organic/inorganic project'.[132] In Marx's materialist dialectic of organic/inorganic relations, one finds neither a narrowly instrumentalist, anthropocentric perspective, nor a flight into mysticism, but rather the core of an ecological critique of capitalist society – a critique that should allow us to translate ecology into revolutionary praxis.

130 Ibid.

131 Bookchin 1990.

132 Luke 1999, p. 44.

CHAPTER 2

The Origins of Ecological Economics: Podolinsky and Marx-Engels

Until recently, as noted in the Introduction to this book, most commentators, including ecological socialists, have assumed that Marx's historical materialism was only marginally ecologically sensitive at best, or even that it was explicitly anti-ecological. However, research over the last decade and a half has demonstrated not only that Marx viewed environmental issues as central to the critique of political economy and to investigations into socialism, but also that his treatment of the coevolution of nature and society was in many ways the most sophisticated to be put forth by any social theorist at least up to the present century. Still, criticisms continue to be levelled at Marx and Engels for their understanding of thermodynamics and the extent to which their work is said to conflict with the core tenets of ecological economics. In this respect, the rejection by Marx and Engels of the pioneering contributions of the Ukrainian socialist Sergei Podolinsky, one of the founders of energetics, has been frequently offered as the chief ecological case against them.

Since the publication of a number of influential writings by J. Martinez-Alier, including his pioneering *Ecological Economics*, Podolinsky has come to occupy a central place in the ecological literature in two fundamental respects: (1) his direct contribution to ecological economics; and (2) the reception that his work received from Marx and Engels, including the role that this played in the subsequent relationship between Marxism and ecological economics. In connection to his direct contribution, Podolinsky is often credited with being 'the first explicitly to scrutinize the economic process from a thermodynamic perspective'.[1] Martinez-Alier's influential history of ecological economics treats Podolinsky as the major nineteenth-century precursor and perhaps even as the founder of the discipline, arguing, for example, that he was the first to develop 'the concept of energy return to energy input in different types of land use'.[2]

With regard to the Marx-Engels-Podolinsky relationship, the standard interpretation is based on Martinez-Alier's claim that Marx and Engels had a negative reaction to Podolinsky's work, which meant a missed chance to connect

1 Cleveland 1999, p. 128.

2 Martinez-Alier 1987, p. 5.

Marxian value-theoretic analysis to Podolinsky's energetics. Marx himself is faulted for his supposed silence on Podolinsky, whereas Engels is criticised for his dismissal of Podolinsky's analysis.[3] Marx's alleged 'silence from 1880 to the end of his life in 1883' has been characterised by Martinez-Alier as implicit evidence that he agreed with Engels in his 'negative reaction to Podolinsky's work'.[4] For James O'Connor, the point is summed up by saying that Marx turned a 'deaf ear' to Podolinsky.[5] Martinez-Alier has further argued that Marx and Engels's negative response to Podolinsky is the root of 'the Marxist neglect of ecology'.[6] This too has become conventional wisdom, to the point where what are widely regarded as the ecological deficiencies of Marxism are often traced by ecological economists to the inadequate response by Marx and Engels to Podolinsky's work – the citation of Martinez-Alier's individual and coauthored writings often being deemed sufficient to establish this connection.[7]

Our scepticism about this conventional wisdom stemmed initially from our investigation into the chronological development of Podolinsky's work as it related to the working lives of Marx and Engels. We discovered that Podolinsky's analysis was published in four different languages over the years 1880–3, and that there were significant differences among the four versions. Importantly, the version of Podolinsky's analysis that Martinez-Alier and Naredo used to criticise Marx (for his supposed neglect of Podolinsky's argument) was published in the German socialist paper *Die Neue Zeit* in 1883, only after Marx's death.[8] As we further studied Podolinsky's work, and Engels's comments on it, it became clear that the conventional interpretation of this entire episode was seriously misleading.

Accordingly, the present chapter and the next advance a thorough anti-critique of the conventional interpretation, together with a radical reassessment of Podolinsky's place in the history of ecological economics. We show that Podolinsky did not establish a plausible thermodynamic basis for the labour theory of value that could have been adopted by Marx and Engels. Moreover, Marx and Engels did not neglect nor abruptly reject Podolinsky's work as is commonly supposed, but took it seriously enough to scrutinise deeply in the

3 Guha and Martinez-Alier 1997, p. 25; Martinez-Alier 1987, p. xviii; Martinez-Alier 2003, p. 11.

4 Martinez-Alier 2003, p. 11; Martinez-Alier and Naredo 1982, p. 209.

5 O'Connor 1998, p. 3.

6 Martinez-Alier 1995, p. 71.

7 See, for example, Bramwell 1989, p. 86; Cleveland 1999, p. 128; Deléage 1994, p. 49; Hayward 1994, p. 226; Hornburg 1998, p. 129; Kaufman 1987, p. 91; Pepper 1996, p. 230; Salleh 1997, p. 155.

8 Martinez-Alier and Naredo 1982. The German version of Podolinsky's article was published in two installments; see Podolinsky 2008 [1883]. See Appendix 2 below.

spirit of critique. Although verifying Podolinsky's rightful place as a forerunner of ecological energetics, our analysis highlights the severe limitations imposed by his tendencies toward energy reductionism and closed-system thinking as compared to Marx and Engels's metabolic and open-system approach to nature and to human production.

Our analysis in the present chapter is informed by: (a) Marx's previously undisclosed notes on Podolinsky, to appear in the forthcoming volume IV/27 of the *Marx-Engels Gesamtausgabe* (known as MEGA); (b) the significant differences between the several (published and unpublished) versions of Podolinsky's analysis and the respective versions that Marx and Engels read and commented on; (c) a critical analysis of the ecological advances and shortcomings contained in Podolinsky's work in the respective forms in which it was read by Marx and Engels; and (d) a detailed consideration of Marx and Engels's correspondence on Podolinsky, including the probable role that Engels's manuscript *The Mark* played in their discussions. In Appendix 1 to this book, we are including an English-language translation by Angelo Di Salvo of the 1881 Italian version of Podolinsky's work (the one read and commented on by Engels), which has been carefully compared against an English-language translation by Mark Hudson of the published French version of 1880.[9] (The latter translation, though employed in a rigorous comparison with the translation from the Italian, is not reproduced here. Only the specific differences with the Italian version are noted). We are also including in Appendix 2 a translation by Peter Thomas of the 1883 *Die Neue Zeit* version of Podolinsky's work in this area.

Our investigation begins with a brief sketch of Podolinsky's life and work, which serves not only to reintroduce an important socialist ecological thinker, but also to establish the ways in which Podolinsky's political and intellectual milieu intersected with and differed from that of Marx and Engels. We then locate the Italian version of Podolinsky's article in the context of the development of his work on energetics through several versions published in four different languages. This genetic textual analysis, which is also informed by a close look at two letters that Podolinsky wrote to Pyotr Lavrovich Lavrov in 1880, helps to pinpoint the specific versions of Podolinsky's piece most likely read by Marx and Engels. It also begins to establish the very limited, in fact nonexistent, extent to which Podolinsky's work addressed value-theoretic questions in a way that could have been adopted by the founders of historical materialism.

We then summarise and criticise what is actually Podolinsky's main analytical theme: his argument that human labour is uniquely gifted in its ability

9 Podolinsky 1880, 2004 [1881]. See Appendix 1 below.

to accumulate energy in useful forms on the earth and that this unique capability implies that the human being fulfils the thermodynamic requirements of a so-called perfect machine as theorised by Sadi Carnot. Here, our critique notes the practical difficulties with the kind of energy-accounting exercises that Podolinsky used to defend his energy accumulation thesis. More important, we uncover the tendency toward energy reductionism (reduction of human production and consumption, and of its historical dynamics, to pure energetics) that is implied by Podolinsky's framework. We also demonstrate how Podolinsky's so-called perfect machine hypothesis falls prey to closed-system thinking (neglecting the resource extraction problem, as in coal, and ignoring the dissipation of energy and material waste into the environment). We argue that Podolinsky's energy reductionism and closed-system thinking greatly limit the socioecological insights obtainable from his analysis as compared to the power of Marx and Engels's metabolic open-system perspective on human production, capitalism, and the environment. As a crucial case in point, we show that Podolinsky's quantitative energetics does not provide a viable physical-scientific basis for a labour theory of value as Marx understood it, namely, as an analysis of the socioeconomic forms taken on by capitalist alienation of both nature and labour vis-à-vis the direct producers. In asserting that Podolinsky provided a potential thermodynamic basis for value analysis, the conventional interpretation implicitly adopts an energy-reductionist, crude materialist, and nondialectical approach to value that is completely alien to Marx's approach. We clarify this point with reference to the critique of naturist value thinking (ascribing value directly to nature) that was developed by Podolinsky's economic mentor, Nikolai Sieber, who was also an economic follower of Marx.

Turning to Marx and Engels's reaction to Podolinsky, we first note that Marx's detailed extracts from a draft of Podolinsky's (1880) *La Revue Socialiste* article contradict the conventional wisdom that he basically ignored or turned a deaf ear to Podolinsky's work. Moreover, from Marx's extracts, it appears that even those portions of the published versions of Podolinsky's analysis that the conventional wisdom (mistakenly) sees as adaptable to Marx's value theory were most likely absent from the manuscript read by Marx.

Next, we investigate the two key letters to Marx in which Engels comments on the Italian version of Podolinsky's work. We show that Engels's letters constitute much more than an abrupt dismissal or negative reaction. Rather, they exhibit a careful reading of Podolinsky in the spirit of critique. Moreover, Engels's comments on Podolinsky were sent to Marx in December of 1882, less than three months before Marx's death, and they were based on the version published in the Italian journal *La Plebe* in 1881 – one that was much less extensive than the *Die Neue Zeit* article of 1883.[10] The *La Plebe* piece itself was more extensive than an earlier version published in the Parisian *La Revue Socialiste* in June 1880.[11] Engels's criticisms focus mainly on those elements of Podolinsky's analysis that suffer the most from energy reductionism and closed-system thinking, especially the treatment of labour as a purely mechanistic (and not metabolic) process and the failure to account for the squandering of energy in the form of coal. In this way, Engels's comments help clarify the metabolic and open-system character of his (and Marx's) historical-materialist perspective on human production and on its development. This clarification jibes with the fact that Engels's letters on Podolinsky grew out of his discussions with Marx on *The Mark*, a work in which Engels addresses various ecological issues raised by the disintegration of communal peasant agriculture in Germany under the twin pressures of landed property and capitalist competition.[12]

Finally, we consider whether the final German, *Neue Zeit*, version of Podolinsky's work includes any additional analyses running counter to our own reinterpretation. We conclude with a brief reconsideration of the relationship between ecological economics and Marx and Engels's theoretical system in light of our analysis. This reinterpretation sheds new light on Podolinsky's real contribution and limitations. It also suggests an important avenue for expanding upon Martinez-Alier's pioneering excavation of the history of ecological economics.[13]

Podolinsky: Life and Work

Sergei Podolinsky (1850–91) was a Ukrainian socialist and physician, who was an acquaintance of Marx and Engels. He was a member of the wealthy landed gentry class.[14] His father had been postmaster general of the southern provinces and later retired to his estates where he was a gentleman poet. His mother's mother had been the daughter of a French ambassador in Napoleon's day. While a student of the natural sciences in Kiev, Podolinsky gravitated toward Nikolai Sieber (1844–88), who was later the first economics teacher at

10 Engels's two letters to Marx are dated 19 December 1882 and 22 December 1882; see Marx and Engels 1975, Vol. 46, pp. 410–14. The Italian version of Podolinsky's work, like the German one, was published in two installments; see Podolinsky 2004 [1881], Appendix 1 below.

11 Podolinsky 1880.

12 Engels 1978a.

13 Martinez-Alier 1987.

14 Serbyn 1982, p. 5.

a university in the Russian Empire, and probably anywhere, to be influenced by Marx.[15] Sieber was Marx's most brilliant economic follower in the 1870s and 1880s and laid the foundation for Marxist economics in Russia and in the Ukraine. In his master's dissertation, 'David Ricardo's Theory of Value and Capital', published in Russian in 1871, Sieber presented Marx's work as a necessary sequel to Ricardo's. Subsequently, Sieber was to write a series of articles explaining and defending Marx's economics.[16] Marx referred favourably to Sieber in the 1873 postface to the second edition of Volume I of *Capital* and again in his 1880 *Notes on Adolph Wagner*.[17]

In 1872, Podolinsky finished his studies in Kiev and travelled to the West, taking up medical studies in Zurich. He met Marx and Engels that summer in London through Pyotor Lavrovich Lavrov (1823–1900), a leading Russian populist-socialist thinker.[18] In September, he attended the Hague Congress of the First International. Podolinsky authored two articles on the history of the International in the first issues of the journal *Vpered!* (*Forward!*), which he, along with Lavrov, helped launch. In 1875, Podolinsky published two pamphlet-size socialist works in the Ukrainian language. One of these was 'The Steam Engine', a socialist utopian story about a rural worker who is severely injured by a threshing machine while working in the fields and who dreams of a socialist future when workers will own the land and its produce and will reap the rewards.[19] The other was titled *On Poverty*. Podolinsky received his doctorate in medicine at Breslau in 1876 under the supervision of Rudolf Peter Heinrich Heidenhain, a physiologist. He also studied in Zurich under the physiologist Ludimar Hermann, author of the *Handbuch der Physiologie* (published in six volumes from 1879 to 1883). Also, while in Paris, Podolinsky could not help being exposed to the energetic-physiological analyses developed by prominent

15 We use the Russian-German form of Nikolai Sieber's name here because this was the form used by Marx and is the one most familiar in the Marxist literature. In most bibliographies and discussions of Ukrainian intellectuals, however, the Ukrainian spelling of his name is used: Mykola Ziber.

16 Sieber 2001. A revised and expanded edition of *David Ricardo's Theory of Value and Capital*, incorporating later work by Sieber on Marx, was published in 1885 under the new title *David Ricardo and Karl Marx in their Socio-Economic Investigations* (White 2001, p. 6).

17 Marx 1975, p. 184; Marx 1976a, p. 99. Sieber met Marx and Engels for the first time in 1880 (the year that Podolinsky sent his manuscript to Marx), when Sieber was a frequent guest at Marx's house. For detailed discussions of Sieber's economics and relation to Marx, see White 1996 pp. 229–38; White 2001; D.N. Smith 2001; Koropeckyj 1984, pp. 203–14; and Koropeckyj 1990, pp. 194–203.

18 Serbyn 1982, p. 6.

19 Martinez-Alier 1987, pp. 54–6.

French writers such as Claude Bernard (1813–78), Gustave Hirn (1815–90), and Étienne Jules Marey (1830–1904). These theorists applied the emerging ideas of thermodynamic science, as well as Carnot's earlier work on the efficiency of steam engines, more or less directly to human labour, thus conceiving of human beings as 'living machines'.[20] Marey, for example, began his *Animal Mechanism* with the words:

> Living beings have been frequently and in every age compared to machines, but it is only in the present day that the bearing and justice of this comparison are fully comprehensible. No doubt, the physiologists of old discerned levers, pulleys, cordage, pumps, and valves in the animal organism, as in the machine. The working of all this machinery is called Animal Mechanics in a great number of standard treatises ... Modern engineers have created machines which are much more legitimately to be compared to animal motors; which, in fact, by means of a little combustible matter which they consume, supply the force requisite to animate a series of organs, and to make them execute the most varied operations.[21]

As Jacques Gleyse observes, it is difficult to ignore the capitalist functionality of this school together with its elite-engineering perspective on social efficiency and reforms:

> The idea of the rationalized energy-producing body ... was perhaps not only developed in part from the technology of the steam engine, but also through the economic need for more and more efficient factory production. At least we can perceive ... a metaphorical dialogue between these two types of language. But in both instances it would seem that a group of pioneers was instigating a system of control over the general population ... In the industrial universe and in the factory environment 'man' became a theoretic entity in accordance with values represented firstly by the steam engine and then by the machine ... A kind of implacable logical cycle was set up: technology gave birth to science and then science, expanding beyond its first field of application, or else being applied (or even misapplied) to other fields, led in turn to the birth of a technology, or sometimes even a technocracy. It was the human body, or more particularly in this case the physical activity associated with it, that was the subject of this technology.

20 Carnot 1977; Gleyse 2002; Papanelopoulou 2003.

21 Marey 1874, p. 1.

> But this technology should not just be considered as such; above all else it was a widespread system of control that organized society, or at least a system that a few influential people wished to promote for the greater good of the masses ... Hirn consolidated this paradigm and applied it to corporal practices as a whole, going beyond the limited field of industrial production.[22]

Exemplifying such positivistic views in physiology, Bernard wrote in 1865 that:

> There is an absolute determination in all of the sciences because, each phenomenon being linked necessarily to physic-chemical conditions, the scientist can modify these conditions to master the phenomenon, that is to say, to hinder or favour its manifestation. In the case of organic bodies, there is no debate on the subject. I would like to prove that it is the same for living bodies, and that, for them also, determination exists.[23]

The treatment of animals (including human beings) as thermodynamic machines was thought to lead to easily quantifiable relations of food (combustible matter), heat and useful work. Hirn, as explained by Marey, carried out experiments in which he 'enclosed the subject in a hermetically closed chamber, and made him turn a wheel which could, at choice, revolve with or without doing work'.[24] The object was to measure the energy efficiency of human labour in ways equivalent to the measurement of the thermodynamic efficiency of a steam engine.

Unfortunately, such energy-reductionist approaches were to leave a deeper imprint on Podolinsky than did the more metabolic, and less mechanical, methods of Hermann. The most direct influence on him in this respect was likely the French 'living machine' school represented by Bernard. One sees in such outlooks the intellectual roots of Podolinsky's attempt to ground value analysis in energy flows and of his vision of socialism as a tightly engineered machine dedicated to the accumulation of energy on the earth (see below).

In 1877, Podolinsky returned for a time to his family home in Kiev, where he married the daughter of a landowner, Maria Andreeva. They settled in exile in Montpellier, France. In 1879, Podolinsky published his long study, *The Life and Health of People in the Ukraine*, using his knowledge as a physician.

22 Gleyse 2002, pp. 8–9.

23 Bernard 2000, p. 320; compare Olmstead and Olmstead 1952, pp. 131–50.

24 Marey 1874, p. 18.

Podolinsky was personally and financially involved in the Ukrainian socialist and nationalist journal *Hromada* (*Community*), then published in Geneva, for which he coauthored a manifesto on Ukrainian national independence and socialism in 1880.[25] He also wrote *Crafts and Factories in Ukraine* (1880), the first economic monograph to be written in the Ukrainian language.[26]

It should be noted that as a Ukrainian socialist political economist, Podolinsky belonged to a tradition of thought that was closer to French than to German socialism. For example, M. Drahomanov, who edited *Hromada* and with whom Podolinsky was closely associated, considered himself a follower of Proudhon.[27] Podolinsky was equally close to (and is often thought of as belonging to) the tradition known as legal Marxism, which emphasised industrial development and economic determinism rather than the class struggle. This group included Sieber and later Mikhail Tugan-Baranovsky. To the end of his life, Podolinsky combined a commitment to socialism with a strong devotion to the Ukrainian nation.[28]

Meanwhile, by the late 1870s, Podolinsky was also working on his study of agricultural energetics, and it is this work that has drawn the most attention from modern-day ecological economists. In March 1880, Podolinsky sent to Marx his 'Human Labor and the Conservation of Energy', written in French. A new version of this work was completed in May 1880 and published in late June in *La Revue Socialiste* under the title 'Socialism and the Unity of Physical Forces' (a much longer version was also published around the same time in the Russian journal *Slovo* [*The Word*]).[29] In addition, in that same year – his most

---

25 Rudnytsky 1952, pp. 206–8, 223–4.

26 Holubnychy 1971, p. 684.

27 Rudnytsky 1987, pp. 206–7, 263.

28 Himka 1993; Holubnychy 1971; Holubnychy 1993, p. 116; Serbyn 1982, pp. 4, 6.

29 Marx's extracts from Podolinsky's 'Human Labor and the Conservation of Energy' are predominantly in French and correspond word for word with much of the 1880 French version of Podolinsky's manuscript published in *La Revue Socialiste*. As discussed later, however, the limited coverage of Marx's extracts compared to the published text, and their different titles, strongly suggest that Marx was reading an earlier and shorter version of the *La Revue Socialiste* article.

The 1880 Russian article in *Slovo* was titled 'Human Labor and its Relation to the Distribution of Energy'. It was very extensive – 70 pages long in small type, with 12 chapters. It has been reprinted in book form, with an introduction by P.G. Kuznetsov (Podolinsky 1991). Its chapter headings (translated into English by Leontina Hormel) are as follows: 1. 'What is Energy? Its Conservation and Distribution'; 2. 'Converting Energy on Earth'; 3. 'Economy of Energy'; 4. 'The Appearance of Organisms. The Meaning of Plants in the Distribution of Energy'; 5. 'The Meaning of Animals and Man in the Distribution

productive as a scholar – Podolinsky published articles in *La Revue Socialiste* on nihilism and on social Darwinism. He was on the editorial board of *La Revue Socialiste* and emerged as a well-known socialist-populist analyst.

A longer version of 'Socialism and the Unity of Physical Forces' was completed around a year later and published in two installments in Italian in 1881 in the journal *La Plebe*. A still more detailed version was published under a different title, 'Human Labor and the Unity of Physical Forces', in *Die Neue Zeit*, the journal of the German Social Democratic Party, in September-October 1883.[30] Unfortunately, in January 1882 (presumably after sending the final manuscript to *Die Neue Zeit*), Podolinsky suffered a mental breakdown from which he never fully recovered. In 1885, his parents obtained special permission to repatriate him, and he returned to Kiev, where he remained until his death in 1891.

Because we know from the Marx-Engels correspondence that the *La Plebe* version is the one read and commented on by Engels, and because the *La Plebe* version subsumes the *La Revue Socialiste* article, a draft of which was likely the version read by Marx (see below), the *La Plebe* version itself demands close attention, and it is accordingly reproduced in Appendix 1.[31] It is presented in full so that the reader can decide for her- or himself whether Podolinsky's analysis contains ecological insights that could and should have been directly adopted by the founders of Marxism, as is so often claimed. We note the major differences in the final German version (itself reproduced in Appendix 2) and address their significance below as well.

---

of Energy. The Concept of Labor'; 6. 'The Origin of the Capacity Toward Work in the Constitution/Organism of Man'; 7. 'Man as Thermal Machine'; 8. 'Labor as the Means for the Satisfaction of Needs'; 9. 'Various Forms of Labor and Their Relation to the Distribution of Energy'; 10. 'Labor, the Tendency Toward the Production of Mechanical Work'; 11. 'The Plunder and Accumulation of Energy'; and 12. 'General Conclusions'.

30 Podolinsky 2004 [1881], Appendix 1 below and 2008 [1883], Appendix 2 below. The German rendition exceeds the Italian version by over three thousand words. The differences between the German and Italian versions are actually quite significant, especially in terms of the former's more insistent championing of a thermal-machine approach to human labour (see the penultimate section of the present chapter).

31 Marx and Engels 1975, Vol. 46, p. 410.

Development of Podolinsky's Project

Podolinsky's 'Socialism and the Unity of Physical Forces' was a product of the revolution in the scientific understanding of energy in the early nineteenth century, beginning with the discovery of the energy conservation principle in the 1840s. Between 1842 and 1847, four European scientists – J.R. Mayer, James P. Joule, L.A. Colding, and H. von Helmholtz – all introduced the hypothesis of energy conservation. An expanded list of the scientists who made this breakthrough in the period, however, would include Sadi Carnot (before 1832), Marc Séguin, Karl Holtzmann, G.A. Hirn, C.F. Mohr, William Grove, Michael Faraday, and Justus von Liebig.[32] Sadi Carnot's work, in particular, was to lead to the rise of thermodynamics, especially the famous second law (the entropy law) as physicists in the 1850s and 1860s tried to determine the laws of efficient energy use based on the steam engine. Energy – a term that came into wide usage among scientists only in the late 1870s – was found to dissipate when used so that the level of entropy (the amount of energy no longer available for human purposes) increased.

Podolinsky tried to use the new thermodynamic perspective to develop an agricultural energetics, combining elements from physics, physiology, and Marxian economics. His goal was to explore the centrality of human labour to the accumulation of energy on Earth.

From the first, Podolinsky saw his work as in a process of development. His plan, as he indicated in letters to Marx and to Lavrov in 1880 and in 1881, was to publish a set of successive versions of his original, preliminary analysis on labour and energy that would appear in various languages, with each new version extending the field of analysis over the previous ones and with each providing further illustrations. (Although he published a very extensive Russian version in 1880, he sought to spread his ideas in the western European context, and hence began publishing his work, or parts of it, in French, Italian, and German versions, presenting it in a more theoretically developed, if less extensive, form). In his letter to Marx on 30 March 1880, he mentioned his intention to expand the work that he had sent to Marx to take account of diverse forms of production and also his intention to provide a more detailed article with further examples.[33]

Setting aside the question of the lengthy Russian version, Podolinsky did in fact extend his work published in western European languages several times.

---

32 Kuhn 1977, pp. 66–8.

33 See Martinez-Alier 1987, p. 62.

The *La Revue Socialiste* article was most likely an extension of a draft on 'Human Labor and the Conservation of Energy' (also in French) that he had earlier sent to Marx.[34] The Italian version, an English-language translation of which is provided in Appendix 1, added 20 new paragraphs not in the *La Revue Socialiste* article, a couple of sentences beyond that, and a number of footnotes.[35] The German article published in *Die Neue Zeit* in 1883 was the final version (see Appendix 2).[36]

In the opening paragraphs of his article, 'Socialism and the Unity of Physical Forces', Podolinsky referred to the conservation of energy and to the need to understand how human labour should be allocated in this respect to best satisfy human needs. He then mentioned that

> according to the theory of production formulated by Marx and accepted by socialists, human labor, expressed in the language of physics, accumulates in its products a greater quantity of energy than that which was expended in the production of the labor power of the workers. Why and how is this accumulation brought about?

Although it seems to be directed at Marx's theory of surplus value and accumulation of capital, Podolinsky's question, posed in terms of physics, is really quite different: it aims at showing how human labour results in the accumulation of solar energy on Earth.

Nonetheless, due principally to this statement at the beginning of his article, in which he said that Marx's theory of production based on human labour could be 'expressed in the language of physics', and because of a letter that he wrote to Marx in April 1880 that mentioned his interest in relating the physics of human labour to the concept of surplus value, Podolinsky's work has often been presented as if that were its main argument. The actual thrust of his analysis, though, was different and had little to do directly with economic value, however much that may have been his ultimate object. His argument took essentially four steps. First, he provided a competent discussion of the general problem of entropy, explaining, following Clausius, that 'the entropy of the universe tends towards a maximum'. Second, he proposed a definition of useful work as that which results in an accumulation of solar energy on the Earth (so that solar energy does not simply radiate back into space). In this context,

---

34 Podolinsky 1880.

35 Podolinsky 2004 [1881]. See Appendix 1 below.

36 Podolinsky 2008 [1883]. See Appendix 2 below.

he provided statistical examples drawn from agriculture to argue that human labour has the power of increasing the amount of energy generated from plants in comparison to uncultivated nature. Third, he attempted on this basis to argue that human beings (and some animals) constitute the perfect machine referred to in Sadi Carnot's and in William Thomson's thermodynamics. As a perfect machine, a human being, in Podolinsky's terms, is able to recycle work back to its own firebox. Fourth, he suggested that this perfect machine could only be properly used in a socialist system of production.

We will not provide a detailed exposition of Podolinsky's discussion of the problem of entropy, which was insightful for its time. In the article published in *La Revue Socialiste*, this aspect was only sketchily developed. Entropy itself was not mentioned. The introductory references to Carnot and to Thomson found in the later Italian version were not included.[37] Still, it is significant that in both versions, Podolinsky showed his close attention to scientific developments, referring, for example, to Thomas Sterry Hunt's observation that, as stated by Podolinsky, 'even free oxygen in the atmosphere, according to certain geological hypotheses, originated in combination with the carbon that now constitutes coal'. Martinez-Alier has taken this reference to Sterry Hunt as evidence that Podolinsky recognised the effects of carbon dioxide on climate change, writing that,

> (he [Podolinsky] added in a footnote) [that] there was a theory which linked climatic changes to concentrations of carbon dioxide in the atmosphere, as Sterry Hunt had explained at a meeting of the British Society for the Advancement of Science in 1878.[38]

Yet, although it is true that Sterry Hunt is referred to in a footnote in the final German version of Podolinsky's argument, no such statement on carbon dioxide concentrations and their relation to climate change is actually to be found in Podolinsky's article (or in the earlier French and Italian versions).

By 1880, when Podolinsky first published his work, the fact that carbon dioxide and other gases could affect global temperature was well established. The experimental laboratory basis for the notion that carbon dioxide helped

37 Although the paragraphs at the beginning of the 1881 Italian version that referred to Carnot and to Thomson were not included in the *La Revue Socialiste* article of June 1880, Marx's notes show that they were included in the work on 'Human Labor and the Conservation of Energy' that Podolinsky had sent to him in March 1880. These paragraphs were probably dropped in the June 1880 version simply because of lack of space.

38 Martinez-Alier 2005, p. 10.

regulate the climate through what is now commonly called the 'greenhouse effect' had been carried out in 1859 by the British physicist John Tyndall, who was the first to theorise this relation. Interestingly, Marx attended some of Tyndall's lectures in this period and was especially intrigued by his experiments on solar radiation. However, we do not know whether he was present when Tyndall delivered his results on the greenhouse effect.

The later global warming hypothesis with regard to carbon dioxide and the tendency of global temperatures to rise secularly was not introduced until 1896 by Swedish climatologist Svante Arrhenius. Arrhenius, like other climatologists of his day, was responding to Louis Agassiz's introduction of his ice age theory in 1837, which by the mid-1860s had become part of the scientific consensus. Since he was primarily concerned about the appearance of another ice age, Arrhenius saw anthropogenic carbon dioxide emissions as having a possible beneficial effect in raising global temperature. It is quite possible that this prevailing ice-age focus may have contributed to Podolinsky's conviction, which pervades his work, that the terrestrial accumulation of solar energy and the rise in temperature that this involved was an unalloyed good.[39] Such a view that a warming of the earth as a whole could only be beneficial to humanity had been presented in the speculations of the utopian socialist Charles Fourier in *The Theory of the Four Movements* (1808).[40]

Accumulation of Energy on Earth

Central to his overall argument was Podolinsky's point that useful work could be defined as work that increased the accumulation of solar energy on Earth. As he said early on in his article, 'We believe ... that to a certain extent, it is within the power of humanity to produce certain modifications in the distribution of solar energy, in such a way as to render a greater portion profitable to humans'. It was this accumulation of usable energy (or, as we would say today, low entropy) that Podolinsky saw as the very purpose of work and as the material-physical basis for civilisation. Although human beings cannot create useful energy (because all such energy derives from the sun), they can, he argued, assist in its accumulation on Earth in forms available for human purposes. They can do so directly, Podolinsky suggested, through agricultural cul-

39 For all this background, see Fleming 1998, pp. 65–82; Weart 2003, pp. 1–11; Imbrie and Imbrie 1979; Scheider and Londer 1984, pp. 34–6; Uranovsky 1935, p. 140.

40 Fourier 1996.

tivation, draining marshes, irrigation, mechanisation of agriculture, protecting plants against natural enemies, and driving away and exterminating animals harmful to vegetation. Such accumulation of energy, moreover, can also occur in nonagricultural activities. The production of shoes, for example, was a way of transforming energy to make it usable for human purposes, even enhancing human labour potential, and thus fell under the definition of useful work.

Podolinsky used French government statistics and other sources to provide calculations on the energy productivity of (domestic animal and human) labour in agricultural production in 1870s France. He showed that the hay that was generated on a natural pasture without the contribution of human labour embodied much less energy (measured in kilocalories) than did the hay produced in sown pastures or the wheat and the straw produced in fields devoted to wheat agriculture. The energy surplus over natural pastures was accounted for by the input of human and of animal labour from which the hourly energy productivity of that labour in kilocalories per hour could be estimated. Of course, any apparent precision in such calculations hinges on specific, more or less restrictive, assumptions regarding not only crop yields and their energy content but also the quantity of (direct and indirect) energy input, including the energy equivalents of the human and the nonhuman labour applied per hectare of land (on which more presently). Podolinsky's assumptions and calculations are shown in Table 1, which is an extended version of the helpful reconstruction in Martinez-Alier.[41]

The essential idea here was the notion that human labour had increased the throughput in energy terms over what would be found in forests or in natural pastures. This (in modern terms) energy subsidy could be expressed in amounts that were multiples of the inputs of human and of animal labour and thereby translated into figures on energetic labour productivity.

Problems with the Quantitative Energy Accumulation Approach

Even in the relatively simple context of nineteenth-century hay and wheat production, on which Podolinsky relied for his examples, a proper calculation of energy throughput and energy productivity of labour was far more complex than he indicated.[42] In comparing the caloric content of agricultural output

41 Martinez-Alier 1987, p. 48.

42 Many of the problems we raise here are recognised by Martinez-Alier, but he gives them a different emphasis by stressing the common elements between Podolinsky's energy cal-

TABLE 1 *Podolinsky's calculation of the energy productivity of (animal and human) labour (per hectare, based on data for 1870s France)*

Sector	Product (kg)	Energy-product (kcal)[a]	Energy input over natural pastures (kcal)	Energy input (kcal)[b]	Hourly energy productivity of labour (kcal/hour)
Natural pastures	2,500 (hay)	6,375,000	–	none	–
Sown pastures	3,100 (hay)	7,905,000	1,530,000	37,450[c]	40.85
Wheat cultivation	800 (wheat) 2,000 (straw)	8,100,000	1,725,000	77,500[d]	22.26

[a] Assuming 2,550 kcal/kg of hay and straw, and 3,750 kcal/kg of wheat.
[b] Assuming 645 kcal per hour of horse labour and 65 kcal per hour of human labour.
[c] Assuming 50 hours of horse labour and 80 hours of human labour per hectare.
[d] Assuming 100 hours of horse labour and 200 hours of human labour per hectare.

per hectare with that of uncultivated ('natural') pastures, Podolinsky implicitly presumed that the latter had not been reduced by various forms of human extractive labour such as hunting and gathering as well as forestry (not to mention disruptions from agricultural and industrial pollution). Given the negative impacts of expanding human production and population on non-domesticated plant and animal species, the treatment of uncultivated lands as simply 'natural' (in other words, exogenous) undoubtedly results in a sizeable overestimate of the relative caloric content of cultivated harvests.

As shown in Table 1, Podolinsky did not subtract from output or include in input the energy associated with fertilisers, including manure and guano. Podolinsky's failure to include fertilisers in his estimates was quite extraordinary in an 1880 context, given the nature of the agricultural crisis that had swept Europe and North America in the mid-nineteenth century, which resulted in the raiding of the battlefields and catacombs of Europe for bones to fertilise the agricultural lands, the importation of guano and nitrates from Peru and Chile, and the beginnings of an industry for the production of fertilisers. Such issues had occupied as central a figure in the chemistry and agriculture of

culations and more recent agricultural energy-flow analyses, especially their recognition of the role of human activity in imparting an energy subsidy to agricultural production (see Martinez-Alier 1987, pp. 48–50).

his time as Liebig, and had been commented on by Marx in *Capital*, which Podolinsky had presumably read.[43]

Nor is it easy to see how Podolinsky could have left coal out of his estimates. He was, of course, aware of the role of coal in both industrial and agricultural production. As we have seen, he emphasised it as one of the main forms in which plants contribute to the accumulation of useful energy on or in the earth. In addition, one of his early socialist propaganda writings, published in 1875, was *The Steam Engine* – a utopian novelette about a rural worker who is severely injured by a threshing machine while working in the fields and who dreams of a socialist future.[44] Nonetheless, Podolinsky did not include coal and other non-labour inputs in the denominators of his energy-productivity measures.[45] Energy input was considered equivalent to work done and was not measured in terms of the total caloric consumption of humans. Energy expended through the human metabolism was therefore not included on the input side of his energy calculations. Also not included (as biologists would note) was the energy expended in the respiration of plants. Solar energy itself was not calculated as an input. (Some of these inputs were characterised by Podolinsky as 'free gifts of nature'). Nor did Podolinsky consider the fact that not all energy inputs and outputs in agriculture (least of all in forests and in natural pastures) can be measured simply in terms of the energy embodied in the desired product, because natural systems, even when simplified by humans, are more complex than that. Taken together, these points raise serious obstacles to the kind of energy calculations that Podolinsky advanced. And these obstacles are even more imposing when more complex and indirect forms of production are considered.

Such energy flow calculations are, however, meaningful up to a point in that they can reveal some of the material preconditions of production, highlighting its environmental dependence. They can thus help reveal the concrete ways in which the first and the second laws of thermodynamics (conservation and dissipation of matter-energy) impose limits on human production. Energy accounting is thus likely to play an important role (together with biochemical-ecological analysis of production and consumption systems) in postcapitalist society – that is, in a socio-economic system dedicated to a sustainable, all-round human development in coevolution with nature, rather than an ecologically unsustainable process of competitive capital accumulation.

43 Foster 2000, pp. 147–63.

44 Serbyn 1982, p. 6; Martínez-Alier 1987, pp. 54–6.

45 See below for further discussion of this point.

That said, any socialism worthy of the name will eschew the kind of energy reductionism arguably built into Podolinsky's analysis. Although energy flows are an important precondition and limiting factor in the production of goods and services by nature and human labour, this production also involves complex physical, chemical, and biological conditions and processes that are hardly reducible to pure energetics. This should be evident in the case considered most closely by Podolinsky, namely, the production of food. As Roy Rappaport states in his classic study of the ecological energetics of Tsembaga gardens in New Guinea, the nutritional use value of harvested plants includes not just 'a supply of calories' but also 'a supply of minerals, vitamins, and proteins', so that 'it should not be assumed that energy-capturing activities are the only necessary subsistence activities'.[46] Even though Podolinsky did not explicitly make the latter kind of assumption, his historical analysis, as we shall see, failed to recognise 'that while the details of energy transactions may illuminate some aspects of ecological and, perhaps, economic relationships, explanations that are restricted to the consideration of energy inputs and outputs will in some cases fail'.[47]

Podolinsky's Analysis as a Basis for Value Theory

Despite the aforementioned problems, it appears, based on the opening paragraphs of his article and his correspondence with Marx, that Podolinsky may have been aiming at an analysis in the language of physics that would provide a firmer basis for the labour theory of value in its Marxian version. This is, at any rate, the way that his energy productivity calculations have been interpreted by many ecological critics of Marx and Engels. Thus, Martinez-Alier, after going over Podolinsky's energy-accounting exercises, specifically says that Podolinsky saw this accounting as giving 'a scientific basis to the labor theory of value, a point that neither Marx nor Engels appreciated'.[48] This criticism of Marx and Engels has become a conventional wisdom used to distance ecological economics from Marxism.[49] It is, however, somewhat of a stretch insofar as Podolinsky himself undertook no value analysis whatsoever in 'Socialism and the Unity of Physical Forces'. And it becomes even more questionable when

46 Rappaport 1984, p. 63.

47 Ibid; cf. Pimentel and Pimentel 1996, pp. 75–6.

48 Martinez-Alier 1987, p. 49.

49 See, for example, Bramwell 1989, pp. 85–6; Hornburg 1998, p. 129; Kaufman 1987, p. 91; O'Connor 1998, p. 3.

one considers the difficulty of trying directly to translate energy inputs and outputs into economic (labour) values as conceived by Marx.

For Podolinsky's work to be seen as a basis for a labour theory of value, one has to assume that (a) human labour is reducible to energy inputs that can be directly compared to the output of production as a whole, and that (b) the resulting energy-flow figures can be translated directly into commodity values. But from the standpoint of Marx's theory, these assumptions run aground on the complex relations among use values (resources and produced goods and services that directly or indirectly serve human needs and human development), labour values (the abstract labour times necessary to produce commodities), and exchange values (the prices fetched by commodities in the market). In Marx's approach, abstract labour values must be objectified in vendible use values, and neither these use values nor the human labour and natural conditions that produce them can be reduced to pure energy terms. There is, moreover, no stable, one-to-one relationship between abstract, value-creating labour and concrete labour expended (even assuming that the latter can be reduced to pure energy) either intertemporally or across different firms and industries.[50] In short, although Podolinsky's attempt to measure the energy productivity of labour was revealing in many respects and was an important contribution, it was a far cry from anything that could potentially constitute an energetic basis for Marx's labour theory of value. Indeed, the attempt to directly translate energy productivities of human labour into economic value categories is extremely problematic in all respects and belongs to a long history of energy reductionism that has been opposed by some of the major figures in ecological economics.[51]

Stated differently, to interpret Podolinsky's work as a scientific foundation for a labour theory of value is to replace Marx's dialectical material-social analysis of value and use value with a crude materialist (specifically energy reductionist) approach. This point merits a brief historical digression.

Value and Nature: Marx and Sieber versus Podolinsky

Ironically, one of the most pertinent commentaries on the difficulty of finding a scientific proof or basis of the labour theory of value in the physical world (e.g. in productive energy flows) can be gleaned from the work of Nikolai Sieber,

50 Saad-Filho 2002, ch. 5; Burkett 2014. pp. 108–12.

51 See Mirowski 1988, pp. 822–5; Burkett 2003, pp. 139–41, 151–3.

who was both Podolinsky's economic mentor and Marx's most formidable early economic follower. Sieber was exceptional in that, in contrast to many Marxist economists even up to the present, he understood the distinction between the quantitative value problem and the qualitative value problem that was at the heart of Marx's divergences from Ricardo.[52] In the early 1870s, Sieber began to publish a series of articles in the journal *Znanie* [*Knowledge*]. In the first of these, Sieber replied to a German review of Marx's *Capital* by Karl Rössler, who had rhetorically asked why 'the food in the stomach of a worker should be the source of surplus value, whereas the food eaten by a horse or an ox should not'. Sieber had replied, quite inadequately from Marx's standpoint, that Marx's *Capital* was concerned with human society and not domestic animals and thus was directed only at the surplus value created by human beings. Marx commented in his published notes that

> the answer, which Sieber does not find, is that because in the one case the food produces human labour power (people), and in the other – not. The value of things is nothing other than the relation in which people are to each other, one which they have as the expression of expended human labour power. Mr. Rössler obviously thinks: if a horse works longer than is necessary for the production of its (labour power) horse power, then it creates value just as a worker who worked 12 hours instead of 6 hours. The same could be said of any machine.[53]

If Sieber did not grasp the essential point at first, he did subsequently. In 1877, Yu. G. Zhukovskii, a follower of Ricardo, criticised Marx for arguing that only human labour created surplus value. Zhukovskii argued, as explained by James D. White, that 'anything which bore fruit, be it a tree, livestock or the earth, all were capable of providing exchange value. For Zhukovskii one of the main sources of value was Nature'.[54] In response, Sieber said that a good Ricardian ought to be able to grasp that human labour was the sole source of exchange value, which reflected the division of labour and the fragmentation of society. In the following year, the classical liberal political economist Boris Chicherin presented essentially the same argument as did Zhukovskii.[55] Here, Sieber's response was unequivocal, cutting into the commodity fetishism basic to the classical liberal view:

---

52 See Sieber 2001, p. 41.

53 Marx as quoted in White 2001, p. 6.

54 White 2001, pp. 6–7.

55 Chicherin 1998, p. 325.

> But to people it *appears* as though things exchange themselves one for another, that things themselves have exchange value, etc. and that the labour embodied in the thing given is reflected in the thing received. Here lies the whole groundlessness of the refutations of Mr. Chicherin, and before him of Mr. Zhukovskii, that neither the one nor the other could understand ... that Marx presents to the reader the whole doctrine of value and its forms ... as the peculiar way people at a given stage of social development necessarily understand their mutual relations based on the social division of labour. In fact, every exchange value, every reflection or expression of it, etc. represents nothing but a myth, while what exists is only socially-divided labour, which by force of the unity of human nature, seeks for itself unification and finds it in the strange and monstrous form of commodities and money.[56]

Podolinsky was most likely aware of these debates on Marxian economics in Russia and in the Ukraine and of the development of Sieber's position. Yet those who see Podolinsky's energy-accounting exercises as a scientific basis for a labour theory of value saddle him with precisely the kind of crude materialist position that Sieber came to condemn. As shown in Table 1, Podolinsky's energy-productivity calculations lump animal labour in with human labour, which, if translated directly into value terms, implies that domestic animals harnessed by human beings are, like human labour, productive of labour values. Thus, to interpret these calculations as energetic labour-value calculations is to contradict Podolinsky's otherwise strong adherence to the notion that all value derives from human labour. From this, it appears that Podolinsky's champions are extrapolating his analysis in ways that he himself might reject.

Near the beginning of his article, Podolinsky does seem to identify surplus value with physical energy, and in a letter to Marx of 8 April 1880, he did describe his work as an 'attempt to bring surplus labour and the current physical theories into harmony'.[57] Such statements suggest that Podolinsky's viewpoint on value was closer to Ricardo's purely quantitative approach than to Marx's quantitative and qualitative treatment, for they seem to imply that value is a thing embodied in a product that has behind it another thing – a physical reality in the form of physical force or energy (e.g. muscular labour), all of which can be quantified. It is, at any rate, easy to see how Marx and Engels's critics might interpret Podolinsky's statements in this way. Those holding such a view

56 Sieber as quoted in White 2001, p. 8.

57 Quoted in Martinez-Alier 1987, p. 62.

might genuinely think of themselves as addressing a perceived conundrum of Marx's theory – the physical basis of value.

However, in Marx's analysis, value, or abstract labour time, is not a natural-physical substance, but rather an alienated material-social relation behind which lies society's reproductive division of labour enmeshed with nature. For Marx, the reduction of value to abstract labour time is not because of a normative and/or empirical presumption that labour in general (let alone muscular labour in particular) is more important or primary than other production inputs. Rather, it is rooted in capitalism's social separation of the workers from necessary conditions of production (starting with the land) and their recombination only in capitalist enterprises that purchase, produce, and sell commodities for a profit. The separation of value (as abstract labour) from production's material conditions reflects workers' class-alienation from the same conditions (i.e. the conversion of labour power into a buyable commodity). Marx always insisted that as far as the production of wealth or use value is concerned, nature is just as primary as labour and that labour is itself a natural force to which supernatural productive powers should never be ascribed.[58]

Podolinsky, as distinct from his economic mentor, Sieber, does not appear to have grasped such an intricate view of economic value. Still, the extent to which Podolinsky really thought economic values could be explained in pure energetic terms remains an open issue. The point here is that to treat his work as a potential scientific basis for a labour theory of value is implicitly to endorse energy reductionism in the realm of economics. The tremendous gulf between such energy reductionism and Marx's dialectical and material-social approach to capitalist value relations is completely fogged over when Marx and Engels are condemned for not adopting Podolinsky's purported insights.

Podolinsky's Perfect Machine Argument

The main thrust of Podolinsky's article was directed not at value theory, but rather at the notion of the human being as constituting the perfect machine. More specifically, it drew from Sadi Carnot's thermodynamic notion of the perfect machine and the way that this notion had later been developed by William Thomson and by others. For Podolinsky, this thermodynamic perspective, when applied to human beings, demonstrated that although machines as

58 Marx 1976a, pp. 133–4, 647–51.

such could never meet the criteria of the perfect machine, human beings could. Here, it is important to quote Podolinsky at length:

> According to Sadi Carnot, in order to be able to judge the degree of perfection of a thermal machine, one needs to know not only its economic coefficient, *but also its capacity to recycle the heat spent at work*. A machine having the capacity to reheat itself, making the heat spent at work rise toward its fire-box, would be a *perfect machine*, and only such a machine could provide a true conception of the transformation of heat and *vice-versa*. Now, no machine constructed by the hands of men possesses this faculty. No machine heats its own fire-box with its own work alone, and no machine works on a *reverse cycle*, that is, the transformation of work into heat is unknown. As a consequence, the true laws of these transformations cannot be found with the aid of inanimate machines. The plant world, producing almost no effective mechanical motion, also cannot even be remotely considered as an example of a *perfect thermal machine*.
>
> But, observing the work of humans, we see in front of us just exactly what Sadi Carnot calls a *perfect machine*. From this perspective, humanity would be a machine that would not only transform heat and other physical forces in work, but that would also produce the *complete reverse cycle*, which converts its work into heat and other forces essential for the satisfaction of its needs, that is to say it would recycle to its fire-box the heat produced by its own labor. In reality, a steam engine, even admitting that it will run an entire year without the intervention of muscular human labor, could never produce all the elements necessary to sustain its work in the following year. The human machine, by contrast, will have created a new crop, will have raised young domestic animals, will have constructed new machines, and will still be able to continue with success its new work in the following year. The reason is evident: the human machine is a perfect machine, whereas an inanimate machine never achieves the conditions of perfection that Sadi Carnot requires.[59]

Podolinsky clearly believed that he had discovered an important principle in his notion of human beings as perfect machines. Human beings, he argued, were unique in this respect. Plants, although they carried out photosynthesis,

59 Podolinsky 2004 [1881], pp. 69–70. See below: Appendix 1, p. 256.

were not machines because they lacked the mechanical motion necessary for work. Animals could achieve some human-like energy accumulation only on a limited instinctual basis or if domesticated by humans. Machines themselves were dependent on human muscular labour to keep them functioning. Only the human machine was adaptable enough to carry out many different kinds of work while reheating its own firebox and thus was able to carry out the reverse cycle required by Sadi Carnot's and by William Thomson's perfect thermodynamic machine.

A key part of this argument focused on what Podolinsky referred to as the economic coefficient of human beings: the quantity of work they perform compared to the energy they consume. He argued that if one compares the quantity of oxygen inhaled during work (or the amount of food energy burned off, assuming this to be proportional to the oxygen processed by the human body) to the quantity of physical work supplied by the muscles, there is a ratio of 5:1. In this case, the economic coefficient of the human machine is 1:5, or one unit of work performed for every five units of energy consumed. Human needs in civilised society are more complex, however, so the economic coefficient should be considered nearer to 1:10 than 1:5, reflecting the growth of per capita consumption.

> This means that the satisfaction of all of our needs, presently considered as indispensable, represents a quantity of work almost ten times greater than the human muscular labor. This surplus must be accounted for by the greater productivity of human muscular labor, guided by intelligence, by the muscular power of domestic animals, or finally, by inanimate forces both natural and artificial.[60]

For Podolinsky, primitive man, who relies almost exclusively on the free gifts of nature, has a higher economic coefficient than does civilised man. The former has an economic coefficient of 1:6 mainly because of less developed needs. Nevertheless, the latter has a higher productivity and is able to accumulate solar energy on Earth in quantities that surpass 'ten times the force of his muscles'. Thus, despite his reliance on greater per capita energy throughput, civilised man is a more perfect machine.

It is the economic coefficient in relation to the accumulation of energy, according to Podolinsky, that sets the limits to human survival:

60 Ibid.

> As long as muscular labor supplied by the human machine is converted into an accumulation of energy necessary for the satisfaction of human needs, which represents a quantity in excess of the sum of the muscular work of the human machine, by as many times as the denominator of the economic coefficient exceeds the numerator – the existence and the possibility of the labor of the human machine are guaranteed.[61]

Shortcomings of the Perfect Machine Perspective

The most problematic part of Podolinsky's analysis is his central point – the claim that the human labourer constitutes the perfect thermodynamic machine in that it is able to carry out the complete reverse cycle, in effect, reheating its own firebox. In the formulations of Sadi Carnot and of William Thomson, the perfect machine is an ideal benchmark for measurement of actual machine efficiency. As Thomson put it, 'a perfect thermodynamic engine is such that, whatever amount of mechanical effect it can derive from a certain thermal agency; if an equal amount be spent in working it backwards, an equal reverse thermal effect will be produced'.[62] In the case of a steam engine, this would mean that the work w produced by the falling of heat from boiler to condenser could be used to raise the heat back up by an equivalent amount so that there would be no net effect.[63] A more than perfect engine, Carnot theorised, would produce work W with a surplus w – producing more than a net effect if the engine were reversed, constituting a perpetual motion machine. But this, as Sadi Carnot argued, would violate the laws of physics.[64]

Yet at times Podolinsky seems to attribute even this *supra* level of perfection to human muscular labour. For example, he suggests, as already quoted, that

> the human machine, by contrast [to a steam engine so efficient that it runs an entire year without the intervention of human muscle power], will have [in the same year] created a new crop, will have raised young

61 Ibid.

62 As quoted in C. Smith 1998, p. 93.

63 Carnot's ideal engine (known as the Carnot engine), if run backward, would consume 'as much motive power as it produced running forward … Joined together but operating in opposite directions two engines [combined into one larger engine] would, therefore, produce no net effect' (Challey 1971, p. 81).

64 C. Smith 1998, pp. 91–3.

> domestic animals, will have constructed new machines, and will still be able to continue with success its new work in the following year.[65]

The extreme difficulty that Podolinsky runs into here stems from his insufficient recognition that the analysis of the steam engine carried out by physicist-engineers like Carnot, Clausius, and Thomson is constructed in terms of a closed system and an ideal, frictionless engine.[66] In contrast, the human economy (like life itself), despite the emphasis of economists on the circular flow, is not a closed system but one that continually draws on its external environment so as to accumulate energy (or low entropy) within its own (open) system while simultaneously dissipating energy and material waste back into its environment. Indeed, the capitalist economy is arguably the most extreme example possible of a system that draws on a resource tap (at ever increasing rates) and dissipates waste into the environmental sink (also at ever increasing rates), in ways that accelerate entropic degradation. The 'human machine' cannot be analysed apart from this open system.

The chief point that Podolinsky underscores – though he is not able to develop this – is that life to some extent goes against entropy (or feeds on low entropy). Here we can turn to the classic study, *What is Life?*, written in 1944 by the great Nobel prize-winning physicist and pioneer in quantum theory, Erwin Schrödinger, who wrote the following:

> How does the living organism avoid decay? The obvious answer is: by eating, drinking, breathing and (in the case of plants) assimilating. The technical term is *metabolism*. The Greek word μεταβάλλειν means change or

65 Podolinsky 2004 [1881], p. 70, Appendix 1 below, p. 256. The notion of the human body as a 'more perfect' machine than the steam engine had been previously expounded by others. For example, nineteenth-century physicist and pioneer of thermodynamics Peter Guthrie Tait quotes James Prescott Joule (one of the discoverers of the first law of thermodynamics) as having stated that 'the animal frame, though destined to fulfill so many other ends, is, as a machine, more perfect than the best contrived steam-engine; that is, capable of more work with the same expenditure of fuel' (Tait 1864, p. 344; compare Martínez-Alier 1987, p. 51). What Podolinsky offered, however, was a much more extreme interpretation of human labour power as a 'perfect machine' in Carnot's strict sense – indeed exceeding Carnot's own notion of what was thermodynamically possible.

66 'Clausius was no more concerned than Carnot with the losses whereby all real engines have an efficiency lower than the ideal value predicted by the theory. His description, like that of Carnot, corresponds to an idealization. It leads to the definition of the limit nature imposes on the yield of thermal engines' (Prigogine and Stengers 1984, p. 114).

> exchange. Exchange of what? Originally the underlying idea is, no doubt, exchange of material. (E.g. the German for metabolism is *Stoffwechsel*). That the exchange of material should be the essential thing is absurd ... What then is that precious something contained in our food which keeps us from death? That is easily answered. Every process, event, happening – call it what you will; in a word, everything that is going on in Nature means an increase of the entropy of the part of the world where it is going on. Thus a living organism continually increases its entropy – or, as you might say, produces positive entropy – and thus tends to approach the dangerous state of maximum entropy, which is death. It can only keep aloof from it, i.e. alive, by continually drawing from its environment negative entropy ... Or, to put it less paradoxically, the essential thing in metabolism is that the organism succeeds in freeing itself from all the entropy it cannot help producing while alive ... Thus the device by which an organism maintains itself stationary at a fairly high level of orderliness (= fairly low level of entropy) really consists in continually sucking orderliness from its environment.[67]

It is this that gives the appearance that the human machine is the perfect (or more than perfect) thermodynamic machine, which can endlessly carry out a reverse cycle and reheat its own firebox. But this appearance is only sustained insofar as the human socio-metabolic system is not a closed, isolated system to which the entropy law then directly applies, but an open, dissipative system. It continually feeds on its environment and is able to defy (or, more precisely, to give the impression of defying) the entropy law in this way. Nevertheless, human beings exist within a limited biosphere. An open, dissipative system that feeds on its environment on an exponentially rising scale through the commodification of production – demanding a continual increase in the use of energy and of materials and dumping ever more wastes into the environment – is a trait that is carried to its zenith under the profit-driven, generalised commodity economy of capitalism. Such an economy must deplete and despoil the natural conditions of human development ('simultaneously undermining the original sources of all wealth – the soil and the worker', as Marx puts it), especially when the scale of its biogeochemical effects begins to rival that of the biosphere itself.[68] Increasingly, as the system's extreme exploitation of the global environment violates the limits to natural wealth of any given qualities,

67 Schrödinger 1944, pp. 71–2, 75.

68 Marx 1976a, p. 638.

it produces a closing circle on human developmental possibilities within the framework of capitalist relations.[69]

By failing to see human life and human systems as metabolic in nature, involving an exchange with the environment (whether conceived in terms of materials and of energy or in terms of low entropy), Podolinsky gets trapped in a mechanistic and reductionist view that is unable to capture the full material and social complexity of the human relation to nature. The notion of the human being as the perfect thermal machine tends to underestimate both the real dependence of human beings on nature and the full vulnerability of the natural world to human action (i.e. the reality of coevolution). The object simply becomes the accumulation of energy stocks and flows for the benefit of the human economy while downplaying the fact that human beings produce only in conjunction with nature. The difference between Podolinsky and Marx in this respect (a subject we will take up in greater detail below) could not be greater. As ecological economist Kenneth Stokes has argued, Marx and Engels's 'model explicitly embodied the open-systems notion of the metabolic interaction of man and nature; the notion that the economic process is embedded in the Biosphere'.[70]

Indeed, the great danger in exaggerating the role of human labour in the production of wealth, and downplaying nature's contribution, had already been stressed by Marx in 1875. As the British political economist and food activist Susan George explains,

> One day in May 1875, Karl Marx received a political platform intended to reconcile two antagonistic factions of the German Workers' Party at the upcoming Party Congress. Exasperated, he dashed off the marginal notes which came to be known as the *Critique of the Gotha Programme* – rather a grand title for a quick, irritated 'will they never get it through their heads?' sort of reaction.
>
> The first sentence of the offending document declared that 'Labour is the source of all wealth and all culture …' Marx shot back witheringly, 'Labor is *not the source* of all wealth. *Nature* is just as much the source of use values … as labour, which itself is only the manifestation of a force of nature'.[71]

69 Commoner 1971.

70 Stokes 1994, p. 64.

71 George 1998, p. ix; cf. Marx 1966, p. 3.

Nothing could be more incongruous for Marx than the reduction of human beings to the status of machines – based on a thermodynamics of closed systems that ignored the larger part of nature's contribution to production. Nor could a theory of value (which had to encompass use values and their relation to the natural conditions of production) for Marx reduce value to a product of muscular labour. Yet Podolinsky's most explicit attempt to provide a 'definition of the value of production' saw it as dependent on muscular work.

As Podolinsky indicated in his June 1880 letter to Lavrov (see note 80 below), he was most uncertain about the last part of his article in which he tried to relate his theory of the perfect thermodynamic machine (at least potentially) to different land uses under different modes of production. 'Primitive man', he argued in that part of his article, is simply a hunter and gatherer who does nothing to prevent the dispersion of solar energy into interstellar space. Slavery requires standing armies and continuous wars and hence is inefficient in the accumulation of energy. The feudal serf's energy productivity is hindered by the obligatory work on the lord's estate. Capitalism's economic crises periodically throw thousands of workers onto the streets. Only socialism, then, can allow human beings to realise their potential as perfect thermal machines able to accumulate energy on the earth.

But such arguments only showed how limited Podolinsky's analysis was in terms of human-nature interactions and how incapable the energy-accumulation model was of capturing the complexities of human land usage and their socioeconomic-ecological characteristics. Socialism became merely the universalisation of a system of efficient muscular labour for the benefit of all. The inherent limitations of Podolinsky's analytical framework allowed him to go no further.

Marx's Notes on Podolinsky

It is crucial to understand from the outset that the ecological criticisms of Marx and Engels's responses (or lack of responses) to Podolinsky are based to a very large extent on misreadings of the available evidence and on claims that cannot be factually supported or logically justified. For example, Martinez-Alier, who has consistently faulted Marx and Engels for their alleged failure to address Podolinsky's work, has stated that 'Marx died in 1883 and apparently never commented on Podolinsky's work beyond a letter of acknowledgment in the first days of April 1880'.[72] This argument, however, relies simply on the extant

72 Martinez-Alier 2003, p. 10.

evidence and assumes that because we do not have certain documents (e.g. letters from Marx to Podolinsky), that they must either have said very little (in the one case that we know about) or 'apparently' never existed (in all further cases). We know that in his letter to Podolinsky in early April 1880, Marx not only acknowledged receipt of the draft manuscript 'Human Labor and the Conservation of Energy' but also conveyed something more, because Podolinsky indicated in a reply that the letter had given him 'deep joy'.[73] Yet not even this letter of Marx's to Podolinsky that we definitely know about has survived. In short, to argue based on the paucity of extant evidence that Marx 'apparently never commented on Podolinsky's work' beyond a bare acknowledgment is logically impermissible.

We now know that Marx took extensive extracts from a very early version of Podolinsky's manuscript carrying the title 'Le Travail humain et la Conservation de l'Energie' ('Human Labor and the Conservation of Energy'), most likely in early April, 1880.[74] The extracts were around 1,800 words long and focused primarily on Podolinsky's thermodynamic argument. Marx's notes are scheduled for publication in MEGA Volume IV/27.[75] Given the existence of these notes, it is quite plausible that Marx wrote back or otherwise passed on comments to Podolinsky. (Podolinsky's economic mentor, Nikolai Sieber, was a guest at Marx's house on a number of occasions in the years 1880 and 1881). Perhaps Marx sent back the manuscript itself (as was customary in

---

73 Quoted in Martinez-Alier 1987, p. 62.

74 This is the title given on Marx's notes themselves and is also the title provided by the editors of the journal *Russian Contemporaries* in a footnote they attached to Podolinsky's 24 March 1880 letter to Lavrov (Sapir 1974, p. 67). Translated into English by Mikhail Balaev (from the Department of Sociology at the University of Oregon in Fall 2003), this letter reads as follows (copied from the journal *Russian Contemporaries*):

Montpellier, 24 March 1880

I sent to you, Petr Lavrovich, my work 'about labor' ['Le travail humain et la conservation de l'energie' (footnote by the editors of the journal *Russian Contemporaries*)] that I have just received. Please, be so kind as to send the address of Marx to me: I want to send it to him as well as the thing is directly related to him and was inspired in my mind by the theory of added labor ...

I shake your hand. S. Podolinskii.

75 We are grateful to Kevin B. Anderson, David Norman Smith, Norair Ter-Akopian, Georgi Bagaturia, and Jürgen Rohan, the editors of the MEGA volume in which Marx's notes on Podolinsky will appear, for allowing us access to these notes for our research. Although contractual and copyright issues prevent us from directly quoting Marx's notes, it is still possible to give a fairly clear idea of what they tell us about both Marx's engagement with Podolinsky's analysis and the respective versions read by Marx and Engels.

those days before copy machines) with marginal notes or accompanying comments. Unfortunately, the original manuscript that Marx read has not yet been found.

That Marx took the time to compile such detailed extracts – with various passages emphasised, indicating an active and engaged reading – certainly runs counter to the notion that he and Engels turned a deaf ear to Podolinsky's work. Nor can it be reasonably argued that either Marx or Engels neglected thermodynamics (much less natural science in general). We know that Marx and Engels both filled multiple notebooks with extracts from, and commentaries on, the leading natural science writers of their time. We also know that these notebooks covered a wide range of scientific fields – physics, chemistry, biology, physiology, geology, and agronomy – in each of which the analysis of energy dynamics occupied an important if not central position.[76] In fact, Marx and Engels were familiar with and in some cases closely studied the works of many of the scientists involved in the development of thermodynamics (both the first and second laws) – including Hermann von Helmholtz, Julius Robert Mayer, James Prescott Joule, Justus von Liebig, Jean-Baptiste Joseph Fourier, Sadi Carnot, Rudolf Clausius, William Thomson, Peter Guthrie Tait, William Grove, James Clark Maxwell, and Ludwig Eduard Boltzmann. In addition, Marx attended numerous public lectures on natural science in the years leading up to and following the publication of *Capital*, Volume I, in 1867, and among these was a series of lectures by the English physicist John Tyndall, author of *Heat Considered as a Mode of Motion*.[77] Marx studied William Robert Grove's *Correlation of Physical Forces* (1846), John Tyndall's *Heat* (1870), and Adolf Fick's work on the forces of nature and on their interactions (*Die Naturkraefte in ihrer Wechselbeziehung*, 1869). Tyndall, a major figure in the developing physics in his own right, was the principal advocate of the ideas of J.R. Mayer – one of the co-discoverers of the conservation of energy (the first law of thermodynamics). Marx followed Tyndall's research on the sun's rays, particularly as it related to heat. Marx and Engels were also close students of the development of knowledge about electricity (which Marx saw as replacing steam as a motive force), including the work of Michael Faraday who invented the first electric motor. The fact that the second law of thermodynamics and the concept of entropy itself are not explicitly addressed in Marx's economic writings is not evidence of a neglect of natural science. Clausius did not introduce the actual term *entropy* in his attempt to bring the first and second laws of thermodynam-

76 Baksi 2001.

77 Tyndall 1863.

ics into harmony until 1865. Tait's *Sketch of Thermodynamics*, the first widely read text referring to the new science in its title (and emphasising both the first and second laws), did not appear until 1868 – after Marx's *Capital* had been published.[78] Even in 1882, near the end of Marx's life, we find him closely following the research of the French physicist Marcel Deprez, which was directed at the distant transmission of electricity. In the same year, Marx also read Édouard Hospitalier's *Principal Applications of Electricity*, on which he took extensive notes.[79]

The main charge commonly directed at Marx in this area is that he failed to exploit Podolinsky's insights into the natural-scientific basis for the labour theory of value that was provided by the new energetics. This criticism, as we have seen, is based on the notion that Podolinsky had developed the key elements of such an analysis, which he made available to Marx, and that Marx could therefore have incorporated these elements into his system. As discussed above, however, Podolinsky's *La Revue Socialiste* and *La Plebe* articles did not develop even the rudiments of an analysis capable of relating energy flows to capitalist value relations as conceived by Marx, and arguments to the contrary are based on a fundamental misunderstanding of the nature and the purposes of Marx's value analysis.

Moreover, even the underlying assumption that Marx had access to the portions of Podolinsky's analysis that have been considered by some to be relevant for value theory proves to be highly questionable. Marx's notes on Podolinsky's 'Human Labor and the Conservation of Energy' are missing almost all the material that in later versions of Podolinsky's argument conceivably relates to this subject: mention of Marx's analysis, empirical estimates on the energy input of human labour in agriculture and of its effect on the output of agriculture, reference to labour values, and the discussion of land use in alternative modes of production (including socialism). If Marx's notes are an accurate reflection of Podolinsky's argument (and it is doubtful that Marx would have failed to note any of these things if presented in the manuscript), then Podolinsky's analysis in 'Human Labor and the Conservation of Energy' was, as the title suggests, a straightforward treatment of thermodynamics along with its abstract application to human beings as the embodiment of Carnot's (and Thomson's) perfect machine. The other parts of the argument were therefore most likely added to the original manuscript on 'Human Labor and the Con-

78 C. Smith 1998, pp. 255–6; Prigogine and Stengers 1984, p. 117.

79 On these and other aspects of Marx and Engels's natural-scientific studies, see Baksi 1996, pp. 261–96; Baksi 2001, pp. 377–90.

servation of Energy' between April and the completion of the final draft of 'Socialism and the Unity of Physical Forces' for *La Revue Socialiste* in May, possibly even as an attempt to respond to comments made by Marx. In a letter to Lavrov on 4 June 1880, Podolinsky said that his work about labour would be published that month 'with some, unfortunately very short, attachments about socialism' (suggesting that the final parts of the manuscript on forms of property and human energy were in fact written and attached later).[80]

Podolinsky may have believed from the start, as he wrote to Marx on 8 April 1880, that his work constituted an 'attempt to bring surplus labor and the current physical theories into harmony'.[81] Nevertheless, the textual evidence from Marx's extracts strongly suggests that 'Human Labor and the Conservation of Energy' was even less explicit in its argument in this respect than what was published under the title 'Socialism and the Unity of Physical Forces' in *La Revue Socialiste* three months later. Indeed, in terms of a contribution to bringing energetics into harmony with the labour theory of value, it appears likely that all Marx saw was Podolinsky's thesis that human beings were perfect machines able to feed their own fireboxes with their muscular labour – which, if it made any sense whatsoever, would only constitute an argument as to why labour was the source of all value (and one that was hardly likely to impress Marx).

80 Podolinsky's 4 June 1880 letter to Lavrov (Sapir 1974, p. 68), translated into English by Mikhail Balaev (from the Department of Sociology at the University of Oregon in Fall 2003), reads as follows:

Montpellier, 4, June 1880 Quartier Mont Maur

Dear Petr Lavrovich!

Please be so kind as, if possible, to send me a copy of 'Intricate Mechanics' [a brochure by Varzar] in Russian. Malone is asking me for it in French, but I have only my Ukrainian version that I consider myself as having no rights to translate in the author's name. I also could not get a Russian original from Geneva.

On 20 June my work about labour ['Le Socialisme et l'Unite des Forces physiques'] will be published in *Revue Socialiste* with some, unfortunately very short, attachments about socialism. Besides that, 'Steam Engine' with attachments and 'Wealth and Poverty' will be published as fiction.

Will you be in Switzerland this summer? We are planning to go there for some three months.

Shaking your hand firmly, S. Pod.

81 As quoted in Martinez-Alier 1987, p. 62.

Engels's Comments on Podolinsky

In December 1882, a year or so after Podolinsky's Italian article appeared, Engels wrote two letters to Marx on Podolinsky's analysis. The core of Engels's comments resides in his reactions to (a) the translation of human work into energy computations, and (b) the translation of such computations into economic values. And it is these reactions that have been the focus of all the charges regarding Marx and Engels's failure to build on Podolinsky's ideas. To develop a further understanding of the Podolinsky-Marx-Engels relationship, it is therefore necessary to examine Engels's letters.

From late October 1882 until early January 1883, Marx was on the Isle of Wight, where he had gone to try to regain his health, while Engels remained in London. Engels had been working on his essay, 'The Mark', which was to be published as an appendix to the German edition of *Socialism: Utopian and Scientific*.[82] 'The Mark' was about the downfall of peasant communal rights to the land and hence about the conditions of primitive accumulation in Germany. It ended with a consideration of ecological-economic factors:

1. the inability of the peasant to raise cattle without rights to common lands;
2. the obstacles to peasants continuing to farm their small plots of land without the manure provided by the cattle;
3. the growth of landed property on a large scale;
4. the threat to European agriculture then posed by United States agriculture, with its production and export of grain on a gigantic scale; and
5. the gross impoverishment of the remaining German peasants resulting from factors 1 through 4.

On 15 December 1882, Engels sent his draft of 'The Mark' to Marx asking him to return it in a few days. Three days later, Marx sent a note (the second to last item of correspondence from Marx to Engels contained in the *Collected Works*) back to Engels saying that 'The Mark' was 'very good' and that he was returning the manuscript. On 19 December, Engels wrote that he had received Marx's note and then the manuscript. In the second paragraph of his letter, Engels then launched immediately into the 'Podolinsky business'.[83] From this, we conclude that it is likely (but by no means certain) that Marx wrote a marginal note on Podolinsky and on the question of labour efficiency in agriculture on Engels's

82 Engels 1978a, pp. 77–93.

83 Marx and Engels 1975a, Vol. 46, p. 410.

copy of 'The Mark' or sent a note to this effect accompanying it. It was, in any case, immediately after referring to Marx's response to 'The Mark' that Engels took up the Podolinsky question.

Engels wrote on 19 December that although he did not have Podolinsky's article at hand, he had read the version published in *La Plebe*.[84] He then proceeded to a critique of its contents. For Engels, Podolinsky's 'real discovery' was 'that human labor is capable of retaining solar energy on the earth's surface and harnessing it for a longer period than would otherwise have been the case'.[85] But this thermodynamic insight does not, Engels argued, translate directly into economics. 'All the economic conclusions' that Podolinsky drew from this insight 'are wrong'.[86] Engels then proceeded to a succinct theoretical discussion of the accounting of human energy and of its relation to work. Engels's discussion was more complete than Podolinsky's in that it accounted directly for (a) the calories human beings consumed, and (b) the fact that the economic labour performed in work in no way corresponds to the reproduction of the calories used up by human beings during the time that they are working (here Engels considers such issues as friction, the loss of calories because of increased heat, and so forth, and human excretions).

In Engels's view, the energetic significance of the labour conducted during a day in which a certain number of calories are consumed – he provides for illustration a hypothetical figure of 10,000 calories – 'consists rather in the stabilisation over a longer or shorter period of the *fresh* cal' (calories) that workers absorb

> from the radiation of the sun, and this is the only connection that the latter have, so far as labour is concerned, with the first 10,000 cal. Now whether the *fresh* cal stabilised by the expenditure of 10,000 cal of daily nourishment amount to 5,000, 10,000, 20,000 or a million is dependent solely upon the level of development of the means of production.[87]

That is, the crucial issue is human productive powers and human needs in their entirety (the historically determined level of subsistence) developed at any given stage of production. Engels clearly felt that the daily expenditure of human energy in economic work is not easily related to the daily amount of energy physically consumed because human beings (a) draw on other sources

84 Ibid.
85 Ibid.
86 Ibid.
87 Marx and Engels 1975a, Vol. 46, pp. 410–11.

besides their own production for energy, (b) consume energy in ways that are related to their basic metabolic processes and not simply in relation to the work they do, and (c) have needs and productive capacities that are a product of the historical development of society and production.

Engels was well aware that Podolinsky's energy calculations had not included fertilisers, and he pointed to the fact that their inclusion was not only necessary but made the quantification of energy input much more complicated. Equally important, he insisted, was the role of coal, which showed how human beings draw on their environment to exploit solar energy from the past. This was something that had not entered directly into Podolinsky's calculations despite his knowledge of the role that the steam engine played in agriculture, particularly in threshing.[88] As Engels wrote to Marx:

> What Podolinski has completely forgotten is that the working individual is not only a stabiliser of *present* but also, and to a far greater extent, a squanderer of *past*, solar heat. As to what we have done in the way of squandering our reserves of energy, our coal, ore, forests, etc., you are better informed than I am. From this point of view, hunting and fishing may be seen not as stabilisers of fresh solar heat but as exhausters and even incipient squanderers of the solar energy that has accumulated from the past.[89]

The difficulty of analysing human production as a whole in terms of energy emanating from human labour presented problems so formidable that Engels contended that they were virtually insurmountable. This was particularly the case in industry. As he further explained in the same letter,

> In industry all calculations come to a full stop; for the most part the labour added to a product does not permit of being expressed in terms of cal. This might be done at a pinch in the case of a pound of yarn by laboriously reproducing its durability and tensile strength in yet another mechanical formula, but even then it would smack of quite useless pedantry and, in the case of a piece of grey cloth, let alone one that has been bleached,

88 On the introduction of steam engines for threshing into Russian agriculture, see Hume 1914, pp. 67–75.

89 Marx and Engels 1975a, Vol., 46, p. 411. Martinez-Alier argues that Engels was wrong to state 'that Podolinsky had forgotten about coal' (Martinez-Alier 1987, p. 222; Martinez-Alier and Naredo 1982, p. 217). But Engels's criticism was focused precisely on Podolinsky's energy-accounting exercises, not on what Podolinsky said about coal in a broader context.

> dyed or printed, would actually become absurd. The energy value conforming to the production costs of a hammer, a screw, a sewing needle, is an impossible quantity ... To express economic conditions in terms of physical measures is, in my view, a sheer impossibility.[90]

Engels's argument here is one that he had developed earlier, in a notebook entry written in 1875 (later published as part of his *Dialectics of Nature*), in which he commented on attempts by scientists to explore the relation between human physiology, energy, and work. In a manner closely resembling what had happened with the Darwinian theory (specifically the competitive struggle for existence), Engels contended, the concept of work had been transferred from political economy to natural science and was in the process of being transferred back from natural science to political economy with absurd results. 'Let someone try to convert any *skilled labor*', he wrote,

> into kilogram-metres [after the physiological experiments of Fick and Wislicenus] and then to determine wages on this basis! Physiologically considered, the human body contains organs which in their totality, *from one aspect*, can be regarded as a thermodynamical machine, where heat is supplied and converted into motion. But even if one presupposes constant conditions as regards the other bodily organs, it is questionable whether physiological work done, even lifting, can be at once fully expressed in kilogram-metres, since within the body *internal* work is performed at the same time which does not appear in the result. For the body is not a steam-engine, which only undergoes friction and wear and tear. Physiological work is only possible with continued chemical changes in the body itself, depending also on the process of respiration and the work of the heart. Along with every muscular contraction or relaxation, chemical changes occur in the nerves and the muscles, and these changes cannot be treated as parallel to those of coal in a steam-engine. One can, of course, compare two instances of physiological work that have taken place under otherwise identical conditions, but one cannot measure the physical work of a man according to the work of a steam-engine, etc; their external results, yes, but not the processes themselves without considerable reservations.[91]

90 Ibid.

91 Marx and Engels 1975a, Vol. 25, pp. 586–7.

This rejection of attempts to reduce human work to the thermodynamic logic of a steam engine reads like a critique, well before the fact, of Podolinsky's treatment of the human being as the perfect machine based on Carnot. It is worth recalling here that Carnot's notion of a perfect machine (and Clausius' thermodynamics) did not address everyday engineering concepts, such as the loss of heat through friction, because Carnot and Clausius were looking at ideal, completely reversible engines, in which all of this was abstracted from. Likewise Podolinsky, in the examples he uses to illustrate his interpretation of human beings as perfect machines, able to reheat their own fireboxes, abstracts from all the ways in which energy is dissipated through the normal workings of the human metabolism as well as from the important path dependencies involved in metabolic processes of human consumption and work. In doing so, Podolinsky ignored Carnot's own warning against the practical application of the concept of perfect energy efficiency even in the case of steam engines:

> We should not expect ever to utilise in practice all the motive power of combustibles. The attempts made to attain this result would be far more hurtful than useful if they caused other important considerations to be neglected. The economy of the combustible is only one of the conditions to be fulfilled in heat-engines. In many cases it is only secondary. It should often give precedence to safety, to strength, to the durability of the engine, to the small space it must occupy, to small cost of installation, etc.[92]

How much more important such material qualifications of the energetic efficiency criterion must be in the case of human labour! The importance of these factors is clarified by the analysis of so-called energy income (consumption of energy sources convertible into work) and so-called energy expenditure (work), as developed by one of Podolinsky's teachers, the great physiologist Ludimar Hermann.

As Hermann emphasised in his *Elements of Human Physiology*, an adequate analysis of energy flows in human labour must recognise that the biochemical compositions of energy income and of expenditure, and their compatibility (or lack thereof) with nutritional and with other metabolic functions, helps determine the sustainability of any particular labour process – that is, its consistency with the healthy reproduction of the labourer.[93] Different kinds of labour require different biochemical forms of energy income, not just different

92 Carnot 1977, p. 59.

93 Hermann 1875, pp. 199–200, 215–25.

amounts of caloric input. Moreover, according to Hermann, the biochemical constituents of the energy flows associated with labour operate interactively with the effects of immediately previous activity (e.g. in terms of whether the labourer is properly warmed up or, at the other extreme, not already exhausted by prior labour) to determine the full, complex amount of energy expenditure and effective work performance achievable from any given caloric income.[94] In short, although path-dependency effects certainly apply to both inanimate and animate (including human) machines, the complications they pose for the calculation of energy productivities are clearly compounded by the metabolic nature of animate (including human) labour.[95] It is ironic that Engels shows greater sensitivity to this crucial metabolic dimension than does Hermann's one-time pupil, Podolinsky.

Accordingly, Engels resists too sharp a distinction between the forms of life – both between plants and animals, and between human beings and other animate species. For example, he argues that one should not treat human beings as accumulators of energy while downplaying the role of other life forms. Thus, Engels writes in his 19 December 1882 letter to Marx,

> Man, by his labour, does deliberately what plants do unconsciously. Plants – and there is nothing new in this of course – are the great absorbers and repositories of solar heat in modified form. Thus man, by his labour, in so far as it stabilises solar energy (which in industry and elsewhere is by no means always the case), succeeds in combining the natural functions of the energy-consuming animal with those of the energy-gathering plant.[96]

If human beings are able, like plants, to accumulate solar power on Earth, they are also squanderers of energy like animals (and, as Engels had already indicated, on a vast scale). By emphasising the human squandering of energy, Engels raised ecological questions that were much deeper than Podolinsky, with his steam engine analogy, was able to grasp. Although emphasising that Podolinsky had made 'a very valuable discovery' in his treatment of human beings as accumulators or stabilisers of solar energy, Engels insisted that Podolinsky's conflation of physics with economics had driven him to false conclusions that oversimplified some of the fundamental problems of human existence.[97] At

94 Hermann 1875, pp. 240–1.
95 Cf. Mirowski 1988, p. 819.
96 Marx and Engels 1975a, Vol. 46, p. 412.
97 Ibid.

the same time, Podolinsky's recognition of the dependence of industry ultimately on agriculture, although not new, was, Engels reiterated in his letter of 22 December to Marx, a crucial element (a 'time-honoured economic fact') in a materialist perspective – even though Engels did not find its translation 'into physical terms ... particularly rewarding' in relation to the furtherance of economic analysis.[98]

Given Engels's careful critique of Podolinsky, it would be a mistake to argue, as some have done, that he simply ignored or carelessly rejected Podolinsky's ideas. Furthermore, it is significant that in all of this Engels never focuses on value questions, reflecting the fact that although Podolinsky raised the issue of value in a single sentence at the outset of his study, he had nothing to say about it directly. For this reason, it is a misnomer to say that 'Engels was uninterested in Podolinsky's attempts to redefine the labour theory of value'.[99] Rather, Engels approached Podolinsky from his strong point by addressing the latter's attempts to present basic economic conditions in physical terms. But even then, Podolinsky's argument was found wanting.

Engels's criticisms of Podolinsky's perfect machine argument may sound familiar to many present-day ecological economists from the reaction generated by Elias L. Khalil's suggestion that 'the economic process should be conceived after the Carnot cycle, and not the entropy law'.[100] Similar to Podolinsky, Khalil argued that insofar as human labour and the Carnot cycle are both 'designed *purposefully*' to produce net work or 'free energy', neither one is limited by 'the non-purposeful, mechanistic entropy law'.[101] Gabriel Lozada aptly described this argument as 'basically an "ultravitalist" attempt to deny that living, purposeful beings are completely subject to all laws of elementary matter such as the entropy law'.[102] As A.G. Williamson pointed out, one should never confuse the possibility that 'a purposeful agency ... may be interposed in an otherwise spontaneous (or natural) process to produce useful work' with the notion that the 'purposeful agency may be of unlimited potency'.[103] The basic problem, as Biancardi et al. observed, was with Khalil's (and, we might add, Podolinsky's) assumption that 'the Carnot cycle has *the same form* as the economic process'.[104] Unlike Carnot's ideal frictionless engine, which was con-

98 Marx and Engels 1975a, Vol. 46, p. 413.
99 Martinez-Alier and Naredo 1982, p. 218.
100 Khalil 1990, p. 171.
101 Khalil 1990, 170, emphasis in original.
102 Lozada 1991, p. 157.
103 Williamson 1993, pp. 70–1.
104 Biancardi, Donati and Ulgiati 1993, p. 9, emphasis added.

ceived as an isolated thermodynamic system (closed to transfers of matter and energy), the human economy is a dissipative system that both draws upon (in fact mines) and dumps waste back into its natural environment. Hence, 'each economic process can be regarded as an irreversible transformation', i.e. one that, ecologically speaking, never 'returns to the starting conditions'.[105] By neglecting this crucial form-divergence, both Khalil and Podolinsky confused the fact that the reproduction of human life feeds upon the (temporary) fixation of low entropy matter-energy in useful forms, with the fantastic notion that this need not involve increasing entropy from the standpoint of the total biospheric system with which the system of human reproduction co-evolves.

Similarly, Engels's comments on Podolinsky not only reject the latter's energy-reductionist conception of human labour, posing a more metabolic alternative, but also emphasise the failure of Podolinsky's energy-productivity calculations to take into account the great extent to which human production has heretofore operated as 'a squanderer of *past* solar heat', especially by 'squandering our reserves of energy, our coal, ore, forests, etc.'.[106] Moreover, Engels points out how Podolinsky's adoption of the perfect machine analogy led him to downplay the temporary nature of the energy stabilisation achieved by human labour in its output (despite his prior endorsement of the entropy law). Neither Podolinsky's energy accounting nor his historical discussion of energy accumulation recognise that labour and its material throughput are subject to entropic dissipation. Engels, by contrast, emphasises that human labour 'is capable of retaining solar energy on the earth's surface', not permanently, but only 'for a longer period than would otherwise have been the case'.[107] 'In stock farming', for example, 'energy is stabilised in as much as the vegetation, that would otherwise rapidly wither, die and decompose, is systematically converted into animal protein, fat, skin, bone, etc., hence *stabilised over a longer period*'.[108] In his follow-up letter to Marx on 22 December 1882, Engels elaborates on this point and extends it to livestock raising and manufacturing production, observing that the

> storage of energy by means of labour takes place strictly speaking only in *arable farming*. In stock farming the energy stored in plants is, in general, merely transferred to the animal, and hence we can only speak

105 Biancardi, Donati and Ulgiati 1993, p. 10.

106 Marx and Engels 1975a, Vol. 46, p. 411, emphasis in original.

107 Marx and Engels 1975a, Vol. 46, p. 410.

108 Marx and Engels 1975a, Vol. 46, p. 411, emphasis added.

> of storage in so far as nutritive plants are put to use which would, in the absence of stock farming, go to waste. In all branches of industry, on the other hand, energy is merely *expended*. The most one can say is that vegetable products such as wood, straw, flax, etc., and animal products in which plant energy is stored, are made available by processing, i.e. are *preserved for a longer space of time* than if they had been allowed to decay naturally.[109]

Despite all this, Martinez-Alier suggests that Engels's position in these letters indicates that he 'saw no limits to the amount of energy which could be harnessed by the work of man'.[110] The reality, as documented here, is quite different. Neither in his comments on Podolinsky nor anywhere else does Engels suggest that there are no physical or energetic limits to the development of the means of production. Rather, as shown above, Engels's open-system metabolic approach emphasises the constraints placed on labour's energy accumulation by finite stocks of nonrenewable resources, as well as by friction, and other forms of dissipation and decay. *In reality, the 'no limits' perspective applies much more accurately to Podolinsky, with his notion that human labour is a more than perfect machine in the Carnotian sense, than to Engels*. Indeed, it is *Podolinsky's* emphasis on the seemingly limitless capacity for human labour to promote accumulation of solar energy that would *appear* to suggest that a continual warming of the earth can provide a solution to all problems of energy and human production.

Marx never answered Engels's letters on Podolinsky. Perhaps they needed no answer. But also Marx was scarcely in a position to do so because he was extremely ill. A few weeks later, on 10 January 1883, Marx wrote his last letter to Engels. On 11 January, Marx's daughter Jenny died, and he left for London immediately. On 13 January, Marx contracted bronchitis and an inflammation of the throat and was unable to swallow. In February 1883, he developed a lung tumour, and on 14 March, he died.

109 Marx and Engels 1975a, Vol. 46, p. 411.

110 Martinez-Alier 1987, p. 222. Here, Martinez-Alier decontextualises and fixates upon Engels's statement (already quoted above in proper context) that the amount of energy temporarily stabilised per hour of human labour 'is dependent solely upon the level of development of the means of production' (Marx and Engels 1975, Vol. 46, p. 411).

Elaborations in *Die Neue Zeit*

The German version of Podolinsky's argument on energetics, which was published in *Die Neue Zeit* months after Marx's death (see Appendix 2), includes the exact same energy-productivity calculations for agricultural labour that were presented in the earlier French and Italian versions of Podolinsky's work. There are, however, five passages in the German article that do not appear in the Italian version which was evaluated by Engels.

First, there is a much more extensive treatment of 'the radiating energy of the sun', and of how this energy takes on 'higher forms on the earth's surface' that are more or less employable for satisfying human needs. Here, Podolinsky provides more information on the potential usefulness and limitations (due to problems of friction and harnessability) of such inorganic energy sources as the earth's rotation and magnetic force, tides and other water currents, winds, and geothermal heat (including hot springs). He also offers some geohistorical conjectures in support of the superior energy-accumulating capabilities of plants compared to animals. In this latter context, Podolinsky gives a somewhat clearer explanation of the process by which coal deposits were formed, as well as some additional data on coal deposits in Great Britain and North America. Nonetheless, the article in *Die Neue Zeit* does not address the squandering of coal which worried Engels; nor does it consider the role of coal and other non-labour inputs in a proper accounting of labour's energy-productivity. Hence, this central element of Engels's critique still stands.

The second insertion in the German version reconsiders 'the boundaries of useful labour' from the standpoint of 'the muscular labour of animals and humans'. Here, Podolinsky argues that the food-seeking movements of a snail or a butterfly do not qualify as labour insofar as they 'do not transform the slightest quantity of solar energy into such higher forms which by their further deployment could increase the store of energy on the earth's surface'.

Far from qualifying Podolinsky's mechanistic energy-reductionism, this discussion solidifies and amplifies his identification of labour (and implicitly of all use value) with purposeful energy accumulation:

> For we should keep in mind that by the word 'labour' must be understood a 'positive act' of the organism, which has a necessary consequence an accumulation of energy ... Viewed from this perspective, we can conclude that the different movements of animals that are self-evidently goalless or have as a goal merely the seeking out of means of nutrition, etc., cannot be counted as labour, precisely because they leave behind no increase of energy accumulation.

Under this definition, any (mental or physical) activity whose goal is to reduce the energy used by each hour of human labour does not qualify as labour unless and insofar as it increases the caloric content of total output and the earth's 'energy accumulation'.

Third, the article in *Die Neue Zeit* has an extended discussion of the economic coefficient (ratio of work performed to energy input) of human labour. More details are provided on Hirn's 'important experiments on the conversion of the heat of the human organism into labour'. By isolating a man in a box, strictly controlling his intake of air and food, and restricting his activity to a series of calorically measurable tasks, Hirn provided the basis for Helmholtz's efforts to calibrate 'the percentage yield of the heat transformed during labour'. The German *Die Neue Zeit* article also contains a more in-depth qualification of the figure of one-fifth for the economic coefficient implied by the work of Hirn and Helmholtz. Together with the role of non-nutritional needs that was addressed in the French and Italian versions, Podolinsky now emphasises certain ways in which 'the human organism is much more complicated than any other thermal machine'. For instance, human workers are able to consciously impede the dispersal of energy from their bodies through the use of clothing, shelter, and heating devices. Compared to machines, human labour has a variety such that its 'mechanical achievements are already so rich and diverse that they are overtaken by a mechanical apparatus only with difficulty'.

None of these qualifications alter Podolinsky's basic conception of labour as conscious energy accumulation, however. He continues to insist that 'we can apply most of the laws of the steam machine or any other thermal machine (set into movement by heat) also to the labouring human'. And he still ascribes 'the increase and development of humanity', compared to other species, to humanity's superior ability, especially through agriculture, 'to employ its mechanical energy in a direction that enabled a general accumulation of energy on the earth's surface'. For Podolinsky, in short, not only human labour but human evolution can be reduced to purposeful energy accumulation. It is the imperative to accumulate energy, as a condition of human growth and development, that drives humanity's evolution toward socialism in Podolinsky's view.

In a fourth addition to the German version, Podolinsky supports his application of thermal-machine analysis to the human labourer with two quotations from Carnot's work, *Reflections on the Motive Power of Fire*. In the quoted passages, Carnot suggests that the efficiency of heat-engines should first be 'considered independently of any particular agent', so as to derive principles 'applicable not only to steam-engines but to all imaginable heat-engines' regardless of how the 'difference of temperature' needed to create an 'impelling power' is generated in each particular case.[111] But the sentences quoted by Podolinsky have to do with the generality of Carnot's analysis across different kinds of mechanical heat engines, not animate ones. They are not meant to suggest that the abstract analysis of inanimate heat engines is directly applicable to human workers. There is certainly nothing in Carnot's discussion that effectively counters Engels's critique of energy-reductionism applied to human labour.

Podolinsky's fifth and final elaboration in *Die Neue Zeit* addresses the apparent contradiction between Quesnay's and Adam Smith's respective conceptions of productive labour. Quesnay, the physiocrat, held that the source of all value is the land, whereas Smith argued that only labour is productive. Podolinsky suggests that they are both right because even though 'labour ... creates no material', it does add 'something to the object that was not created by labour', namely energy. His entire discussion remains on the level of use value, unconnected to the social relations of production. As such, it has no clear implications for Marxian value analysis, according to which productive labour *under capitalism* is that which 'creates *surplus-value* directly, i.e ... is directly *consumed* in the course of production for the valorisation of capital'.[112] As Marx says, *apropos* Podolinsky:

> Only the bourgeoisie can confuse the questions: what is productive labour? and what is a productive worker from the standpoint of capitalism? with the question: what is *productive* labour as such? And they alone would rest content with the tautological answer that all labour is productive if it produces, if it results in a product or some other use-value ...[113]

Marx's critique applies doubly to Podolinsky's tautological conception of useful labour as energy accumulation.

All in all, the additions in Podolinsky's *Die Neue Zeit* article do nothing to correct the shortcomings highlighted by Engels's notes on the *La Plebe* version. In fact, the energy-reductionist notion of human labour apparent in Podolinsky's earlier articles emerges even more clearly from the more detailed discourse and illustrations in his final *Die Neue Zeit* article. For example, Podolinsky now defines labour more overtly as physical (muscular) activity, an activity that purposefully accumulates energy on the earth. In general, the German version reduces use value to pure energy quantities even more insistently than

111 Carnot 1977, pp. 6, 8.

112 Marx 1976a, p. 1038, emphases in orginal.

113 Marx 1976a, p. 1039, emphases in orginal.

does the Italian version, rendering more palpable the divergence of Podolinsky's approach from Marx and Engels's metabolic conception of human labour and human wealth. At the same time, the German rendition carries over the anti-ecological features that Engels found in the Italian version, namely, the calculation of energetic labour productivities without taking account of non-labour inputs (including coal), the failure to deal seriously with the role of friction and biochemical process in human labour and production, and the closed-system interpretation of the human labourer as a 'perfect machine'.[114]

Unfortunately, by the time the piece in *Die Neue Zeit* appeared in print, not only was Marx dead (the second instalment of Podolinsky's article was preceded by an obituary for Marx), but Podolinsky's own career as an intellectual and activist was over. In January 1882, he had suffered his mental collapse from which he never recovered.

Stoffwechsel

In the same year (1880) that Podolinsky sent his first small work on labour and energy to Marx, the latter wrote his *Notes on Adolph Wagner*. In these notes, Marx reiterated his conception of value accumulation as necessarily accompanied by real material exchanges – involving the human metabolism with nature – that limit and constrain it. It is this understanding of capitalist society as alienated from itself and from nature that constitutes the heart of Marx's critique and the core of his ecological vision. Referring to the method used in his political-economic works, Marx wrote, 'I have employed the word [*Stoffwechsel*] for the "natural" process of production as the material exchange … between man and nature'.[115] *Stoffwechsel* translates as *metabolism*, and it was essential to Marx's conception of human labour as a metabolic relation between human beings and nature occurring in and through society, i.e. as a material-social metabolism.[116] In this conception, the relation between human

114 Nonetheless, given the significant additions to the German version, which were not in the Italian rendition read by Engels (much less in the original, still shorter manuscript read by Marx), it should be noted that Martinez-Alier and Naredo were incorrect when they stated that 'Engels's comments are addressed to the article published in *Die Neue Zeit*, which is most probably the same one that Podolinsky sent to Marx in 1880'. They said that even after noting that 'Engels refers to the Italian version of Podolinsky's article' (Martinez-Alier and Naredo 1982, pp. 216, 222).

115 Marx 1975, p. 209.

116 Marx 1976a, pp. 207, 283.

beings and nature could not be reduced to the closed thermodynamic model of nineteenth-century physics, but had to be seen in terms of an open, dissipative system in which the human metabolic relation fed upon nature – and not simply in terms of quantifiable energy but also more qualitative elements, such as specific soil nutrients. Marx's analysis highlighted the emergence of a metabolic rift as human beings robbed the environment that constituted the basis for human production, undermining the conditions of sustainability.[117] The analysis of living systems, including human society, as metabolic systems was to be the key to the development later on of ecosystem ecology, which was never reducible to pure energetics.

Marx's view, which refused to conflate physics with value, is entirely compatible with ecological economics as long as this does not take the form of energy reductionism. As Martinez-Alier has pointed out, Nicholas Georgescu-Roegen (1906–2004) – unquestionably the greatest of modern ecological economists – 'knew [Engels's] *Dialectics of Nature* to the core' and liked 'to highlight Engels's anticipation of arguments against an absurd theory of energy-value when he [Engels] wrote in 1875 that "no-one could convert specialized work into kilogrametres and determine salary differences based on that criteria"'.[118] Georgescu-Roegen shared some of the same criticisms as Engels of attempts to reduce economic values to physical energy, siding in this respect with Engels rather than Podolinsky.[119]

The important thing, as Georgescu-Roegen would have said, is not that the human economy can be reduced to energy (or to low entropy), but rather that it feeds on it and must attempt to conserve it.[120] Looked at from this perspective, Podolinsky, inspired by Marx, presented what Engels called 'a valuable discovery' in laying out some of the physical (and ecological) conditions of human production – a discovery that helped open the way to a more developed critical analysis and to a more complete ecological view.

Yet, our analysis has shown that the conventional wisdom that finds in the 'Podolinsky business' evidence that Marx and Engels were indifferent or even hostile to ecological concerns is entirely fallacious. If they had objections to Podolinsky's analysis, it had to do with his energy reductionism together with his poor energy accounting. Nevertheless, such questions bring to light the seriousness with which Marx and Engels approached issues of energy, metabolism, open thermodynamic systems, and the ecology of production in

117 Foster 2000, pp. 141–77.
118 Martinez-Alier 1997, p. 231.
119 Georgescu-Roegen 1986, pp. 8–9.
120 Georgescu-Roegen 1971, p. 277.

general. This suggests that Martinez-Alier's pioneering survey should be built upon in a way that incorporates Marx and Engels as pioneers of ecological economics, as well as incorporating other important socio-ecological thinkers (including Marxists) whose work was less strictly centred on energy issues.[121] Such an extension would merely be following Martinez-Alier's own recognition that 'to seek out the writers who have counted calories' is 'a somewhat one-sided' approach to the history of ecological economics.[122]

---

121 Cf. Foster 2000, pp. 226–56.

122 Martinez-Alier 1987, pp. 1–2. We must, however, reject Martinez-Alier's accompanying assertion that 'not much is lost analytically by focusing on the use of energy as the central point in ecological economics' (1987, p. 2). This assertion seems to us to represent an unjustified narrowing of the discipline's subject matter.

CHAPTER 3

Classical Marxism and Energetics

Introduction

Chapter 2 established that Marx and Engels did not ignore or coldly dismiss Podolinsky's early attempt to introduce thermodynamics into political economy; rather they engaged Podolinsky's ideas seriously and offered specific, well-developed criticisms of Podolinsky's energy-accounting and energy-reductionism. It remains to explain how thermodynamics entered into Marx's *Capital* and subsequent writings. Although Marx was writing at what was still an early stage in the development of thermodynamics, he paid close attention to contemporary work in physics and other natural sciences and sought to ensure that his analysis was consistent with the most up-to-date science of the time. Marx's critique of political economy was always aimed at capturing the dynamic forces creating the potential for revolutionary socio-economic transformation. His materialist and dialectical approach arguably made Marx's *Capital* a crucial starting point for nearly all of the early work in ecological economics.

In *Capital*, Marx was engaged primarily in a critique of capitalist industrialisation and not of course in the development of an ecological economics as constituted today. His contributions to the latter therefore have to be seen in the way in which he incorporated thermodynamic conceptions into the deep structure of his work. We can see this by examining Marx's treatments of: (1) labour power and its value; (2) energy and surplus value; (3) capitalist industrialisation and thermodynamics; and (4) entropy and ecological crisis, including the metabolic rift. The virtue of Marx's analysis from a socio-ecological viewpoint is that at every point capitalist production is seen as both an economic and an ecological regime, i.e. a form of production that represents specific mediations with the physical environment.

It is simply incorrect to say, as some have suggested, that Marx ignored energy flows and thermodynamic conceptions. Thus in his polemic against Marx, Martinez-Alier goes so far as to assert that 'Marx does not seem to have considered the metabolic energy flow, so he could not trace the distinction ... between endosomatic use of energy in nutrition and the exosomatic use of energy by tools'.[1] In conflict with this statement, however, is the now

1 Martinez-Alier 2005, p. 3.

well-known fact, which Martinez-Alier recognises elsewhere, that it was Marx who introduced the concept of social metabolism and metabolic exchange of materials and energy into social science. Moreover, this is at the core of the analysis of the labour and production process in *Capital*. Far from ignoring the distinction between the endosomatic (bodily) and exosomatic (tool-based) uses of energy, Marx (as we showed in Chapter 1 and as we will demonstrate with respect to his approach to machinery in the present chapter), built his analysis around this distinction, which derived from ancient Greek materialism. It was subsequently to contribute to Engels's powerful analysis of human evolution, focusing on the human-species as a tool-making animal, drawing on Marx's insights and in line with Darwin's analysis.[2]

Indeed, a growing body of research has demonstrated the enormous extent to which Marx built thermodynamic concepts and other elements of contemporary physics into his critique of political economy.[3] Thermodynamics is to be found in the very pores of an analysis that to the superficial reader addresses the capitalist laws of exchange value alone.

This is because Marx's dialectical conception of value gives it from the very start a twofold character, both use value and exchange value, which together constitute commodity relations. Use value incorporates the conditions of production and in particular the natural-material properties embodied in production that are universal prerequisites. Exchange value, in contrast, is concerned with the enhancement of economic surplus value for the capitalist – a specific social form of production. Marx's method is never to ignore either part of this dialectic but to analyse their developing relations and contradictions together. Hence, every chapter of *Capital* addresses conditions related to physics and economics.[4]

It is no wonder then that Marx incorporated into *Capital* and other works Liebig's understanding of the metabolic exchange underlying agriculture, Ludimar Hermann's biochemical physiology, Grove's studies of the correlation of physical forces and of electricity, Charles Babbage's treatment of machinery and power, Robert Willis's studies of the mechanics of energy transfer, and so on.[5]

---

2 See Foster 2000, pp. 196–207; Winder, McIntosh and Jeffrey 2005, pp. 351, 354–5.

3 Rabinbach 1990; Wendling 2009.

4 Rabinbach 1990, p. 76.

5 Wendling 2009, pp. 185–7.

Labour Power and its Value

Marx defines 'labour-power, or labour-capacity' as 'the aggregate of those mental and physical capabilities existing in the physical form, the living personality, of a human being, capabilities which he sets in motion whenever he produces a use-value of any kind'.[6] Labour power 'is a natural object, a thing, although a living, conscious thing'.[7] It is, 'above all else, the material of nature transposed into a human organism'.[8] The metabolic-energetic content of Marx's conception is evident not just in his choice of the term labour *power*, but also in an alternative (and more descriptive) translation of the definition just quoted: 'Labour-power itself is energy transferred to a human organism by means of nourishing matter'.[9] Capitalist exploitation is not a process in which workers create something out of nothing. To emphasise this point, Marx tells us: 'What Lucretius says is self-evident: "*nil posse creari de nihilo*", out of nothing, nothing can be created'.[10] 'All the phenomena of the universe', Marx quotes the eighteenth-century Italian economist Pietro Verri as saying, 'whether produced by the hand of man or indeed by the universal laws of physics, are not to be conceived of as acts of creation but solely as a reordering of matter'.[11]

Energy considerations are accordingly central to Marx's analysis of the value of labour power. As is well known, Marx equates labour power's value with the value of the commodities entering into the consumption of workers and their families. Two aspects of this consumption are distinguished: a physical subsistence component and 'a historical and moral element'.[12] Our main concern here is with the physical subsistence element. This begins, of course, with the worker's 'natural needs, such as food, clothing, fuel and housing' – needs which 'vary according to the climatic and other physical peculiarities of his country'.[13] Even at this basic level, Marx recognises both the role of matter-energy dissipation, as well as the energy requirements for the individual worker's reproduction. Precisely because 'labour-power exists only as a capacity of the living individual', it is by nature (regardless of what happens in the labor-process) sub-

6 Marx 1976a, p. 270.
7 Marx 1976a, p. 310.
8 Marx 1976a, p. 323.
9 Marx 1967, p. 215.
10 Marx 1976a, p. 323.
11 Marx 1976a, p. 133.
12 Marx 1976a, p. 275.
13 Ibid.

ject to 'wear and tear ... and death'.[14] 'The owner of labour-power is mortal', and must therefore 'perpetuate himself by procreation'.[15] Hence, the value of labour power includes the value of commodities 'necessary for the worker's replacements, i.e. his children, in order that this race of peculiar commodity-owners may perpetuate its presence on the market'.[16]

It should perhaps not surprise us that Marx, in addressing the physiological and energetic requirements of production, was always aware of the arrow of time. The publication of more and more of Marx and Engels's voluminous notes on the sciences – chemistry, physics, mechanical engineering, biology, geology, agronomy, cosmology, anthropology, mathematics, philosophy of science, and so on – allows us to see how this was concretely accomplished. Marx's notes on thermodynamics and energetics begin as early as 1851 with his reading of Büchner and Liebig.[17] The very concept of labour power was first introduced in Germany by Helmholtz.[18]

The fact that Marx adopted the labour power category and used it both in its material-energetic sense and in relation to economic value analysis (i.e. the way labour power was translated into a commodity that generated surplus value for the capitalist) has led such analysts as Anson Rabinbach and Amy Wendling to refer to the 'marriage of Marx and Helmholtz' both in Marx's work and in particular in Engels's.[19] Rabinbach claims that Marx always emphasised the energetic basis of labour power and saw it connected to thermodynamics because labour generally involved mechanical work. Marx's confrontation with thermodynamics in his critique of political economy, Wendling concludes, was such that it caused him 'to superimpose a thermodynamic model of labor over the ontological model of labor he inherits from Hegel'.[20]

The study of the role that thermodynamics played in the development of Marx's theory of labour power and machinery led Rabinbach to contend, 'The

14 Marx 1976a, p. 274.

15 Marx 1976a, p. 275.

16 Ibid. The physical requirements of reproduction not only of the individual labourer but also whole families supported by, and supporting, the worker are always explicit in Marx. While 'a certain mass of necessaries must be consumed by a man to grow up and maintain his life ... another amount' is required 'to bring up a certain quota of children'. In order 'to maintain and reproduce itself, to perpetuate its physical existence, the working class must receive the necessaries absolutely indispensable for living and multiplying' (Marx 1976b, pp. 39, 57).

17 Wendling 2009, p. 69.

18 Rabinbach 1990, pp. 55–61; Wendling 2009, pp. 82–4.

19 Rabinbach 1990, pp. 72–4; Wendling 2009, pp. 83–4.

20 Wendling 2009, p. 59.

most important nineteenth-century thinker to absorb the insights of thermodynamics was Marx, whose work was influenced and perhaps even decisively shaped by the new image of work as "labor power"'.[21] Wendling, for her part, observes, 'Marx's concept of labor-power [Arbeitskraft], as distinct from labor [Arbeit], is an energeticist notion'.[22]

The interpretation of Rabinbach and Wendling is, unfortunately, limited by its failure to delve fully into Marx's political-economic, value-theoretic analysis and the metabolic basis of Marx's treatment of human labour. Nonetheless, the evidence that they provide on the role of thermodynamics in Marx's basic concept of labour power and its employment in mechanised production is too firmly rooted to be questioned.

Indeed, the metabolic dimension certainly looms large in Marx's consideration of the connections between the worker's labouring activity and labour power's value. 'The use of labour-power is ... labour itself', and 'the purchaser of labour-power consumes it by setting the seller of it to work'.[23] This is true whether labour is considered to be production of use values or production of values. Even though the substance of value is abstract labour ('homogenous human labour, ... human labour-power expended without regard to the form of its expenditure'), the 'creation of value' still requires 'the transposition of labour-power into labour', i.e. 'a productive expenditure of human brains, muscles, nerves, hands, etc., of the labour-power possessed in his bodily organism by every ordinary man'.[24] Conservation of labour's value-creating power therefore imposes additional maintenance requirements on the worker:

> However, labour-power becomes a reality only by being expressed; it is activated only through labour. But in the course of this activity, i.e. labour, a definite quantity of human muscle, nerve, brain, etc. is expended, and these things have to be replaced. *Since more is expended, more must be received.* If the owner of labour-power works today, tomorrow he must again be able to repeat the same process in the same conditions as regards health and strength. His means of subsistence must therefore be sufficient to maintain him in his normal state as a working individual.[25]

21 Rabinbach 1990, pp. 69–70.

22 Wendling 2009, p. 81.

23 Marx 1976a, p. 283.

24 Marx 1976a, pp. 128, 323, 134–5. Marx was always very careful to explain that there was no new materiality being created. Rather, matter-energy takes a new form as a result of labour. See, for example, his footnote to Lucretius (Marx 1976a, p. 323).

25 Marx 1976a, pp. 274–5, emphasis added.

An alternative translation of the italicised sentence is: 'This increased expenditure demands a larger income'.[26] Here, Marx is employing an 'energy income and expenditure' framework adapted from the work of the great German energy physiologist Ludimar Hermann. We know that Marx studied Hermann's *Elements of Human Physiology*, which treats energy flows in human labour from a biochemical standpoint.[27] In Hermann's analysis, 'energy income' connotes consumption of energy sources convertible into work, while 'energy expenditure' refers to the labourer's loss of energy when work is performed (see Chapter 2). Marx evidently found Hermann's approach quite useful for determining the 'ultimate or minimum limit of the value of labour-power', i.e. 'the value of the commodities which have to be supplied every day to the bearer of labour-power … so that he can renew his life-process' in something more than 'a crippled state'.[28] In addition, Marx was undoubtedly aware of Liebig's discussion of the application of thermodynamics to physiology in the last chapter of his *Familiar Letters on Chemistry*, entitled 'The Connection and Equivalence of Forces'.[29]

Marx follows Hermann and Liebig in declining to reduce the content of energy income and expenditure to pure energetic terms. For Hermann, the biochemical processes of energy income and expenditure, and their degree of compatibility with nutritional and other metabolic functions, help determine whether any given work situation is consistent with the healthy reproduction of the labourer.[30] In short, the living worker cannot be treated like a steam engine that will just keep running as long as adequate coal is shovelled in. Marx applies this aspect of Hermann's approach when discussing the value of labour power in terms of the length of daily worktime:

> When the working day is prolonged, the price of labour-power may fall below its value, although that price nominally remains unchanged, or even rises. The value of a day's labour-power is estimated … on the basis of its normal average duration, or the normal duration of the life of a worker, and on the basis of the appropriate normal standard of conversion of living substances into motion as it applies to the nature of man. Up to a certain point, the increased deterioration of labour-power inseparable from a lengthening of the working day may be compensated for by

26 Marx 1967, p. 171.

27 Hermann 1875; Baksi 2001, p. 378.

28 Marx 1976a, pp. 276–7.

29 Liebig 1864, pp. 387–97.

30 Hermann 1875, pp. 199–200, 215–25.

> making amends in the form of higher wages. But beyond this point deterioration increases in geometrical progression, and all the requirements for the normal reproduction and functioning of labour-power cease to be fulfilled. The price of labour-power and the degree of its exploitation cease to be commensurable quantities.[31]

In a footnote to the passage just cited, Marx provides a quotation from a work by the 'father of the fuel cell' – the English jurist and physical chemist Sir William Robert Grove – entitled *On the Correlation of Physical Forces*, which states: 'The amount of labour which a man had undergone in the course of 24 hours might be approximately arrived at by an examination of the chemical changes which had taken place in his body, changed forms in matter indicating the anterior exercise of dynamic force'.[32] Marx and Engels had, in fact, read Grove's book with deep interest as early as 1864–5, as part of their studies of the mechanical theory of heat and the convertibility of different forms of energy.[33] They were familiar with the fourth edition of Grove's work, published in 1862, in which Grove had already provided a detailed discussion of the second law of thermodynamics.[34] Marx obviously found these studies directly relevant to his analysis of the value of labour power.[35]

Marx's analysis of the value of labour power clearly incorporates the conservation of energy as well as the inevitability of matter-energy dissipation. That Marx does not use the terms 'entropy', 'thermodynamics', or 'first and second laws', is explained by the fact that these terms were only then being introduced into physics and thus were not used widely even within the scientific community at the time of Marx's *Capital*. (Clausius introduced the term 'entropy' – from a Greek construction meaning 'transformation' – in 1865, two years before

---

31 Marx 1976a, p. 664.

32 Ibid.

33 In a letter to Lion Philips, written on 17 August 1864, Marx reports: 'I recently had an opportunity of looking at a very important scientific work, Grove's *Correlation of Physical Forces*. He demonstrates that mechanical motive force, heat, light, electricity, magnetism and CHEMICAL AFFINITY are all in effect simply modifications of the same force, and mutually generate, replace, merge into each other, etc.' (Marx and Engels 1985, p. 551, capitalisation in original). Marx reaffirmed his excitement with Grove's work two weeks later in a letter to Engels, suggesting that Grove 'is beyond doubt the most philosophical of the English (and indeed German!) natural scientists' (Marx and Engels 1975a, Vol. 41, p. 553). Marx did not dispense this kind of praise very often.

34 Grove 1864, pp. 1–208; Marx and Engels 1975b, p. 162.

35 Cf. Stokes 1994, pp. 52–3; Baksi 2001, p. 385.

the publication of *Capital*, while Clausius's *Mechanical Theory of Heat* appeared in 1867, the same year as *Capital*. The first use of the term 'thermodynamics' in the title of a book was in 1868 in Tait's *Thermodynamics*).[36]

Energy and Surplus Value

As shown in Chapter 2, Marx's approach to energy and value did not align with that of Podolinsky (who in any case made only suggestive comments in this regard). But what, then, was the specific nature of Marx's argument? More precisely, how is the creation of surplus value consistent with the first law of thermodynamics?

At several points in *Capital* and its preparatory works, Marx considers the creation of surplus value in terms of the difference between: (1) the energy equivalent of the value of labour power, as determined by the labour required to produce the means of subsistence purchased with the wage; and (2) the energy expended by labour power, insofar as it corresponds to the energy content of the commodities in which value is objectified. But given the inability of the commodity (value) form to adhere to the metabolic-energetic requirements of labour power and the work it performs, it is as incorrect to identify the energy equivalent of labour power's value with *all* the energy that enters into the reproduction of labour power as it is to identify the energy content of commodity values with *all* the energy entering into their production. Podolinsky's energy reductionist notion that an excess of energy-product over the energy 'expended in the production of the labour power of the workers' holds the key to labour value, is thus full of misapprehensions insofar as it is meant to refer to Marx's theory.[37] For Marx, moreover, the production of surplus value is a social and material effect specific to capitalism; it is not susceptible to a purely natural scientific proof. Nonetheless, Marx's qualified application of the energy income and expenditure approach to surplus value demonstrates the thermodynamic consistency of his theory.[38]

---

36 C. Smith 1998, p. 255; Lindley 2004, p. 110.

37 Podolinsky 2004, p. 61, Appendix 1, below, p. 243.

38 Throughout this discussion we follow Marx's assumptions, in Volume I of *Capital*, that commodity prices = commodity values, and that competition among firms has converted all concrete labours into abstract labour simultaneous with the formation of commodity prices (Saad-Filho 2002, Chapter 5). Our discussion of the energetics of surplus value builds upon the work of Elmar Altvater. See Altvater 1990, pp. 20–5; 1993, pp. 188–92; 1994, pp. 86–8.

For Marx, the possibility of surplus value stems from labour power's 'specific use-value ... of being a source not only of value, but of more value than it has itself'.[39] This use value has two important characteristics. First, given capitalism's reduction of 'value' to abstract labour time, 'the use value of labour capacity, as value, is itself the value-creating force; the substance of value, and the value-increasing substance'.[40] Second, 'the past labour embodied in the labour-power and the living labour it can perform, and the daily cost of maintaining labour-power and its daily expenditure in work, are two totally different things'.[41] While the value of labour power is determined by the value of workers' commodified means of subsistence,

> The *use* of that labouring power is only limited by the active energies and physical strength of the labourer. The daily or weekly *value* of the labouring power is quite distinct from the daily or weekly exercise of that power, the same as the food a horse wants and the time it can carry the horseman are quite distinct. The quantity of labour by which the *value* of the workman's labouring power is limited forms by no means a limit to the quantity of labour which his labouring power is apt to perform.[42]

In energy terms, 'What the free worker sells is always nothing more than a specific, particular measure of force-expenditure'; but 'labour capacity as a totality is greater than every particular expenditure'.[43] 'In this exchange, then, the worker ... sells himself as an effect', and 'is absorbed into the body of capital as a cause, as activity'.[44] The result is an energy subsidy for the capitalist who appropriates and sells the commodities produced during the portion of the workday over and above that required to produce the means of subsistence represented by the wage. The apparently equal exchange of the worker's labour power for its value thus 'turns into its opposite ... the dispossession of his labour'.[45] Marx develops this point in terms of the distinction between surplus labour and the 'necessary labour' objectified in workers' commodified means of subsistence:

39 Marx 1976a, p. 301.
40 Marx 1973, p. 674.
41 Marx 1976a, p. 300.
42 Marx 1976b, p. 41, emphases in original.
43 Marx 1973, p. 464.
44 Marx 1973, p. 674.
45 Ibid.

> During the second period of the labour process, that in which his labour is no longer necessary labour, the worker does indeed expend labour-power, he does work, but his labour is no longer necessary labour, and he creates no value for himself. He creates surplus-value which, for the capitalist, has all the charms of something created out of nothing.[46]

Of course, this value (energy) surplus is not really created out of nothing. Rather, it represents capital's appropriation of a portion of the *potential* work embodied in labour power through its metabolic regeneration largely during non-worktime. And this is only possible insofar as the regeneration of labour power, in both energy and biochemical terms, involves not just consumption of calories from the commodities purchased with the wage, but also fresh air, solar heat, sleep, relaxation, and various domestic activities necessary for the hygiene, feeding, clothing, and housing of the worker. Insofar as capitalism forces the worker to labour beyond necessary labour time, it encroaches on the time required for all these regenerative activities. As Marx observes,

> But *time* is IN FACT the active existence of the human being. It is not only the measure of human life. It is the space for its development. And the ENCROACHMENT OF CAPITAL OVER the TIME OF LABOUR is the appropriation of the *life*, the mental and physical life, of the worker.[47]

Viewed in this way, Marx's metabolic-energetic analysis of surplus value is an essential foundation for his analysis of capitalism's tendency 'to go beyond the natural limits of labour-time' – a tendency 'that forcibly compels even the society which rests on capitalist production ... to restrict the normal working

---

46 Marx 1976a, p. 325. 'The matter can also be expressed in this way: If the worker needs only half a working day in order to live a whole day, then, in order to keep alive as a worker, he needs to work only half a day. The second half of the labour day is forced labour, surplus-labour ... One half a day's work is objectified in his labouring capacity – to the extent that it exists in him as someone ALIVE or as a LIVING instrument of labour. The worker's entire living day (day of life) is the static result, the objectification of half a day's work. By appropriating the entire day's work and then consuming it in the production process with the materials of which his capital consists, but by giving in exchange only the labour objectified in the worker – i.e. half a day's work – the capitalist creates the surplus value of his capital; in this case, half a day of objectified labour' (Marx 1973, pp. 324, 334, capitalisations in original). Note that Marx here looks at the labour time required to reproduce the worker from the point of view of the capitalist, i.e. as identical to the labour-time equivalent of the commodities purchasable with the wage.

47 Marx 1991, p. 493, emphases and capitalisations in original.

day within firmly fixed limits'.[48] Unless forcibly constrained from doing so, capitalist production encroaches not just on the time the worker needs 'to satisfy his intellectual and social requirements', but also on 'the physical limits to labour-power':

> Within the 24 hours of the natural day a man can only expend a certain quantity of his vital force. Similarly, a horse can work regularly for only 8 hours a day. During part of the day the vital force must rest, sleep; during another part the man has to satisfy other physical needs, to feed, wash and clothe himself ... But what is a working day? At all events, it is less than a natural day. How much less? The capitalist has his own views of this point of no return, the necessary limit of the working day. As a capitalist, he is only capital personified. His soul is the soul of capital. But capital has one sole driving force, the drive to valorize itself, to create surplus-value, to ... absorb the greatest possible amount of surplus labour.[49]

Capitalism's inherent drive to extend working time beyond labour power's metabolic-energetic limits is, in fact, one of the major themes in Volume I of *Capital*. It is precisely capitalism's attempt to convert labour power into a surplus-labour machine that threatens the worker's metabolic reproduction:

> But in its blind and measureless drive, its insatiable appetite for surplus labour, capital oversteps not only the moral but even the merely physical limits of the working day. It usurps the time for growth, development and healthy maintenance of the body. It steals the time required for the consumption of fresh air and sunlight. It haggles over the meal-times, where possible incorporating them into the production process itself, *so that food is added to the worker as to a mere means of production, as coal is supplied to the boiler, and grease and oil to the machinery*. It reduces the sound sleep needed for the restoration, renewal and refreshment of the vital forces to the exact amount of torpor essential to the revival of an absolutely exhausted organism. It is not the normal maintenance of labour-power which determines the limits of the working day here, but rather the greatest possible daily expenditure of labour-power, no matter how diseased, compulsory and painful it may be ...[50]

48 Marx 1991, p. 386.

49 Marx 1976a, pp. 341–2.

50 Marx 1976a, pp. 375–6, emphasis added.

It is worth noting that Marx's use of metabolic-energetic analysis led him to a direct comparison between the overextension of worktime and the overexploitation of land. After all, he closely studied the works of the leading agronomists of his time, including Justus von Liebig and James Johnston – works emphasising the biochemical recycling processes required to maintain soil fertility.[51] In Marx's view, capitalism's incessant pressure to produce as much surplus value as possible within any given time period caused it to violate the metabolic conditions for sustaining the productive vigour of both land and labour power.[52] Referring directly to the work of Johnston, Marx argued in *Capital* that

> The way that the cultivation of particular crops depends on fluctuations in market prices and the constant changes in cultivation with these price fluctuations – the entire spirit of capitalist production, which is oriented towards the most immediate monetary profit – stands in contradiction to agriculture, which has to concern itself with the whole gamut of permanent conditions of life required by the chain of human generations.[53]

Similarly, in the case of forestry, Marx suggested that:

> The long production time (which includes a relatively slight amount of working time), and the consequent length of the turnover period, makes forest culture a line of business unsuited to private and hence to capitalist production … The development of civilization and industry in general has always shown itself so active in the destruction of forests that everything that has been done for their conservation and production is completely insignificant in comparison.[54]

The common element in capitalism's tendencies to overexploit land and labour power is the failure to provide sufficient time (and biochemical energy inputs) for the restoration of productive power. In both cases, this productive power winds up being depleted insofar as free competition reigns:

> Capital asks no question about the length of life of labour-power. What interests it is purely and simply the maximum of labour-power that can

51 Krohn and Schäfer 1983, pp. 32–9; Baksi 1996, pp. 272–4; Baksi 2001, pp. 380–2; Foster 2000, pp. 149–54.

52 Mayumi 2001, pp. 81–4; Burkett 2014, p. 138.

53 Marx 1981, p. 754.

54 Marx 1978, pp. 321–2.

> be set in motion in a working day. It attains this objective by shortening the life of labour-power, in the same way as a greedy farmer snatches more produce from the soil by robbing it of its fertility.[55]

Hence, when considering the forces behind the English Factory Acts, which placed a cap on worktime, Marx suggested that:

> Apart from the daily more threatening advance of the working-class movement, the limiting of factory labour was dictated by the same necessity as forced the manuring of English fields with guano. The same blind desire for profit that in the one case exhausted the soil had in the other case seized hold of the vital force of the nation at its roots.[56]

That this analogy was underpinned by the energy income and expenditure framework is clear from the following passage in *Theories of Surplus Value*, written just a few years before the publication of *Capital*, Volume I:

> Anticipation of the future – real anticipation – occurs in the production of wealth in relation to the worker and to the land. The future can indeed be anticipated and ruined in both cases by premature overexertion and exhaustion, and by the disturbance of the balance between expenditure and income. In capitalist production this happens to both the worker and the land ... What is shortened here exists as power and the life span of this power is shortened as a result of accelerated expenditure.[57]

Given this parallel, it is not surprising that Marx developed a full-blown ecological critique of capitalism – one that synthesised his metabolic-energetic analyses of capital's exploitation of labour and of the land. But an essential place in this synthesis was occupied by the capitalist mechanisation of production.

---

55 Marx 1976a, p. 376.

56 Marx 1976a, p. 348. In the same passage (Marx 1976a, pp. 348–9), Marx points to 'the diminishing military standard of height in France and Germany' as evidence of labour power's deterioration under the duress of capitalist exploitation – citing data compiled in Liebig 1865, pp. 117–18.

57 Marx 1971, pp. 309–10.

Capitalist Industrialisation and Thermodynamics in Marx's *Capital*

Thermodynamic considerations – the conservation of energy, its entropic dissipation through friction in particular, and the correlation of physical forces – play a crucial role in Marx's analysis of 'Machinery and Large-Scale Industry' in Chapter 15 of *Capital*, Volume 1. This chapter represents the core of Marx's analysis of industrial development under capitalism.

Marx depicts the Industrial Revolution using a model of machinery systems consisting of 'three essentially different parts, the motor mechanism, the transmitting mechanism and finally the tool or working machine'.[58] He perceives machine-based production as a transfer of force from one part of the system to another – starting from the motor mechanism which 'acts as the driving force of the mechanism as a whole', on through the transmission mechanism which 'regulates the motion, changes its form where necessary, and divides and distributes it among the working machines', and finally to the working machine which 'using this motion ... seizes on the object of labour and modifies it as desired'.[59] This entire framework is clearly informed by an extensive theoretical and practical study of both energy conservation and the mechanics of energy transfer.[60]

In an 1863 letter to Engels outlining his research for 'the section on machinery', Marx wrote that he had not only 're-read all [of his] note-books (excerpts) on technology', but was 'also attending a practical (purely experimental) course for working men given by Prof. Willis'.[61] The lecturer he referred to was the Reverend Robert Willis (1800–75), the brilliant British architect and mechanical engineer (and, from 1837 onward, Jacksonian Professor of Natural and Experimental Philosophy at the University of Cambridge). That the mechanics of energy transmission were a central theme in these lectures is clear from the working models that Willis used – models he had himself designed and integrated into an instructional system.[62] As described by technology-educator Eric Parkinson:

58 Marx 1976a, p. 494.

59 Ibid.

60 Baksi 1996, pp. 274–8. Wendling (2007, p. 255) reproduces a page from Marx's unpublished notebooks that shows a meticulous drawing by Marx of a machine and its various motor mechanisms for the transmission of energy.

61 Marx to Engels, 28 January 1863, in Marx and Engels 1975a, Vol. 41, p. 449.

62 Willis 1851.

> Willis developed a special construction kit which could be used as a means of demonstrating principles of mechanisms to his students. It was devised so that mechanical components could be added, removed, or re-positioned with speed and accuracy during a lecture-demonstration.[63]

When combined with Marx's theoretical and historical studies, such practical instruction led him to argue that the Industrial Revolution started not with the motor mechanism and its energy sources, but rather with the tool or working machine – specifically with the mechanisation of the portion of labour that incorporated directly the principal material(s). As explained in *Capital*,

> The entire machine is only a more or less altered mechanical edition of the old handicraft tool ... The machine, therefore, is a mechanism that, after being set in motion, performs with its tools the same operations as the worker formerly did with similar tools. Whether the motive power is derived from man, or in turn from a machine, makes no difference here.[64]

This argument 'establish[ed] a connection between human social relations and the development of these material modes of production'.[65] After all, the ability of the capitalist to separate the tool from the worker and install it in the machine – and the subsequent application of science to the technical improvement of machinery on the capitalist's profit-making behalf – presumed that the worker had already been socially separated from control over the means of production.[66] But this historical primacy of social relations, and corresponding primacy of machine-tools over energy sources and mechanisms, hardly prevented Marx from emphasising the crucial enabling role of power supply and transmission in the Industrial Revolution. For one thing, the mechanisation of tools means they are freed from the limitations of the individual worker's labour power as the direct motive force. 'Now assuming that [the worker] is acting simply as a motor, that a machine has replaced the tool he is using, it

63 Parkinson 1999, p. 67. Parkinson adds that Willis's model-based approach 'was something of a benchmark in education in mechanics. Willis was a clear leader in his field, established a novel, practically-based teaching mode, and communicated his ideas to an influential cadre of future engineers' (Parkinson 1999, p. 67).

64 Marx 1976a, pp. 494–5.

65 Marx to Engels, 28 January 1863, in Marx and Engels 1975a, Vol. 41, p. 450.

66 For details on Marx's analysis of capitalism's development and application of science as a form of workers' alienation from the means of production, see Burkett 2014, pp. 158–63.

is evident that he can also be replaced as a motor by natural forces'.[67] Once installed in machines, tools may be driven by a greater variety of power sources and on a much larger energy-scale. Indeed, the growing scale of machinery itself precludes the continued use of labour power as motive force:

> An increase in the size of the machine and the number of its working tools calls for a more massive mechanism to drive it; and this mechanism, in order to overcome its own inertia, requires a mightier moving power than that of man, quite apart from the fact that man is a very imperfect instrument for producing uniform and continuous motion.[68]

The replacement of labour power with other motive forces starts with 'a call for the application of animals, water and wind as motive powers', but it soon graduates to the development of coal-driven steam engines and eventually (as Marx projected) electric power mechanisms.[69] It is here, with the development of motor mechanisms and their power sources in response to the energy demands of increasingly complex and large-scale machine-tool systems, that Marx emphasises the role of friction as a fundamental entropic process.[70] Hence, in explaining that the 'increase in the size of the machine and its working tools calls for a more massive mechanism' and motor force to drive it, Marx observes that the question of force (or energy) became critical when water power, which in Britain had hitherto been the main source of power, no longer seemed adequate: 'the use of water-power preponderated even during the period of manufacture. In the seventeenth century attempts had already been made to turn two pairs of millstones with a single waterwheel. But the increased size of the transmitting mechanism came into conflict with the water-power, which was now insufficient, and this was one of the factors which gave the impulse for a more accurate investigation of the laws of friction'.[71]

---

67 Marx 1976a, p. 497.

68 Ibid.

69 Marx 1976a, p. 496.

70 That Engels also had a keen interest in friction, but on a more theoretical level, is clear from the numerous passages on this subject in *Dialectics of Nature* (1964), e.g. pp. 95–6, 108, 110, 228–9, 252, 258–60, 284, 297. This may help explain why Georgescu-Roegen seems to have very much liked the book (see Martinez-Alier 1997, p. 231). It is harder to explain how Georgescu missed the more practical discussions of friction in Marx's *Capital*.

71 Marx 1976a, pp. 497–8.

Here Marx demonstrates an acute understanding of the way in which water and steam, as contemporaneous power technologies, affected the early history of industrialisation. Although the 'take-off' associated with the Industrial Revolution is usually seen as occurring around 1760 or 1780, water power remained the principal motive force for industry in Britain until well into the nineteenth century. In the eighteenth and early nineteenth centuries, scientist-engineers such as Parent, Smeaton, Deparcieux, and Lazare Carnot explored the efficiency requirements of water power, the problem of friction, and, in Lazare Carnot's case, the maximum efficiency under ideal conditions from a given fall of water. At this time, despite the improvements of Watt's steam engine, the water wheel provided far more motive power. The steam engine was thus commonly used as a supplement to water power. However, the increasing efficiency of the steam engine, coupled with its greater versatility (the areas of serviceable water power in Britain – principally Scotland and the North – were already in use) led to its steady displacement of water power as the nineteenth century progressed.[72]

Not only do Marx's comments seem to be cognisant of these developments, but his point here may reflect awareness of the fact that the Scottish physicists James Thomson and his brother, William Thomson (the future Lord Kelvin), were initially drawn to their research into thermodynamics by their practical explorations into fluid friction.[73] It was William Thomson who rediscovered and promoted Sadi Carnot's 1824 work on thermodynamics, which had hitherto fallen on deaf ears, and who introduced the term 'thermodynamics' (referring initially to the laws of heat as a source of power) in 1849.

In any event, despite common misinterpretations regarding Marx's polemic with Proudhon, in which Marx glibly stated that 'the hand-mill gives you society with the feudal lord, the steam engine society with the industrial capitalist', Marx clearly did not adopt the view that the steam engine literally generated either the capitalist or industrialisation.[74] He recognised that water power not only dominated in the early manufacturing/mercantilist period preceding industrialisation, but even led the way in the initial phase of industrialisation proper (the age of 'machinofacture'). In fact, his analysis emphasises that steam power only displaced water power as the entire mechanism of capitalist production began to demand increasingly large concentrations, and more versatile forms, of energy.

---

72 Cardwell 1971, pp. 67–88; Lindley 2004, pp. 64–5.

73 C. Smith 1998, pp. 39, 48.

74 Marx 1963a, pp. 109–10; Foster 2000, p. 280.

More specifically, with respect to the 'power source' of industry, Marx observed that it is the versatility of steam power, which allowed the right amount of power to be applied at the right time in the right location, which gives it an advantage over water power. With 'tools ... converted from being manual implements of man into the parts of a mechanical apparatus', it becomes possible to reduce 'the individual machine to a mere element in production by machinery'; but this presumes that the motive mechanism is 'able to drive many machines at once'.[75] Thus, the required 'motor mechanism grows with the number of the machines that are turned simultaneously, and the transmitting mechanism becomes an extensive apparatus'.[76] Insofar as 'the object of labour goes through a connected series of graduated processes carried out by a chain of mutually complementary machines of various kinds', the power source must meet demanding scale, flexibility and transmission requirements.[77] In industries using machines to produce precision machines, especially, an 'essential condition ... was a prime mover capable of exerting any amount of force, while retaining perfect control'.[78] The material nature of water power precluded its use for such purposes beyond a certain level and locality, given problems of friction, containment, storability and transportability:

> The flow of water could not be increased at will, it failed at certain seasons of the year, and above all it was essentially local. Not till the invention of Watt's second and so-called double-acting steam-engine was a prime mover found which drew its own motive power from the consumption of coal and water, was entirely under man's control, was mobile and a means of locomotion, ... and, finally, was of universal technical application and little affected in its choice of residence by local circumstances.[79]

The victory of steam power over water power was thus a product of the fact that it allowed for the location of industry near major population centres, magnified by the fact that, with the further development of industry, the entire mechanism of production (in both scale and complexity) demanded increasingly large concentrations, and more flexible, controllable, transportable, and storable forms of energy.

---

75 Marx 1976a, p. 499.
76 Ibid.
77 Marx 1976a, p. 501.
78 Marx 1976a, p. 506.
79 Marx 1976a, p. 499.

From this it is clear that 'matter matters, too', in *Capital*'s analysis of the energetics of capitalist industrialisation. The concreteness of Marx's analysis in terms of energy and its integration with his economics sets his work apart from other economic treatises of his day. One cannot help but be astonished by the close attention that Marx paid to the physical wear and tear of machinery. In the chapter on machinery and large-scale industry, we are told that:

> The physical deterioration of the machine is of two kinds. The one arises from use, as coins wear away by circulating, the other from lack of use, as a sword rusts when left in its scabbard. Deterioration of the first kind is more or less directly proportional, and that of the second kind to a certain extent inversely proportional, to the use of the machine.[80]

Such physical deterioration is central to the analysis of the costs of fixed capital replacement and repair in Volume II, Chapter 8, of *Capital*, where Marx again distinguishes between wear and tear from 'actual use', and 'that caused by natural forces', showing through various real-world examples how the labour necessitated by each type enters into the values of commodities.[81]

Aside from friction, another reason why Marx eschewed energy-reductionism in his analysis of industry was his awareness that capitalism's 'development of the social powers of labour' involved not just machines and their motive forces, but also 'the appliance of chemical and other natural agencies' in a way that is not reducible to pure energy-transmission.[82] This is most evident from Marx's analysis of capitalist agriculture, where the 'conscious, technological application of science', in the service of profit-making, confronts a barrier in 'the fertility of the soil', with its necessary basis in 'the metabolic interaction between man and the earth'.[83] But there is an irreducible biochemical element to any kind of production wherein something is 'added to the raw material to produce some physical modification of it, as chlorine is added to unbleached linen, coal to iron, dye to wool'.[84] 'In all these cases', as Marx puts it when considering their effect on value accumulation, 'the production time of the capital advanced consists of two periods: a period in which the capital exists in the labour process, and a second period in which its form of existence – that of an unfinished product – is handed over to the sway of natural processes, without

80 Marx 1976a, p. 528; cf. Marx 1976a, pp. 289–90.
81 Marx 1978, pp. 248–61.
82 Marx 1976b, p. 34.
83 Marx 1976a, pp. 637–8.
84 Marx 1976a, p. 288.

being involved in the labour process'.[85] Such biochemical production processes obviously reduce the relevance of analyses anchored solely in energetics.[86]

An important connection between Marx's analysis and ecological economics – specifically the entropy school – involves the latter's view that human production became unsustainable when it 'broke the budget constraint of living on solar income'.[87] However, although Daly limits this post-solar income regime to 'the last 200 years', neither he nor Georgescu-Roegen venture a structural explanation for it – that is, an explanation combining specific social production relations with the development of specific technologies relying on fossil fuels and other 'geological capital'.[88] Marx's analysis of machinery and large-scale industry (and industrialised agriculture) under capitalism provides just such an explanation for the growing industrial mechanism's voracity for materials and energy. Apart from the standard interpretation of the Podolinsky debate (see Chapter 2), perhaps what has bolstered ecological economists' misperceptions of Marx's views are passages such as the following, extracted from its proper context:

> In the first place, in machinery the motion and the activity of the instrument of labour asserts its independence *vis-á-vis* the worker. The instrument of labour now becomes an industrial form of perpetual motion. It would go on producing for ever if it did not come up against certain natural limits in the shape of the weak bodies and the strong wills of its human assistants.[89]

The 'perpetual motion' of which Marx speaks here, replaced in its proper context, concerns the entire social mechanism behind the instrument of production, as perceived from the standpoint of the individual worker alienated from

85 Marx 1978, p. 317.

86 These kinds of processes have been termed 'eco-regulated' by Benton 1989, pp. 51–86. For a detailed rebuttal of Benton's claim that Marx's analysis failed to take such processes into account, see Burkett 1998, pp. 125–33; Burkett 2014, pp. 41–7. It should be noted in relation to biochemical and energetic processes that the more sophisticated purely energetic approaches do not deny the qualitative aspects of biochemical processes but nonetheless attempt to subsume them under a kind of energetic reductionism. For a contemporary example, see Smil 1991. Marx's rejection of quantitative energy reductionism is consistent with that of many later ecological economists. Compare, for example, Georgescu-Roegen 1972 and Daly 1981. For further discussion see Burkett 2003, pp. 140–1.

87 Daly 1992, p. 23.

88 Ibid.; Burkett 2005a, pp. 117–52.

89 Marx 1976a, p. 526.

the means of production. This 'perpetual motion' is that of a material-social class relation; it is not an inherent physical property, a matter only referred to metaphorically and hence inviolate of the laws of thermodynamics. Marx's main point involves how the machine-system 'confronts the worker as a pre-existing material condition of production':[90]

> An organized system of machines to which motion is communicated by the transmitting mechanism from an automatic centre is the most developed form of production by machinery. Here we have, in place of the isolated machine, a mechanical monster whose body fills whole factories, and whose demonic power, at first hidden by the slow and measured motions of its gigantic members, finally bursts forth in the fast and feverish whirl of its countless working organs.[91]

As discussed in Chapter 1, Marx's reference to the 'working organs' of this machine monstrosity goes back to the original Greek term *organon*, which refers both to tools and to bodily organs, in what amounts to a theory of natural technology. At any rate, one can certainly imagine from the above-quoted passages how Marx must have felt about Podolinsky's designation of workers as 'perfect machines', i.e. idealised steam engines. Indeed, the main way in which 'Podolinsky went astray', as Engels put it in his 19 December 1882 letter to Marx, was to bypass the alienated character of real-world machinery and mechanised labour under capitalism.[92] Instead, Podolinsky 'sought to find in the field of natural science fresh proof of the rightness of socialism', and thus 'confused the physical with the economic'.[93] Although contemporary ecological economics does not (for the most part) champion a crude mechanistic vision of socialism, as Podolinsky did, it nonetheless suffers from the same tendency, as he did, to confuse the physical with the economic – stemming from its failure to grapple with the deep material-social contradictions of capitalist production and monetary valuation.[94]

In addressing matter-energy throughput more broadly, Marx emphasises that capitalism's development of 'the productive powers of labour' is dependent upon 'the *natural* conditions of labour, such as fertility of soil, mines, and

90 Marx 1976a, p. 508.
91 Marx 1976a, p. 503.
92 Marx and Engels 1975a, Vol. 46, p. 412.
93 Ibid.
94 Burkett 2009.

so forth'.[95] Capitalist industrialisation is a process in which 'science presses natural agencies into the service of labour' under the pressures of private profit-making and competition.[96] Nature provides capitalist enterprise with use values that act not only as bearers of value, but also as 'free natural productive power[s] of labour'.[97] Both functions are evident in Marx's analysis of raw materials in the capital accumulation process.

Marx's main theme here is that capitalism's development of machine-based production, and of a complex division of labour among competing enterprises, generates an unprecedented increase in labour productivity that necessarily corresponds to an unprecedented demand for raw materials. As he says, 'the increasing productivity of labour is expressed precisely in the proportion in which a greater quantity of raw material absorbs a certain amount of labour, i.e. in the increasing mass of raw material that is transformed into products, worked up into commodities, in an hour, for example'.[98] 'The growth of machinery and of the division of labour has the consequence that in a shorter time far more can be produced', so that 'the part of capital transformed into raw materials necessarily increases'.[99] As labour productivity grows, so grows the quantity of materials that capital must appropriate and process in order to achieve any given expansion of value.

As has been shown, Marx was also well aware of the crucial importance of power supplies for capitalist industry. Accordingly, he includes energy sources in capital's growing demand for 'auxiliary' or 'ancillary' materials, defined as those materials which, while not forming part of 'the principal substance of the product', are nonetheless required 'as an accessory' of its production.[100] They provide heat, light, chemical and other necessary conditions of production distinct from the direct processing of principal materials by labour and its instruments. Obviously, consumption of energy sources ('coal by a steam-engine ... hay by draft-horses', or 'materials for heating and lighting workshops') is a large part of such ancillaries' usage.[101] As Marx observes, 'After the capitalist

95 Marx 1976b, p. 34, emphasis in original.
96 Ibid.
97 Marx 1981, p. 879; Burkett 2014, Chapter 6.
98 Marx 1981, p. 203.
99 Marx 1976c, p. 431.
100 Marx 1976a, p. 288; see also Marx 1976a, p. 311. 'Under raw material we also include the ancillary materials such as indigo, coal, gas, etc. ... Even in branches of industry that do not use any specific raw material of their own, there is still raw material in the form of ancillary material or the components of the machinery, etc.' (Marx 1981, p. 201).
101 Marx 1976a, p. 288.

has put a larger capital into machinery, he is compelled to spend a larger capital on the purchase of raw materials *and the fuels required to drive the machines*'.[102] In short, capitalist industrialisation results in 'more raw material worked up in the same time, and therefore a greater mass of raw material *and auxiliary substances* enters into the labour process'.[103]

This is not to say that the goal of capitalist production is simply to maximise matter-energy throughput. Capitalism is a competitive system in which individual enterprises feel a constant pressure to lower costs. Hence, in its own historically limited way, capitalism does penalise waste of materials and energy. As Marx observes, 'value is not measured by the labour-time that [an] article costs the producer in each individual case, but by the labour-time socially required for its production'.[104] Competition thus penalises excessive matter-energy throughputs by not recognising the labour time objectified in them as necessary, value-creating labour. In this sense, 'all wasteful consumption of raw material or instruments of labour is strictly forbidden, because what is wasted in this way represents a superfluous expenditure of quantities of objectified labour, labour that does not count in the product or enter into its value'.[105] Such waste also includes any 'refuse' that could have been 'further employed as a means in the production of new and independent use values' – at least insofar as competitors are able to implement the necessary recycling operations.[106] 'As the capitalist mode of production extends', Marx argues, 'so also does the utilization of the refuse left behind by production'.[107]

Nonetheless, such competitive economisation and recycling of materials only operates along a path of rising labour productivity, i.e. of the processing of matter-energy into commodities on an ever-growing scale. The main 'motive for each individual capitalist' is 'to cheapen his commodities by increasing

102 Marx 1976c, p. 431, emphasis added. The ancillary materials category also helped Marx analyse situations, mentioned earlier, where biochemical processes make up an essential phase of production. See Burkett 2014, pp. 42–3.

103 Marx 1976a, p. 773, emphasis added. Similarly, when specifying capitalism's inventory requirements, Marx includes 'material for labour at the most varied stages of elaboration, *as well as ancillary materials*. As the scale of production grows, and the productive power of labour grows through cooperation, division of labour, machinery, etc., so does *the mass of raw material, ancillaries, etc.* that go into the daily reproduction process' (Marx 1978, pp. 218–19, emphases added).

104 Marx 1976a, p. 434.

105 Marx 1976a, p. 303.

106 Marx 1976a, p. 313.

107 Marx 1981, p. 195.

productivity of labour'.[108] By lowering cost per commodity produced, such productivity gains enable manufacturers to reap surplus profits and/or an increased market share. Although they still feel pressure to keep throughput at or below the normal level, this level is itself a positive function of the constant drive to boost output per labour hour.

Moreover, capitalism's competitive enforcement of its own standards of matter-energy use does nothing to counter the throughput produced by the 'moral depreciation' of fixed capital precipitated by the development of more advanced machinery and structures, or by rising labour productivity in the industries producing them.[109] Through such moral depreciation (loss of capital values objectified in machinery and buildings), 'competition forces the replacement of old means of labour by new ones before their natural demise' – a clear acceleration of material throughput resulting in environmental degradation.[110] The constant threat of moral depreciation also compels individual enterprises to speed up the turnover of their fixed capital stocks by prolonging work-time and intensifying labour processes, further magnifying the system's normal matter-energy throughput.[111] Advanced capitalism's extension of such accelerated turnover to consumer 'durables' (personal computers, televisions, audio equipment, kitchen appliances, etc.) only worsens these entropic dynamics.[112]

Entropy and the Metabolic Rift

Given this background, one can better understand Engels's critique of Podolinsky's attempt to calculate the energy productivity of agricultural labour (see Chapter 2). In Marx's view, capitalist development of productive forces translates into a growing throughput of matter and energy per labour hour. The amount of energy that each hour of labour (temporarily) stabilises depends on the total amount of matter-energy processed per hour as well as the amount of ancillary energy used per unit of output – both of which correlate to the development of production. Given that the increase in labour productivity under capitalism is generally accompanied by increases in material throughput, Podolinsky's failure to include non-labour inputs in his calculations is a

108 Marx 1976a, p. 435.

109 Marx 1976a, p. 528; Marx 1978, pp. 208–9.

110 Marx 1978, p. 250; for details, see Horton 1997, pp. 127–39.

111 Marx 1981, pp. 208–9.

112 Huws 1999, pp. 29–55; Strasser 1999.

serious omission indeed, seeing as how 'the energy value of auxiliary materials, fertilisers, etc., must ... be taken into consideration' – and increasingly so.[113] The general lesson, Engels tells his life-long comrade (in a statement already referred to but worth repeating), 'is that the working individual is not only a stabiliser of *present* but also, and to a far greater extent, a squanderer of *past*, solar heat. As to what we have done in the way of squandering our reserves of energy, our coal, ore, forests, etc., you are better informed than I am'.[114]

Engels's critique of Podolinsky's energy-reductionist framework is thus fully consistent with Marx's sophisticated metabolic-energetic approach to wage-labour and industrial capital accumulation. For Marx, the capitalist economy is an open system reliant on environmental inputs of labour power and non-human matter-energy. Marx emphasises capitalism's tendency to deplete and despoil the land, while exploiting the worker. Stated differently, Marx argues that the metabolic systems that reproduce the productive powers of labour and the land are susceptible to adverse shocks from the system of industrial capital accumulation to which they are conjoined.[115]

It is thus no accident that Marx chooses the final section of his chapter on machinery and large-scale industry as the place to develop an initial synthesis of capitalism's tendency to 'simultaneously [undermine] the original sources of all wealth – the soil and the worker'.[116] This was for Marx a major consequence

113 Engels to Marx, 19 December 1882, in Marx and Engels 1975, Vol. 46, p. 411.

114 Ibid., emphases in original. Far from dismissing energetic considerations, Engels's comments – informed by Marx's analysis of capitalist productivity growth – show a healthy awareness of how a faulty specification of the relevant dimensions of energy use can generate misleading results. As two leading energy analysts emphasise, one cannot overestimate 'the importance of the choice of space and time boundaries' for any 'assessment of the energetic requirement of human labor' (Giampietro and Pimentel 1991, p. 119). Engels's approach to energy accounting, unlike Podolinsky's, encompasses 'all the energy consumed at societal level to raise the workers and to support their dependents' (Ibid.).

115 Although acknowledging recent research that 'rediscovered Marx's "metabolism"', and its relation to energetics and ecology, Martinez-Alier (2007, p. 224) claims that 'in his published work Marx did not refer to the flow of energy as metabolism'. This criticism is overstated, however, because Marx defined the labour process itself as a metabolic process and analysed this in physical terms as well as in terms of the physiological transfer of energy (even if he did not count calories). It was precisely Marx's definition of labour power in terms of metabolism – and the fact that because of this 'when Marx speaks as he frequently does of the "life process of society" he is not thinking in metaphors' – which in Hannah Arendt's view made him 'the greatest of modern labor theorists' (Arendt 1958, pp. 93, 99, 106–8).

116 Marx 1976a, p. 638.

of the industrialisation of agriculture, which led to the systematic and intensive robbing of the soil, as well as exploitation of the worker. Here, Marx invokes Liebig's theory of biochemical reproductive cycles to argue that capitalism 'disturbs the metabolic interaction between man and the earth'.[117] Specifically, capitalism concentrates population and manufacturing industry in urban centres in a way that 'prevents the return to the soil of its constituent elements consumed by man in the form of food and clothing; hence it hinders the operation of the eternal natural condition for the lasting fertility of the soil'.[118] In short, the capitalist division of town and country disrupts the soil's reproductive cycle, and this disruption is accentuated by the tendency of industrial capitalist agriculture towards 'robbing the soil' and 'ruining the more long-lasting sources of [its] fertility'.[119]

Marx made it clear, while he was preparing the manuscripts for *Capital*, that he viewed the issue of Liebig and agricultural chemistry as crucial.[120] He continually returned to his critique of the metabolic rift associated with capitalist industrialisation, in the process of analysing the origins of agricultural land rent.[121] In the third volume of *Capital*, he wrote:

> Large landed property reduces the agricultural population to an ever decreasing minimum and confronts it with an ever growing industrial population crammed together in large towns; in this way it produces conditions that provoke an irreparable rift in the interdependent process of social metabolism, a metabolism prescribed by the natural laws of life itself. The result of this is a squandering of the vitality of the soil, which is carried by trade far beyond the bounds of a single country.[122]

---

117 Marx 1976a, p. 637.

118 Ibid. 'The natural human waste products, remains of clothing in the form of rags, etc. are the refuse of consumption. The latter are of the greatest importance for agriculture. But there is a colossal wastage in the capitalist economy in proportion to their actual use' (Marx 1981, p. 195).

119 Marx 1976a, p. 638.

120 Marx and Engels 1975a, Vol. 42, pp. 507–8; Henderson 1976, Vol. 1, pp. 262–71; Stanley 2002, pp. 46–51.

121 Although the manuscripts on which Volumes II and III of *Capital* were based were drafted before the publication of Volume I of *Capital*, it is clear that these parts of the original manuscript continued to be revised – particularly with respect to agricultural science (see Saito 2014).

122 Marx 1981, p. 949.

The metabolic rift between town and country created by the industrial capitalist system vitiates the reproduction both of labour power and the land, two things that in reality constitute a unified metabolic system, however much capitalism may treat them merely as separable external conditions. To quote Marx once again,

> Large landed property undermines labour-power in the final sphere to which its indigenous energy flees, and where it is stored up as a reserve fund for renewing the vital power of the nation, on the land itself. Large-scale industry and industrially pursued large-scale agriculture have the same effect. If they are originally distinguished by the fact that the former lays waste and ruins labour-power and thus the natural power of man, whereas the latter does the same to the natural power of the soil, they link up in the later course of development, since the industrial system applied to agriculture also enervates the workers there, while industry and trade for their part provide agriculture with the means of exhausting the soil.[123]

Marx's analysis is fully consistent with the central concept of Liebig's agricultural chemistry paradigm: 'the cycle of processes constitutive for the reproduction of organic structures'.[124] This concept is not energy-reductionist, but it does abide by the first and second laws of thermodynamics. As Krohn and Schäfer describe it, 'plant and animal life, together with meteorological processes, jointly circulate certain "substances"; apart from the irreversible transformation of energy into heat, living processes do not "use up" nature, but reproduce the conditions for their continued existence'.[125]

Capitalism's assault on the biochemical processes necessary to sustain the human-land system does not create or destroy matter-energy, but it does degrade its metabolic reproductive capabilities. This degradation can clearly be seen as a form of entropic matter-energy dissipation. In Marx's view, this phenomenon – to some extent inherent in production – is dramatically worsened by capitalism's specific form of industry, which is based on the social separation of the producers from the land and other necessary conditions of production. Hence it is possible for society to achieve a 'systematic restoration' of its reproductive metabolism with the land 'as a regulative law of social production, and in a form adequate to the full development of the human race'.[126] But

123 Marx 1981, pp. 949–50.
124 Krohn and Schäfer 1983, p. 32.
125 Ibid.
126 Marx 1976a, p. 638.

this requires 'co-operation and the possession in common of the land and the means of production', based on 'the transformation of capitalist private property ... into social property'.[127]

The power of Marx and Engels's metabolic-energetic approach, uniting their discussions of production in general, is an indispensable part of their scientific outlook. As Kenneth Stokes observed, *Capital*'s 'surprisingly contemporary thermodynamic vision of the economic process is a clear departure from the circular flow concept; for it is suggestive of the modern open-systems theoretical perspective'. Marx and Engels's 'model explicitly embodied ... the metabolic interaction of man and nature; the notion that the economic process is embedded in the biosphere', and it treated 'social change' as 'an endogenous dialectical process in which the nature-society nexus displays reciprocal and complex interpenetrations'.[128]

127 Marx 1976a, pp. 929–30.

128 Stokes, 1994, p. 64.

CHAPTER 4

Engels, Entropy, and the Heat Death Hypothesis

Introduction

Ever since Nicholas Georgescu-Roegen wrote his magnum opus, *The Entropy Law and the Economic Process*, the entropy law (or the second law of thermodynamics) has been viewed as a *sine qua non* of ecological economics.[1] Georgescu-Roegen argued strongly that both the entropy law and the first law of thermodynamics (conservation of matter-energy) were incompatible with orthodox neoclassical economics. The relation of ecological economics to Marxian economics, however, was much more ambiguous. Attempts to explore the history of ecological-economic ideas, following Georgescu-Roegen's contributions, immediately brought to the fore the close relationship between those thinkers who had pioneered in ecological-economic thinking and classical Marxism.

Georgescu-Roegen himself pointed, although not uncritically, to Marx and Engels's discussions of energetics and thermodynamic principles. It was, after all, as he noted, the 'first pillar' of historical materialism that 'the economic process is not an isolated system'.[2] He also indicated his support for Engels's critique of energy reductionism.[3] Both Marx and Engels were well versed in the scientific literature on thermodynamics. As even their most persistent ecological-economics critic, J. Martinez-Alier, has acknowledged, 'Engels ... had read everything on the fundamental studies on thermodynamics'.[4] Anson Rabinbach claimed in his important study of nineteenth-century applications of thermodynamics to human labour that 'the most important 19th-century thinker to absorb the insights of thermodynamics was Marx, whose later work was influenced and perhaps even decisively shaped by the new image of work as "labour power"'.[5] Early contributors to ecological-economics thinking, such as Sergei Podolinsky and Frederick Soddy, were inspired by Marx.[6]

---

1 Georgescu-Roegen 1971.

2 Georgescu-Roegen 1971, p. 316.

3 Georgescu-Roegen 1986, p. 9; Georgescu-Roegen's criticisms of classical Marxism focused on the alleged ecological inadequacies of Marx's labour theory of value and Marx's reproduction schemas. These criticisms have been rebutted by Burkett (2004, and 2014, Chapters 6–8).

4 Quoted in Ravaioli 1995, p. 130.

5 Rabinbach 1990, pp. 69–70.

6 See Podolinsky 1995, pp. 127–9, 138; Podolinsky 2004, p. 61; Soddy 1922, pp. 12–13.

Ironically, it is perhaps because of the strong *prima facie* case for a link between classical Marxism and thermodynamic conceptions that the argument is so fervently advanced that Marx neglected thermodynamics. Some of the leading figures in ecological economics have gone to extraordinary lengths to separate at birth the Marxian and ecological critiques and then to deny any direct relationship through a series of disconnects: (a) Marx and Engels's own integration of thermodynamic concepts into their analysis (admittedly not given strong emphasis or even understood in later Marxist thought) is simply ignored; (b) circumstantial evidence is offered to suggest that Marx and Engels actively *rejected* some of the crucial discoveries in thermodynamics in their day; (c) it is alleged that Engels went so far as to cast doubt on the entropy law itself; and (d) the fact that early developments in ecological economics occupied the same intellectual universe as Marxism, which led to much cross-fertilisation of thought, is downplayed if not deliberately obfuscated.

The leading role in criticising Marx and Engels for neglecting and/or misunderstanding thermodynamics has been taken by Martinez-Alier, not only in his very influential book *Ecological Economics* but also in other, frequently cited, writings, appearing in such high-profile journals as *Ecological Economics*, *New Left Review*, and *Socialist Register*. Recent analyses by ecosocialists have strongly challenged these arguments with respect to Podolinsky and Marx-Engels, demonstrating that Podolinsky's perfect-human-machine model of ecological economics was fundamentally flawed from the standpoint of thermodynamics itself (a fact that the founders of historical materialism clearly recognised at the time).[7]

It is perhaps not surprising therefore that greater emphasis has been placed of late on the criticism that Engels (and by imputation Marx) rejected the second law of thermodynamics itself. Thus, in an article in the 2007 *Socialist Register*, Martinez-Alier underscored Engels's alleged 'unwillingness to accept that the First and Second Laws of thermodynamics could apply together' – a claim that was often presented previously as simply 'another interesting point'.[8]

Martinez-Alier's current reputation as the foremost historian of ecological economics makes his criticism in this regard particularly important. Nevertheless, it should be noted, he is not the only one to issue such charges. Much earlier, the renowned social theorist Daniel Bell suggested: 'He [Engels] attacked the formulation of the second law of thermodynamics, as set forth by

7 See the discussion in Chapters 2 and 3 above.

8 Martinez-Alier 2006, pp. 275–6; also see Martinez-Alier 1995, p. 71; Martinez-Alier 2005, p. 5; Martinez-Alier 2007, p. 224.

Clausius in 1867, because of its implicit argument that matter is creatable and destructible'.[9] It appears as though Bell based this claim largely on the discussion in Gustav Wetter.[10]

Benedictine priest and distinguished professor of physics Stanley Jaki authored an attack on Engels's *Dialectics of Nature* in which he contended that, for Engels,

> there could be no mercy for Clausius of entropy fame. In Engels' eyes Clausius was a bogeyman scientist whom he tried to discredit, ridicule or dismiss whenever opportunity arose ... Clausius, entropy, and the heat-death of the universe meant one thing for Engels. They represented the most palpable threat to the materialistic pantheism of the Hegelian left for which the *material* universe was and still is the ultimate, ever active reality. Engels made no secret about the fact that the idea of a universe returning cyclically to the same configuration was a pivotal proposition within the conceptual framework of Marxist dialectic. He saw the whole course of science reaching in Darwin's theory of evolution the final vindication of the perennial recurrence of all, as first advocated by the founders of Greek philosophy ... Such a contention depended, of course, on the ability of dissipated energy to reconcentrate itself. This question, an insoluble enigma to the best minds in physics, represented no problem for Engels. While he admitted that radiating heat disappeared, so to speak, into infinite space, he felt sure that the cold bodies of defunct stars must, sooner or later, collide with one another.[11]

More recently, French Marxist Daniel Bensaïd also presented such criticisms in his *Marx for Our Times*. Bensaïd contended that Engels 'adhered to the first principle (conservation of energy), while rejecting the second (its progressive dissipation)'.[12] Engels was said to have done so on 'ideological' grounds; specifically, Engels objected to 'religious extrapolations from the theory of entropy as to a "thermic death sentence on the universe"'.[13] Likewise, Danish professor of the history of science Helge Kragh dismissed Engels for his criticism of 'the idea of an ever-increasing entropy and its consequence, the heat death' –

9 Bell 1966, p. 84.

10 Wetter 1958, pp. 302–3.

11 Jaki 1974, pp. 312–13.

12 Bensaïd 2002.

13 Bensaïd 2002, pp. 330–2.

simply because he saw it as 'ideologically dangerous'.[14] George Steiner referred to the 'strident rejecting of entropy by Engels'.[15] Ecological political economist Kenneth Stokes argued in his *Man and the Biosphere* that 'Engels's understanding of the second law of thermodynamics was clearly partial' and implied 'that dialectical materialism ... can contravene the second law'.[16] In his book *Pulse*, former *Audubon* contributing editor Robert Frenay recently repeated Martinez-Alier's basic charges, according to which Podolinsky urged Marx and Engels

> to consider the central role of energy flows, and with that the effect of the 2nd Law. To their discredit they refused, and proceeded down a road that pointedly ignored the 2nd Law (Engels misunderstood it, thinking it contradicted the 1st) and environmental considerations in general.[17]

More pointedly, Leszek Kołakowski claimed in his *Main Currents of Marxism* that 'the second law of thermodynamics ... appeared to Engels an absurdity, as it posited an over-all diminution of energy in the universe'.[18] Kołakowski went on to disparage what he referred to as 'Engels's statement that the energy dispersed in the universe must also be concentrated somewhere'.[19] This, Kołakowski claimed, was nothing less than an attempt to dispose of the second law of thermodynamics.

Kołakowski raised his objection to Engels's notes related to the second law as part of a much broader attack on Marx's historical materialism and especially on Engels's dialectics of nature. Similarly, Martinez-Alier declared that Engels's '"dialectics of nature" failed him there', that is, in the analysis of the first and second laws of thermodynamics.[20] The implication here is that the 'dialectics of nature' associated with classical historical materialism, and especially with Engels, is itself thrown into doubt by Engels's supposed rejection of the entropy law. More important, however – both for Martinez-Alier and for us here – is the contention that, by allegedly scorning the second law of thermodynamics, Engels (and by implication Marx) severed any possible connection between classical Marxism and ecological economics.

---

14 Kragh 2004, p. 58.

15 Steiner 1975, p. 162.

16 Stokes 1994, p. 246.

17 Frenay 2006, p. 364.

18 Kołakowski 1978, Vol. 1, p. 395.

19 Kołakowski 1978, Vol. 3, p. 150.

20 Martinez-Alier 2006, p. 275.

Bensaïd's argument can be viewed as somewhat distinct because he defended some of Engels's contributions to ecological economics while contending that Engels was led astray by cosmological speculations on the heat death of the universe, which produced conclusions that were more 'ideological' (having to do with materialist philosophy) than scientific. Thus, Bensaïd contended that 'the law of entropy seemed to [Engels] manifestly to be a breach through which religion could make a return. This is a leitmotiv of the notes on physics in the *Dialectics of Nature*'.[21]

All of the above complaints against the founders of historical materialism have only served to feed the widespread myth that classical Marxism was estranged from thermodynamics.[22]

We believe that these issues, particularly the allegation that Engels (and by imputation Marx as well) rejected the second law of thermodynamics, can be decided purely on the evidence. Accordingly, we first examine in considerable detail Engels's notes on thermodynamics and the heat death hypothesis, which have been presented as evidence for the above-mentioned ecological-economic critique of classical Marxism (and for arguments against Engels's dialectics of nature). We compare these preliminary notes to Engels's more developed view in his draft introduction to *The Dialectics of Nature* and in *Anti-Dühring*, both of which the critics generally (with the partial exceptions of Jaki and Bensaïd) ignore. We then locate Engels's notes on the heat death hypothesis in the historical context of nineteenth-century science and explain that this hypothesis was problematic within that context and is even more questionable from the standpoint of present-day cosmology and astrophysics. Although Engels's critics see his rejection of the heat death hypothesis as a violation of basic physics, in reality this hypothesis was looked at sceptically by many leading physicists in Engels's day (including pioneers in thermodynamics such as Mayer, Rankine, Grove, Boltzmann, and, in his later writings, Helmholtz – plus even at one point William Thomson [after 1892 Lord Kelvin]). Engels's notes questioning the heat death hypothesis in fact directly rely on the later criticisms of that hypothesis by none other than Helmholtz, the figure who is usually credited with introducing it!

Following this treatment of Engels and the heat death controversy, we go on to re-examine very briefly (because this has been dealt with at length above) the contention that Marx and Engels rejected the fundamental principles of ecological economics in distancing themselves somewhat from Podolinsky's

21 Bensaïd 2002, p. 332.

22 For example, Faber and Grossman 2000.

perfect-human-machine model. We conclude by explaining how Marx's incorporation of energetics into his open-system analysis of human labour, capitalist machine-driven production and environmental crisis was an outgrowth of his broader materialist and dialectical conception of human-natural history.

The Second Law and the Heat Death of the Universe

The second law of thermodynamics says that in an isolated system, entropy (levels of disorganisation or unutilisable energy) will expand to a maximum.[23] The second law (together with the first law, which stipulates that matter-energy can be neither created nor destroyed) is fundamental to understanding practical problems in the utilisation of energy. Indeed, it was largely the development of the steam engine that germinated the science of thermodynamics. Although the entropy law was extrapolated to the cosmological level by some of the founders of thermodynamics to form the notion of the 'heat death of the universe', the latter notion was questioned in its day and is now generally considered problematic and misleading, obscuring the real complexity of the evolution of the universe.[24] Yet it is precisely the notion of the heat death of the universe as a guaranteed final end that Engels opposed in what has wrongly been called his rejection on that basis of thermodynamics.

It was arguably the failure to recognise the distinction between the second law and the heat death theory that led Martinez-Alier and Naredo to assert 34 years ago that Engels 'studied Clausius' Second Law, but dismissed it in unequivocal terms as being contradictory of the First Law'.[25] In his pathbreaking *Ecological Economics*, Martinez-Alier went on to claim:

---

23 Charles Perrings explains, in relation to ecological economics (but having a larger applicability), that 'under the Second Law an isolated system (not receiving energy from its environment) is characterized by the fact that its entropy will increase up to the point at which it is in thermodynamic equilibrium and energy flows cease. The entropy of an isolated system cannot decrease. On the other hand, a closed system (receiving energy across its boundaries) will still experience the same irreversible increase in the entropy of its mass, but will be able to avoid the oubliette of the thermodynamic equilibrium by tapping the energy flowing into the system from outside' (Perrings 1987, p. 148).

24 Schneider and Sagan 2005, p. 6; Toulmin 1982, pp. 38–9.

25 Martinez-Alier and Naredo 1982, p. 209; see also Martinez-Alier 1995, p. 71; Martinez-Alier 2006, p. 275.

> The second law was mentioned by Engels in some notes written in 1875 which became, posthumously, famous passages of the *Dialectics of Nature*. Engels referred to Clausius' entropy law, found it contradictory to the law of conservation of energy, and expressed the hope that a way would be found to re-use the heat irradiated into space. Engels was understandably worried by the religious interpretation of the second law.[26]

Furthermore, in his article in the 2007 *Socialist Register*, Martinez-Alier stated:

> One intriguing point arises from Engels' unwillingness to accept that the First and Second Laws of thermodynamics could apply together: the 'dialectics of Nature' failed him there. As Engels became aware of Clausius' concept of entropy, he wrote to Marx: 'In Germany the conversion of the natural forces, for instance, heat into mechanical energy, etc., has given rise to a very absurd theory – that the world is becoming steadily colder ... and that, in the end, a moment will come when all life will be impossible ... I am simply waiting for the moment when the clerics seize upon this theory'.[27]

The notion that Engels would have been guilty of either a neglect of thermodynamics or a fundamental misunderstanding of the second law is rather implausible in light of the extensive natural science research of both Marx and Engels. We know from their notes and letters that from the early 1850s onward they studied the works, and/or attended public lectures, of many of the scientists involved in the development of the first and second laws – including not only Clausius and Thomson but also Hermann von Helmholtz, Julius Robert Mayer, John Tyndall, William Robert Grove, James Clark Maxwell, James Prescott Joule, Justus von Liebig, Adolph Fick, Jean-Baptiste Joseph Fourier, Sadi Carnot, Peter Guthrie Tait, Ludwig Boltzmann, and Ludwig Büchner.[28] Marx and Engels kept abreast of the natural-scientific literature and did not dispute the conclusions of natural-scientific research where there was an actual scientific consensus – although they did raise questions about what appeared to be incomplete, inconclusive, partial, and contradictory results.

Virtually the entire case levelled against Engels (and by implication Marx) for questioning the laws of thermodynamics – and the entropy law in partic-

26 Martinez-Alier 1987, p. 221.

27 Martinez-Alier 2006, pp. 275–6.

28 Baksi 1996, 2001; Burkett and Foster 2006; Foster 2000, Chapters 5 and 6.

ular – is based on four paragraphs in his work: a single paragraph in a letter that he wrote to Marx in 1869 and three paragraphs separately written in 1874 or 1875 and included in his *Dialectics of Nature*. All of these paragraphs are directed not at the entropy law but at its extrapolation into a theory of the heat death of the universe. Because these four paragraphs constitute the primary (and, for Martinez-Alier, the sole) basis on which it is claimed that Engels rejected the second law of thermodynamics, they will all be quoted in full below.

The first of these four paragraph-long notes is from a letter that Engels wrote to Marx on 21 March 1869. The boldface is added to highlight the parts of this paragraph that are quoted by Martinez-Alier in his criticism of Engels.[29] Capitalised words are those excluded by Martinez-Alier from his quote *without the appropriate ellipses marking their removal*:

> **In Germany the conversion of the natural forces, for instance, heat into mechanical energy, etc., has given rise to a very absurd theory**, WHICH INCIDENTALLY FOLLOWS LAPLACE'S OLD HYPOTHESIS, BUT IS NOW DISPLAYED, AS IT WERE, WITH MATHEMATICAL PROOFS: **that the world is becoming steadily colder**, that the temperature in the universe is leveling down **and that, in the end, a moment will come when all life will be impossible** and the entire world will consist of frozen spheres rotating round one another.[30] **I am simply waiting for the moment when the clerics seize upon this theory** as the last word in materialism. It is impossible to imagine anything more stupid. Since, according to this theory, in the existing world, more heat must always be converted into other energy than can be obtained by converting other energy into heat, so the original *hot state*, out of which things have cooled, is obviously inexplicable, even contradictory, and thus presumes a god. Newton's first impulse is thus converted into a first heating. Nevertheless, the theory is regarded as the finest and highest perfection of materialism; these gentlemen prefer to construct a world that begins in nonsense and ends in nonsense, instead of regarding these nonsensical consequences as proof

29 Martinez-Alier 2006, pp. 275–6 (see above); Martinez-Alier 2007, p. 224.

30 It was common among scientists in the nineteenth century and even in the early twentieth century to employ the words *universe* and *world* somewhat interchangeably, sometimes using the term *world* for solar system or even universe. For examples, see Rankine (1852, p. 360) and the Friedmann-Einstein correspondence quoted in O'Connor and Robertson (1997). Engels uses *world* in both senses in this paragraph (also using *universe*). Yet the context is the cosmological order. (Note by the present authors).

> that what they call natural law is, to date, only half-known to them. But this theory is all the dreadful rage in Germany.[31]

Martinez-Alier, as we have seen, has repeatedly offered this paragraph from Engels's March 1869 letter to Marx as direct 'evidence' that Engels rejected the second law of thermodynamics. However, a close examination of the entire paragraph shows that Engels's criticism is not levelled at the second law of thermodynamics itself but at two controversial hypotheses that were commonly extrapolated from the second law: the steady cooling down of the earth and the heat death of the universe. The reference to 'Laplace's old hypothesis', which Martinez-Alier removed from the quote without ellipses, is a clear indication (along with other lines later on not included in Martinez-Alier's quotation from Engels) that the argument is primarily about cosmology, that is, the heat death of the universe, and that by 'world' in the first sentence Engels was referring not simply to the fate of the earth itself but to the universe. In short, the context makes it clear that Engels is not concerned with the second law of thermodynamics here as much as with the questionable cosmology that was being built on it.

In an earlier reference to this same letter in his *Ecological Economics*, Martinez-Alier wrote as follows: 'In a letter to Marx of 21 March 1869, when he became aware of the second law, [Engels] complained about William Thomson's attempts to mix God and physics'.[32] But we know that Engels read Grove's *The Correlation of Physical Forces* by 1865 – shortly after Marx. Grove's work included a detailed treatment of the second law, and there is no possibility that Engels or Marx – both of whom frequently praised Grove's book – missed this discussion.[33] Moreover, as Engels was undoubtedly a regular reader of the British *Philosophical Magazine* (the key scientific outlet for British natural philosopher-physicists), he was almost certainly aware of Clausius's concept of 'entropy' from the moment it was introduced to British readers in 1868 through the translation of Clausius's 1867 'On the Second Fundamental Theorem of the Mechanical Theory of Heat'.[34] From the same source he would have encountered the early work of Thomson, Tait, Rankine and others.

31 Marx and Engels 1975a, Vol. 43, p. 246; boldface indicates those words in the quote from Martinez-Alier; capitalised words are those words deleted by Martinez-Alier without ellipses; italics are those in Engels's original.

32 Martinez-Alier 1987, p. 221.

33 Draper 1986, p. 83; Marx and Engels 1975a, Vol. 25, p. 325; Marx and Engels 1975b, p. 162.

34 C. Smith 1998, pp. 256, 361.

Nor is there any mention, contrary to what Martinez-Alier claims here, of Thomson in the above-quoted letter. Indeed, Thomson was a *Scottish* physicist, whereas Engels's letter refers only to the heat death theory as promulgated *in Germany*.

In his 1869 letter, Engels sees a possible contradiction between *the heat death theory* (specifically its requirement for an exogenous 'first heating') and the conservation of energy. Engels sees this 'nonsensical consequence' (the heat death hypothesis) as a puzzle for a consistently materialist philosophy of science – one that can be solved only through future scientific research that deepens our knowledge of what is currently 'to date, only half-known'. Clearly, Engels feels that to accept the heat death theory as 'the finest and highest perfection of materialism' would be to hold back the progress of this scientific research. Far from contravening the second law in this March 1869 letter, as Martinez-Alier suggests, Engels does not even mention the entropy law in his letter, which is directed instead against the heat death hypothesis.[35]

In the section of Engels's *The Dialectics of Nature* titled 'Notes on Physics', there are three paragraph-long notes written in 1874 or 1875 on the heat death of the universe hypothesis that have been cited but not usually quoted – and never quoted in full – by Engels's critics.[36] These paragraphs have been interpreted by Martinez-Alier, Kołakowski, and others as offering further evidence of Engels's rejection of the second law of thermodynamics. They are therefore included in their entirety below. As in Marx and Engels's *Collected Works*, the paragraphs are separated from each other, indicating that they are distinct, if related, notes. All three, as a close reading will show, are unmistakably directed against the heat death hypothesis. The first paragraph establishes the substance of Engels's objections and the following two his sense that the heat death hypothesis creates all sorts of theoretical difficulties (even absurdities) for any consistent materialist interpretation of cosmological developments:

> *Radiation of heat into universal space*. All the hypotheses cited by Lavrov of the renewal of extinct heavenly bodies (p. 109) *involve loss of motion*. The heat once radiated, i.e., the infinitely greater part of the original motion, is and remains lost. Helmholtz says, up to 453/454.[37] Hence one finally

35 Martinez-Alier 1987, p. 221.

36 On the dating of the three paragraphs from the *Dialectics of Nature*, see Marx and Engels 1975a, Vol. 25, p. 697.

37 Engels is referring here in his notes to Helmholtz's *Populäre wissenschaftliche Vorträge* – a work he utilised extensively in his notes in *The Dialectics of Nature*. Helmholtz included here possible counters to the heat death hypothesis on the lines that Rankine had devel-

arrives after all at the exhaustion and cessation of motion. The question is only finally solved when it has been shown how the heat radiated into universal space becomes *utilisable* again. The theory of the transformation of motion puts this question categorically, and it cannot be got over by postponing the answer or by evasion. That, however, with the posing of the question the conditions for its solution are simultaneously given – *c'est autre chose* [that is quite another thing]. The transformation of motion and its indestructibility were first discovered hardly thirty years ago, and it is only quite recently that the consequences have been further elaborated and worked out. The question as to what becomes of the apparently lost heat has, as it were, only been *nettement posée* [clearly posed] since 1867 (Clausius). No wonder that it has not yet been solved; it may still be a long time before we arrive at a solution with our small means. But it will be solved, just as surely as it is certain that there are no miracles in nature and that the original heat of the nebular ball is not communicated to it miraculously from outside the universe. The general assertion that the *total amount* (die *Masse*) *of motion is infinite*, and hence inexhaustible, is of equally little assistance in overcoming the difficulties of each individual case; it too does not suffice for the revival of extinct universes, except in the cases provided for in the above hypotheses, which are always bound up with loss of force and therefore only temporary cases. The cycle has not been traced and will not be until the possibility of the re-utilisation of the radiated heat is discovered.[38]

∴

Clausius – if correct – proves that the universe has been created, *ergo* that matter is creatable, *ergo* that it is destructible, *ergo* that also force, or motion, is creatable and destructible, *ergo* that the whole theory of the 'conservation of force' is nonsense, *ergo* that all his conclusions from it are also nonsense.[39]

---

oped and that Engels was to recapitulate in part in his notes here (Marx and Engels 1975a, Vol. 25, p. 562; Sternberger, 1977, pp. 37–8). (Note by the present authors).

38 Marx and Engels 1975a, Vol. 25, pp. 561–2; Engels refers at this point to Volume 1 of Pyotr Lavrovich Lavrov's *Attempt at a History of Thought*, published in St. Petersburg in 1875, in which, in a chapter titled 'The Cosmic Basis of the History of Thought', Lavrov addresses the question of extinct suns and their systems of planets and how these cannot be revived but may become the material for accelerating the formation of new worlds (Marx and Engels 1975a, Vol. 25, p. 682). See the discussion of Lavrov's ideas below.

39 Marx and Engels 1975a, Vol. 25, p. 562.

> *Clausius' second law*, etc., however it may be formulated, shows energy as lost, qualitatively if not quantitatively. *Entropy cannot be destroyed by natural means but it can certainly be created.* The world clock has to be wound up, then it goes on running until it arrives at a state of equilibrium from which only a miracle can set it going again. The energy expended in winding has disappeared, at least qualitatively, and can only be restored by an *impulse from outside*. Hence, an impulse from outside was necessary at the beginning also, hence, the quantity of motion, or energy, existing in the universe was not always the same, hence, energy must have been created, i.e., it must be creatable, and therefore destructible. *Ad absurdum!*[40]

Martinez-Alier and the other critics of Marx do not generally draw extensively from the foregoing passages from *The Dialectics of Nature*, but instead confine themselves to quoting a sentence here or there. We are thus offered on flimsy evidence broad and seemingly authoritative interpretations that the reader unfamiliar with the texts, issues, or historical context has no reason to doubt. As Martinez-Alier puts it in *Ecological Economics* (in a passage quoted more fully above),

> The second law was mentioned by Engels in some notes written in 1875 which became, posthumously, famous passages of *The Dialectics of Nature*. Engels referred to Clausius' entropy law, found it contradictory to the law of conservation of energy, and expressed the hope that a way would be found to re-use the heat irradiated into space.[41]

Here, the use of the words *utilisable* and *re-utilisation* with respect to energy in space by Engels is interpreted by Martinez-Alier in *direct human utilitarian terms*, as if such energy could actually be directly 're-used' by human beings. The wording employed by Martinez-Alier – the substitution of *re-use* for *re-utilisation* – doubtless makes it sound to most readers as if this heat dissipated into space is (quite absurdly) to be put back into use by human beings, opening up Engels to ridicule. Yet this fails to acknowledge here that terms such as *work* and *utilisable energy* as employed in physics refer to *physical forces* independent of human action. The notion of utilisable energy is seen as the potential of a system to do 'work'. As we shall see, nowhere in *The Dialectics of Nature*

---

40 Marx and Engels 1975a, Vol. 25, p. 563.

41 Martinez-Alier 1987, p. 221; see also Martinez-Alier 2006, p. 276; Martinez-Alier 2007, p. 224.

can be found any suggestion whatsoever that such a purposeful recovery and reuse of irradiated energy on the terrestrial level might be possible. Rather, Engels simply raised the question of a natural, nonpurposeful reconcentration (or 'reutilisation') of 'the heat radiated into universal space' as a countertendency or offset to *universal* heat death on the level of the entire cosmos.[42]

In other words, Engels's discussion was limited to the cosmological space timeframe and in no way implied a rejection of the second law as applied to terrestrial dimensions and timeframes. Although Engels's notes raised questions about the ultimate effects of the second law of thermodynamics, they did so not in terms of the physics of the earth or even the solar system, but rather in terms of its extrapolation into a theory of the universe, the laws of which, as he indicated, were little understood. In this respect, the real nature of Engels's argument and its connection to similar arguments by natural scientists in his time (though it is acknowledged that Engels was not entirely alone in his views in this respect) are never analysed by Martinez-Alier.

Bensaïd, for his part, quoted three discontinuous sentences from the first of these paragraphs and one sentence from the third paragraph as *prima facie* evidence that Engels rejected the heat death theory and thus the entropy law.[43] But all that was really conveyed was Engels's scepticism regarding the heat death hypothesis, extrapolated from the second law. Bensaïd, interestingly, did note that Engels was not alone in raising these questions and that 'for a long time to come, physicists would question whether reconcentration of the enormous quantity of energy radiated in all directions was possible'.[44]

---

42 Marx and Engels 1975a, Vol. 25, p. 562.

43 Bensaïd 2002, p. 332.

44 Bensaïd 2002, p. 331; Bensaïd quotes a further stand-alone paragraph (in addition to the other three paragraphs from the *Dialectics of Nature* already quoted in the text above) that was taken from the section on 'Mechanics and Heat' and written as early as 1873 (Marx and Engels 1975a, Vol. 25, pp. 551, 695). This fragmentary paragraph (not even a complete sentence) says: 'Newtonian attraction and centrifugal force – an example of metaphysical thinking: the problem not solved but only *posed*, and this preached as the solution. – Ditto Clausius' dissipation of heat'. Engels's note here refers to Clausius's lecture 'On the Second Fundamental Theorem of the Mechanical Theory of Heat', delivered to the German Scientific Association in Frankfurt-am-Main on 23 September 1867, in which Clausius claimed that the tendency for entropy to reach a maximum level meant that eventually 'the universe would be in a state of unchanging death' (quoted in Brush 1978, p. 61; see also Smith 1998, p. 256). Given this, and the fragmentary nature of Engels's note (clearly meant for himself), it can hardly be said to provide unequivocal support for the conclusion that Bensaïd imposes on it, namely that Engels 'obstinately declined to accept Clausius' principles', meaning the second law of thermodynamics (Bensaïd 2002, p. 332).

Kołakowski, as we have seen, charged in relation to the above passages that Engels wrongly stated 'that the energy dispersed in the universe must also be concentrated somewhere' – as Engels sought to refute an 'over-all diminution of energy in the universe'. This was interpreted by Kołakowski as an attempt to 'dispose' of the second law of thermodynamics.[45] In Kołakowski, in contrast to Martinez-Alier, however, the real issue of the *reconcentration* of energy, as it was raised by leading physicists, including many of the major figures in the development of thermodynamics in Engels's day, was at least acknowledged.

It cannot be repeated too frequently that a close scrutiny of the above passages from Engels's 1869 letter and his notes in *The Dialectics of Nature* reveals that Engels was not challenging the second law of thermodynamics at all, but rather its much more dubious cosmological extrapolation or extension in the form of the heat death of the universe conception, which was being used by physicists such as Thomson and Tait to promote a Christian eschatology. Engels was particularly disturbed by the notion that the universe was simply winding down (like a clock) and would eventually descend into some sort of motionless equilibrium. He was also sceptical regarding the obvious implication that this pointed to a moment of creation, seemingly contradicting the first law. As he had indicated in his March 1869 letter to Marx, theories of the universe were being propounded on the basis of natural laws 'only half known'. A certain dialectical scepticism was therefore to be maintained.

In order to reconstruct more fully Engels's views reflected in the notes from the *Dialectics of Nature* quoted above and shed some light on the evolution of his analysis, it is necessary to look at his subsequent 'Introduction' to the *Dialectics of Nature* and his later work *Anti-Dühring*, both of which the critics (aside from Bensaïd and Jaki) have generally ignored. Engels's 'Introduction' was written in 1875–6, *directly after* his notes on the second law and the heat death hypothesis were jotted down and some *six or seven years after* his 1869 letter to Marx on the subjects of the earth cooling and heat death hypotheses. Furthermore, Engels also explored these issues in *Anti-Dühring*, published in 1877–8, that is, not long after he drafted the 'Introduction' to the *Dialectics of Nature*. In contrast to his initial scattered notes, these writings constitute Engels's more developed understanding of these questions.

Engels's 1875–6 draft introduction to the *Dialectics of Nature*, which in contrast to his earlier hurried jottings on the heat death hypothesis was clearly intended for publication (though he did not publish it in his lifetime), shows the care and circumspection with which he actually approached these issues.

45 Kołakowski 1978, Vol. 3, p. 150, Vol. 1, p. 359.

He began by explaining the superiority of ancient Greek natural philosophy, by which he meant primarily Heraclitus, Aristotle, and Epicurus, to the mechanistic natural science of the Enlightenment. 'For the Greek philosophers the world was essentially something that had emerged from chaos, something that had developed, that had come into being'. In contrast, Newtonian mechanism 'everywhere … sought and found its ultimate resort in an impulse from outside [God] that was not to be explained from nature itself'. Only in the late eighteenth century, he argued, did this begin to break down with the nebular hypothesis on the origins of the solar system introduced by Kant and Laplace. For Engels, a dialectical approach was grounded in nature's (including the universe's) evolution and could not rely on the 'first movers' and 'final causes' characteristic of religion. He argued, moreover, that materialist science invariably developed such an evolutionary approach as its overall analysis was extended.[46]

This set the theme for Engels's entire introduction in which he recounted the development of science and ended by raising questions about the cooling of the earth and the heat death hypothesis. Engels closed his discussion in the last five pages with a brief narrative of the inevitable death of the solar system because of the eventual cooling of the sun. In 1862, Thomson had published two articles – 'On the Secular Cooling of the Earth' and 'On the Age of the Sun's Heat' (the latter article also questioned the heat death hypothesis) – in which he postulated the cooling of the sun over as little as a few million years. Thomson's results were accepted by Engels and knowledgeable scientific observers at the time. Later scientific discoveries in radioactivity and nuclear physics, however, were to show that this was based on a faulty notion of the source of the sun's energy and that the sun's cooling was far slower than had been supposed, on the order of billions of years.[47]

Engels even presented the heat death hypothesis as a major conclusion of science, the truth of which he did not directly deny:

> Millions of years may elapse,[48] hundreds of thousands of generations be born and die, but inexorably the time will come when the declining warmth of the sun will no longer suffice to melt the ice thrusting itself

46 Marx and Engels 1975a, Vol. 25, pp. 322–4.

47 Brush 1978, pp. 34–5; Eiseley 1958, pp. 233–53; Thomson 1862, 1891.

48 Here again, Engels's estimate was based on William Thomson's calculations on the age of the earth and of the sun (considered to be the best estimates of the time), which were later shown to be faulty with the discovery of radioactivity and nuclear energy. See Eiseley 1958, pp. 233–53.

> forward from the poles; when the human race, crowding more and more about the equator, will finally no longer find even there enough heat for life; when gradually even the last trace of organic life will vanish; and the earth, an extinct frozen globe like the moon, will circle in deepest darkness and in an ever narrower orbit about the equally extinct sun, and at last fall into it. Other planets will have preceded it, others will follow it; instead of the bright, warm solar system with its harmonious arrangement of members, only a cold, dead space will pursue its lonely path through universal space. And what will happen to our solar system will happen sooner or later to all the other systems of our island universe; it will happen to all the other innumerable island universes, even to those the light of which will never reach the earth while there is a living human eye to receive it.[49]

But then Engels asked a pregnant question (indicating that this was even more speculative): 'And when such a solar system has completed its life history and succumbs to the fate of all that is finite, death, what then? Will the sun's corpse roll on for all eternity through infinite space?'[50]

Engels made it clear that he viewed the solar system as part of a larger 'island universe' – a term introduced by Kant in 1755 in his *Universal Natural History and the Theory of the Heavens* to describe what we would now call 'galaxies' – within a broader perspective that includes other island universes beyond our empirically discernible knowledge.[51] In this view, island universes (or galaxies)

---

49 Marx and Engels 1975a, Vol. 25, pp. 331–2. Ironically, given all of the criticisms of Engels for supposedly rejecting the second law on the basis of his denial of the heat death hypothesis, Jonathan Hughes quotes the above passage in his *Ecology and Historical Materialism* to demonstrate *that Engels did support the heat death hypothesis* (2000, p. 61). He is seemingly unaware of both Engels's notes challenging that hypothesis and the wider controversy. Likewise, Nicholas Churchich, citing the same passage from Engels, attacks Engels for his 'very pessimistic and apocalyptic vision' – his view of 'dialectical regression' – resulting from his *adoption* of the heat death of the universe hypothesis rooted in the entropy law (Churchich 1990, p. 216).

50 Marx and Engels 1975a, Vol. 25, p. 332. In Newtonian physics, to quote Einstein, 'the stellar universe' was usually thought 'to be a finite island in the infinite ocean of space' (2006, p. 98).

51 Fraser 2006, pp. 87–8; Gribben 1998, p. 28. The island-universe theory introduced by Kant claimed that the distant spiral nebulae were other 'island-universes' like the Milky Way. This was controversial in the nineteenth century but was embraced by Engels and by some nineteenth-century astronomers and scientists. The theory was to triumph in the twentieth century with the discoveries of Hubble (see Fraser 2006, pp. 87–8). The

rather than stars were the unit of analysis in astronomy, a viewpoint that later triumphed in the early twentieth century. This raised the issue of the stellar universe beyond the solar system and even beyond the Milky Way.

Here Engels partly relied on the argument of Pyotr Lavrovich Lavrov in his *Attempt at a History of Thought*, published anonymously in St. Petersburg in 1875 and sent by the author to Engels that same year, which contained a chapter titled 'The Cosmic Basis of the History of Thought'. In this work, Lavrov had argued that

> isolated island masses ... gravitate towards one another and move under influence of this gravitation, which thus constitutes the most general cosmic phenomenon accessible to us ... We scarcely know even one island universe in immeasurable space, viz. the one to which we ourselves belong. By means of thought we can convince ourselves of the *probability* of the existence of other island universes beyond its boundaries, of the reality of which mankind will never be certain; but everything that we *know* of the *universe* is restricted to *our* single island universe.[52]

Engels's dynamic conception of the 'island universe' (and of 'island universes', which we, in our expanding concept of the universe – also seen as expanding – now call galaxies) was based not only on the work of Lavrov but also on that of the Italian astronomer Pierro Angelo Secchi and the German astronomer Johann Heinrich von Mädler.[53] In this conception, the death of one star, and one solar system (and indeed one 'island universe'), could possibly become the basis for the formation and evolution of others under the force of gravitation. This theory did not contradict the entropy law because it was conceptualised in open terms, that is, in relation to the interaction of solar systems and island universes (plural), not isolated systems.[54]

---

term *island-universes* is still sometimes used today to refer to other 'pocket universes' in inflationary theories of the cosmos. See, for example, Vilenkin 2006, pp. 81–2, 203–4.

52 Quoted in Engels 1940, pp. 352–3.

53 Marx and Engels 1975a, Vol. 25, p. 328.

54 Johann Heinrich von Mädler (1794–1874) was a distinguished German astronomer. He is most famous for his early attempts to map the moon and his 'central sun hypothesis' on the location of the centre of the galaxy (which proved wrong). He published a general two volume *History of Descriptive Astronomy* in 1873. In the nineteenth century, the Milky Way, at least a billion times larger than the solar system, was thought to be the entire universe. Mädler, along with others, hypothesised that other cosmological bodies then described as nebulae were similar masses of stars, like our Milky Way galaxy. Likewise, his view of

As Lavrov explained more fully,

> Dead suns with their dead systems of planets and satellites continue their motion in space as long as they do not fall into a new nebula in the process of formation. Then the remains of the dead world become material for hastening the process of formation of the new world ... [A] world long since dead obtains the possibility of entering in the process of formation of a new solar system, there a world in formation which has come close to rigid masses is disintegrated into comets and falling stars. Violent death threatens worlds just as easily as inevitable natural extinction. But eternal motion does not cease, and new worlds eternally develop in place of former ones.[55]

Similarly, Engels himself wrote,

> The sudden flaring up of new stars, and the equally sudden increase in brightness of familiar ones, of which we are informed by astronomy, are most easily explained by such collisions. Moreover, not only does our group of planets move about the sun, and our sun within our island universe, but our whole island universe also moves in temporary, relative equilibrium with the other island universes, for even the relative equilibrium of freely floating bodies can only exist where the motion is reciprocally determined; and it is assumed by many that the temperature in space is not everywhere the same.[56]

The dynamic analysis of the universe or universes presented here by Lavrov and Engels, building on the work of astronomers Mädler and Secchi, obviously did not contradict the entropy law yet raised questions related to the heat death hypothesis.

Engels significantly quoted an 1872 work by Secchi, which similarly asked 'are there forces in nature which can reconvert the dead system into its original

---

extinct suns is now confirmed (see J.B.S. Haldane's note in Engels 1940, p. 14; also see Coles 2001, p. 7; Royal Astronomical Society 1875). Angelo Secchi (1818–78) was a famous Italian astronomer (pioneering in the classification of stars), an early astrophysicist, and director of the Rome Observatory ('Angelo Secchi' 1913, pp. 669–71).

55 Quoted in Engels 1940, p. 353.

56 Marx and Engels 1975a, Vol. 25, pp. 333–4; Engels's discussion of the movement of island universes and the way in which this related to temperatures in the stellar universe as a whole was closely related to the argument of Grove.

state of glowing nebula and re-awaken it to new life?' Secchi's answer was simply, 'We do not know'.[57] After redescribing heat death as a situation where 'all existing mechanical motion will be converted into heat and the latter radiated into space, so that ... all motion in general would have ceased', Engels conjectured that 'in some way, which it will later be the task of scientific research to demonstrate, it must be possible for the heat radiated into space to be transformed into another form of motion, in which it can once more be stored up and become active'.[58]

Engels even suggested, based on the Mädler-Secchi-Lavrov argument on the death and formation of stellar systems, that within infinite space there is the possibility of 'an eternal cycle' of universal entropic dissipation, reconcentration, and redissipation of energy, operative over 'periods of time for which our terrestrial year is no adequate measure'.[59] Significantly, Engels in these carefully written passages intended for publication did not employ the easily misunderstood term *re-utilisation* in relation to energy radiated into space, which he had jotted down in his preliminary notes.

Nothing in Engels's discussion of the limits of the heat death hypothesis can be viewed as conflicting with the entropy law precisely because Engels's viewpoint, rooted in the astronomical theories of his time (and not simply on extrapolations from thermodynamics), suggests that the universe is in fact an open, dynamic system.

Engels, as we have seen, had an additional reason for questioning the heat death hypothesis related to his dialectical conception that the universe was a natural-material system removed from any supernatural causes. The heat death hypothesis implicitly relied on some initial exogenous source of motion (usually conceived as emanating from God, as the First Mover). Such reliance

57 Marx and Engels 1975a, Vol. 25, p. 332.

58 Marx and Engels 1975a, Vol. 25, p. 334.

59 Marx and Engels 1975a, Vol. 25, p. 334; Haldane included a note in the 1940 English edition of Engels's *Dialectics of Nature* in which he pointed to the fact that cosmologists were then divided between heat death, cyclical, and historical (infinite with respect to both past and future) theories of the universe and that Engels would have been attracted to either of the latter two (see Engels 1940, p. 24). Jaki (1974, pp. 312–13) is sharply critical of Engels for speculatively raising the issue of an 'eternal cycle' in the universe. Yet he does so in a book that is dedicated almost entirely to criticising notions of an 'oscillating universe' that not only were present in nineteenth-century physics but also were to achieve a prominent place in twentieth-century (and now in twenty-first-century) physics. Most recently, the notion of a cyclic universe – an endless cycle of expansion and contraction – was revived in 2002 by physicists at Cambridge University (see Vilenkin 2006, p. 171).

could be interpreted, Engels argued, as a seeming contradiction between the first and second laws of thermodynamics, but only in the sense of an abrogation of the first law (matter can be neither created nor destroyed) to extrapolate the heat death hypothesis from the second. For Engels, any materialist-scientific theory of the universe's evolution had to be free of initial conditions provided by supernatural creative acts (aka 'intelligent design').[60]

Bensaïd, as noted, argued, based to a considerable extent on Engels's draft introduction to *The Dialectics of Nature*, that 'Engels rejected the second principle of thermodynamics on account of its possible theological consequences'.[61] But it would be better to say that Engels believed that theological concerns sparked the premature extrapolation of the second law into a hypothesis of the inevitable heat death of the universe – a hypothesis whose validity seemed highly doubtful. Overlooking this possibility, Bensaïd instead selectively quoted from Engels's argument to suggest that it was based on a dialectical-materialist 'ideology' and metaphysics that was allowed to override his science. In the process, however, Bensaïd largely passed over Engels's actual scientific arguments, including his reliance on Mädler, Secchi, and Lavrov and the fact that the 'reconcentration' hypothesis arose *from within* thermodynamics in the work of Rankine, Helmholtz, and others (see below). To reduce Engels's argument simply to an irrational 'profession of faith' was therefore a serious error.[62]

The discussion in Engels's 1877–8 *Anti-Dühring* – the central text on historical materialism published during Marx and Engels's lifetimes – shows still further how misleading it is to describe all of Engels's writings on the heat death theory as 'hasty private notes on the second law of thermodynamics' while ignoring his more considered analysis in this area.[63] In Chapter 6 of *Anti-Dühring*, in a section concerned with Dühring's cosmology, Engels adopted the term *motion* for energy (the latter term was only then coming into use), arguing that Dühring arbitrarily 'reduces motion to mechanical force as its supposed basic form, and thereby makes it impossible for himself to understand the real connection between matter and motion'.[64] For Engels, this real connection

60 See the third of the paragraphs from the *Dialectics of Nature* provided as exhibits above (Marx and Engels 1975a, Vol. 25, p. 563).

61 Bensaïd 2002, p. 332.

62 Ibid.

63 Martinez-Alier 1995, p. 71.

64 Marx and Engels 1975a, Vol. 25, p. 55.

> is simple enough. *Motion is the mode of existence of matter*. Never anywhere has there been matter without motion, nor can there be. Motion in cosmic space, mechanical motion of smaller masses on the various celestial bodies, the vibration of molecules as heat or as electrical or magnetic currents, chemical disintegration and combination, organic life – at each given moment each individual atom of matter in the world is in one or other of these forms of motion, or in several forms at once ... Matter without motion is just as inconceivable as motion without matter. Motion is therefore as uncreatable and indestructible as matter itself.[65]

Note how Engels related the different forms of motion or energy to the *different forms of matter* with which energy is bound up in reality. For Engels, the crucial implication of the intrinsic unity and indestructibility of matter and energy is that all apparent cessations of motion represent only states of equilibrium *relative to* the ceaseless motion inherent to the universe as a *qualitatively variegated* material system:

> All rest, all equilibrium, is only relative, only has meaning in relation to one or other definite form of motion. On the earth, for example, a body may be in mechanical equilibrium, may be mechanically at rest; but this in no way prevents it from participating in the motion of the earth and in that of the whole solar system, just as little as it prevents its most minute physical particles from carrying out the vibrations determined by its temperature, or its atoms from passing through a chemical process.[66]

From this perspective, there could be no universal absolute equilibrium in which all motion ceases. Clearly alluding to the heat death theory, Engels did not shy from this conclusion:

> A motionless state of matter is therefore one of the most empty and nonsensical of ideas ... In order to arrive at such an idea it is necessary to conceive the relative mechanical equilibrium, a state in which a body on the earth may be, as [at] absolute rest, and then to extend this equilibrium over the whole universe ... This conception is nonsensical,

65 Marx and Engels 1975a, Vol. 25, pp. 55–6.

66 Marx and Engels 1975a, Vol. 25, p. 56.

> because it transfers to the entire universe a state as absolute, which by its nature is relative and therefore can only affect a *part* of matter at any one time.[67]

In all of his formulations in his draft introduction to *The Dialectics of Nature* and in *Anti-Dühring*, and in his early notes on the heat death hypothesis in *The Dialectics of Nature*, Engels remained consistent throughout with the second law of thermodynamics. He objected only to the extrapolation of the second law into the heat death of the universe hypothesis and to the seeming contradictions that this created within thermodynamics and cosmology as theoretical systems rooted in a consistent materialist outlook.

The Heat Death Hypothesis and Nineteenth-Century Physics

The heat death of the universe hypothesis was promoted by such leading figures in thermodynamics as Helmholtz, Clausius, Thomson, and Tait. Nevertheless, a scientific consensus was lacking. Extrapolation of the heat death hypothesis from the entropy law was opposed in various ways by some of the pioneers in thermodynamics, including Mayer, Rankine, Grove, and Boltzmann, whereas Helmholtz and Thomson, who had both played leading parts in the development of the hypothesis, were later to express serious reservations. As Garber, Brush, and Everitt wrote in their study of Maxwell on heat, 'Scientists' reactions to the idea of heat death were mixed. Acceptance or rejection of the idea followed no disciplinary or national boundaries'.[68] Thomson and Tait appear to have been especially attracted to the notion for religious reasons.[69] Others resisted it because of its apparent conflict with materialism. From the mere fact that there were leading figures in thermodynamics on both sides of the controversy – with Helmholtz and Thomson seemingly lending support at different times to both sides – it stands to reason that the mere rejection of the heat death hypothesis did not entail the abandonment of thermodynamics in general or the entropy law more specifically.

What became known as the 'heat death of the universe' hypothesis was first suggested by Thomson in reference to the solar system and fate of the earth as

67 Ibid.

68 Garber, Brush and Everitt 1995, p. 56.

69 C. Smith 1998, pp. 111, 119–20, 253–4.

early as 1851–2.[70] But its origin is conventionally, if somewhat mistakenly, traced to a lecture, 'On the Interaction of Natural Forces', delivered in Königsberg in 1854 by Helmholtz (one of the co-discoverers of the first law of thermodynamics). Helmholtz made it explicit that this hypothesis was to apply to the entire universe, stating,

> If the universe be delivered over to the undisturbed action of its physical processes, all force will finally pass into the form of heat, and all heat come into a state of equilibrium. Then all possibility of a further change will be at an end, and the complete cessation of all natural processes must set in. The life of men, animals, and plants, could not of course continue if the sun had lost its high temperature, and with it his light ... In short, the universe from that time forward would be condemned to a state of eternal rest ... At all events we must admire the sagacity of Thomson, who ... was able to discern consequences which threatened the universe, though certainly after an infinite period of time, with eternal death.[71]

Clausius was later to argue in 1867 that as '*the entropy of the universe tends toward a maximum* ... and supposing this condition to be at last completely attained, no further change could evermore take place, and the universe would be in a state of unchanging death'.[72]

Nevertheless, even before Helmholtz delivered his 1854 lecture, the heat death hypothesis had come under strong attack within thermodynamics. After Thomson presented his early version of the theory under the title 'The Universal Tendency in Nature to the Dissipation of Mechanical Energy' to the Royal Society of Edinburgh in April 1852, William Rankine issued a rebuttal, 'On the Reconcentration of the Mechanical Energy of the Universe', before the Belfast meetings of the British Association for the Advancement of Science in September of that same year. This is significant because Rankine was one of the foremost contributors to thermodynamics and formulated the basic entropy idea even before Thomson and Clausius.[73] He now argued that although there is a tendency toward 'an end of all physical phenomena', it was still

> conceivable that, at some indefinitely distant period, an opposite condition of the world may take place, in which the energy which is now being

70 Lindley 2004, pp. 108–9; C. Smith 1998, p. 111.
71 Helmholtz 1873, pp. 228–9.
72 Quoted in Brush 1978, p. 61.
73 Lindley 2004, p. 110.

> diffused may be reconcentrated into foci, and stores of chemical power again produced from the inert compounds which are now being continually formed.

More specifically, Rankine supposed that there might be a boundary around what he called 'the visible world' (by which he probably meant the solar system or at most the Milky Way; i.e. the extent of the visible universe) and that 'on reaching those bounds the radiant heat of the world would be totally reflected, and will ultimately be reconcentrated into foci'. Thus, despite the second law of thermodynamics, 'the world, as now created, may possibly be provided with the means of reconcentrating its physical energies, and renewing its activity and life'.[74] As Crosbie Smith portrayed Rankine's argument,

> While not disputing Thomson's claim to 'represent truly the present condition of the universe, as we know it', Rankine refused to accept the pessimistic conclusion. He therefore speculated that radiant heat – 'the ultimate form to which all physical energy tends' – might be totally reflected at the boundaries of the very interstellar medium through which the radiation had been transmitted and diffused. The energy might then be 'ultimately re-concentrated into foci; at one of which, if an extinct star arrives, it will be resolved into its elements, and a store of energy reproduced'.[75]

This disagreement about the heat death hypothesis was clearly a dispute not about entropy itself but about whether there were other physical processes within the universe at large (or beyond the boundaries of the known universe) that could produce an opposite effect. Questioning the heat death theory in this way did not imply any rejection of the second law as such.

Although Engels doubtless paid close attention to articles published in the *Philosophical Magazine*, he did not cite Rankine's article in his notes on the subject (though he did cite Helmholtz's own discussion of this following Rankine). But Rankine's specific conjecture of the 'reflection' of radiant heat from the boundaries of the universe (and Helmholtz's later version of this) continued to crop up in nineteenth-century physics and astronomy and was central to questions raised about the heat death hypothesis in the scientific literature of the day.

---

74 Harman 1982, p. 68; Martinez-Alier 1987, p. 61; Rankine 1852, pp. 359–60.

75 C. Smith 1998, p. 142.

Although Thomson is generally seen to be an originator and strong supporter of the heat death theory, which appealed to him on religious grounds, he nonetheless expressed some reservations in the 1860s. Thus, in his famous 1862 lecture, 'On the Age of the Sun's Heat', he argued, in direct opposition to his own 1852 essay on the universal tendency toward the dissipation of energy, that there were metaphysical reasons for doubting that this universal tendency extended to the universe itself:

> The result [of the entropy law] would inevitably be a state of universal rest and death, if the universe were finite and left to obey existing laws. But it is impossible to conceive a limit to the extent of matter in the universe; and therefore science points rather to endless progress, through an endless space, of action involving the transformation of potential energy into palpable motion and thence into heat, than to a single finite mechanism, running down like a clock and stopping forever.[76]

Helmholtz also raised questions about the heat death hypothesis that he had helped introduce. Helmholtz was to declare, in line with Rankine,

> The heat [radiating out into space] was lost for our solar system, but not for the universe. It radiated out and is still moving out into unending spaces, and we do not know whether the medium carrying the vibrations of light and heat has any frontier where the rays must turn back or whether they will continue their journey to infinity forever.[77]

What is important is that Helmholtz, often credited with having introduced the heat death hypothesis, suggested that Rankine could be right on the 'reconcentration' of energy and frankly admitted that 'we do not know' regarding the supposed heat death of the universe.

Significantly, Engels was taking notes from this very same page of Helmholtz's *Populäre Wissenschaftliche Vorträge* (*Popular Scientific Lectures*) when he raised questions about the reconcentration ('re-utilisation') of energy radiated out into space. He was therefore clearly influenced by Helmholtz's thinking in this regard.[78] Engels's brief allusion to the possible 're-utilisation' of

76 Kragh 2004, p. 46; Thomson 1891, pp. 356–7.

77 Helmholtz 1876, p. 121, translation as quoted in Sternberger 1977, pp. 37–8; see also Harman 1982, p. 68; Martinez-Alier 1987, p. 61; Rabinbach 1990, p. 62.

78 Marx and Engels 1975a, Vol. 25, p. 562.

energy dissipated into space – for which he has been criticised by Martinez-Alier and others – therefore occurred only in notes (not intended for publication) on Helmholtz, who was himself raising the question of the 'reconcentration' of energy in response to the heat death hypothesis. In fact, the context of Engels's notes make it difficult to discern to what extent he was simply relating his version of Helmholtz's views and to what extent he was stating his own.

Helmholtz, in the 1870s, in passages with which Engels was familiar, made it even clearer that he was inclined toward a metaphysical-cosmological view that went against any absolute heat death notion:

> The flame ... may become extinct, but the heat which it produces continues to exist – indestructible, imperishable, as an invisible motion, now agitating the molecules of ponderable matter, and then radiating into boundless space as the vibration of an ether. Even there it retains the characteristics peculiar of its origin, and it reveals its history to the inquirer who questions it by the spectroscope. United afresh, these rays may ignite a new flame, and thus, as it were acquire a new bodily existence.[79]

At about the same time as Helmholtz, Mayer, also one of the co-discoverers of the conservation of energy, stated his doubts that 'the entire machine of creation must eventually come to a standstill'.[80]

The famed English jurist and physical chemist Sir William Robert Grove, whose work was greatly admired by Marx and Engels, combined an acute understanding of the laws of thermodynamics with a deep scepticism about universal heat death. Grove is known as 'the father of the fuel cell' for his development of the platinum-zinc voltaic battery. His theoretical and practical researches greatly enhanced our understanding of the conservation of energy and of the limits to conversion of energy into work. In his main published work, *The Correlation of Physical Forces*, he – much like Engels – questioned

79 Helmholtz 1908, p. 194.

80 Mayer 1870, pp. 566–7; see also Garber, Brush and Everitt 1995, p. 56. The preceding discussion does not exhaust the list of important physicists who rejected the heat death hypothesis in Engels's day. Two American physicists, H.F. Wailing and Pliny Earl Chase, also rejected it and (as Garber, Brush and Everitt explain) 'looked to the reconcentration of energy or rotational break up of dead stars to go on forever creating new worlds, thus preserving the idea of cosmic evolution against the tendency to degradation' (1995, pp. 55–6). Maxwell was apparently motivated to introduce his famous demon hypothesis in examining the second law of thermodynamics as a statistical law to find a way around the divisions in scientific circles generated by the heat death hypothesis.

the heat death theory on the grounds that 'we know not the original source of terrestrial heat; still less that of the solar heat'.[81] And he also – again comparably to Engels – pointed out the likely presence of disequilibrating counters to an entropic 'evening out' of energy on a universal scale, suggesting,

> We know not whether or not systems of planets may be so constituted as to communicate forces, *inter se*, so that forces which have hitherto escaped detection may be in a continuous or recurring state of inter change.
>
> The movements produced by mutual gravitation may be the means of calling into existence molecular forces within the substances of the planets themselves. As neither from observation, nor from deduction, can we fix or conjecture any boundary to the universe of stellar orbs, as each advance in telescopic power gives us a new shell, so to speak, of stars, we may regard our globe, in the limit, as surrounded by a sphere of matter radiating heat, light, and possibly other, forces.
>
> Such stellar radiations would not, from the evidence we have at present, appear sufficient to supply the loss of heat by terrestrial radiations; but it is quite conceivable that the whole solar system may pass through portions of space having different temperatures, as was suggested, I believe, by Poisson; that as we have a terrestrial summer and winter, so there may be a solar or systematic summer and winter, in which case the heat lost during the latter period might be restored during the former. The amount of the radiations of the celestial bodies may again, from changes in their positions, vary through epochs which are of enormous duration as regards the existence of the human species.[82]

In short, Grove based his argument against universal heat death on the indestructibility and non-homogeneity of material forces and on the general difficulty of treating the universe as a finite isolated system in equilibrium. The issue raised by Grove of the limits of the universe as then known (e.g. 'each advance in telescopic power gives us a new shell, so to speak') pointed to the fact that there were too many unknown aspects of cosmological phenomena to arrive at conclusions on the thermal death of the universe based on the second

81 Grove 1873, p. 80.

82 Grove 1873, pp. 80–1.

law of thermodynamics alone. Grove clearly believed that the answers were to be found in the expansion of astronomical data and that the role of gravitation was key. His general approach conformed to principles of scientific inference long understood: that where there are multiple conceivable explanations and possibilities, scientific conclusions – especially when they conflict with known principles – must 'await confirmation'.[83]

Another leading thermodynamic theorist and contemporary of Engels who developed a critique of the heat death theory was the Austrian materialist and physicist Ludwig Boltzmann. Engels was familiar with Boltzmann's work, both directly and via the detailed discussions of Boltzmann in Maxwell's *Theory of Heat*, a book of which Engels made extensive use in *The Dialectics of Nature*.[84] David Lindley describes the state of Boltzmann's thinking on heat death by the years 1895 and 1896:

> In thinking of the universe as a whole, which was generally presumed at that time to be eternal, it might seem that everything would have to settle down into a perfectly uniform, perfectly stable equilibrium – clearly not the heterogeneous universe of stars and planets and empty space that astronomers were beginning to map out. The notion of an inexorable winding down of the universe into a featureless stasis had been pointed out by Clausius, who called it the 'heat death'. Boltzmann now suggested that even in such a state, there would be pockets that, strictly for reasons of chance, ran temporarily away from the general equilibrium and then fell back again. The corner of the universe currently occupied by humanity, he suggested, must be just such a place, where entropy happened to have hit a temporary low and was increasing again. Elsewhere there would be pockets of the universe where entropy was running down, and in such places, Boltzmann speculated, it might appear that time itself was running backward.[85]

Thinkers such as Boltzmann, Grove, Thomson, and Engels argued that a 'basic error' of the heat death hypothesis lay 'in the fact that laws holding for finite

83 See the discussion of ancient Epicurean rules of scientific inference under conditions of 'multiple explanations' in Asmis (1984, pp. 320–30). The same point from Epicurus is emphasised in Marx's dissertation and Engels's commentary on that dissertation (Marx and Engels 1975a, Vol. 1, p. 69; Voden n.d., p. 332).

84 Marx and Engels 1975a, Vol. 25, pp. 389–90, 407–8, 466, 565.

85 Lindley 2001, p. 144. Because Engels died in 1895, he did not have the benefit of Boltzmann's more developed views.

[isolated] systems cannot be transferred without further ado to a universe postulated to be infinite'.[86]

Biologist Ernst Haeckel, famous as the leading promoter of Darwinian ideas in Germany in the nineteenth century and for coining the word *ecology*, also rejected the heat death hypothesis on grounds similar to those of Rankine and Helmholtz, though with a less adequate grasp of the physics involved.[87]

Not only natural scientists but also leading cross-disciplinary thinkers concerned with merging the natural and social sciences propounded the possibility of the 'reconcentration' of energy within the universe as a result of still unknown forces, thereby questioning the heat death hypothesis but not the entropy law. Herbert Spencer was a major populariser of the notion of 'universal death' associated with the heat death controversy. Nevertheless, in an analysis that repeatedly referred to Helmholtz's 1854 paper presenting the heat death theory, he asserted, on the basis of gravitational tendencies and astronomical developments suggesting the collision of stars, that concentration of energy in the universe would proceed to be followed later by its diffusion in an *eternal cycle* of attraction and repulsion. As Spencer put it in his *First Principles*,

> Apparently the universally co-existent forces of attraction and repulsion, which, as we have seen, necessitate rhythm in all minor changes through the universe also necessitate rhythm in the totality of its changes, produce now an immeasurable period during which the attractive forces predominating cause universal concentration and then an immeasurable period during which the repulsive forces predominating cause universal diffusion – alternate eras of evolution and dissolution.[88]

More significantly, economist William Stanley Jevons included in his 1874 *Principles of Science* a section titled 'Speculations on the Reconcentration of Energy' in which he stated that we 'cannot deny the possible truth of ... Rankine's hypothesis' opposing the heat death theory. It nevertheless remained 'practically incapable of verification by observation, and almost free from restrictions afforded by present knowledge'. Jevons argued that Rankine's hypothesis meant

86 Wetter 1958, p. 436.

87 Haeckel 1929, pp. 202–3. Engels was a close reader of all of Haeckel's work, but Haeckel's critique of the heat death theory in *The Riddle of History* was written three years after Engels's death in 1895. Martinez-Alier mentions Haeckel a number of times in his *Ecological Economics*, but takes no note of his critique of the heat death hypothesis.

88 Spencer 1880, pp. 458, 465; see also Brush 1978, pp. 63–4.

that we had to admit to 'the finiteness of the portion of the medium [within the universe] in which we exist', whereas the heat death hypothesis required assumptions about 'the finiteness of [the] past duration of the world' since 'progressing from some act of creation, or some discontinuity of existence'. In either case, he argued, the unity of our physical view is interfered with and 'we become involved in metaphysical and mechanical difficulties surpassing our mental powers'.[89] Jevons's doubts about the universe 'progressing from some act of creation', and his clear sympathy for Rankine's argument on 'reconcentration' – although he emphasised we have no practical way of ascertaining this one way or another – reflect considerations similar to those of Engels.

In this context, it is very significant that Martinez-Alier acknowledged both Rankine's criticism of the heat death hypothesis and even Helmholtz's own criticism of it, in the context of justifying Jevons's criticisms based on Rankine – indicating at the same time that Jevons was fully cognisant of the second law of thermodynamics.[90] His very favourable treatment of Jevons was part of an attempt to present him as one of the founders of ecological economics. The fact that Engels had used some of the same arguments against the heat death hypothesis and developed them far more fully than Jevons, was, however, in *Engels's case*, repeatedly presented by Martinez-Alier as *prima facie* evidence that Engels had rejected the second law of thermodynamics itself. Such intellectual double standards are difficult to fathom.

Today, the Newtonian world of classical physics has been replaced by a much more complex view of the universe. As Peter Coveney and Roger Highfield write in *The Arrow of Time*, the original heat death argument is widely viewed as simplistic or

> flawed because it ignores the role of gravity (and black holes): when gravity is included, it turns out that the universe must go further and further away from the uniform distribution of matter envisaged in the Heat Death ... [Moreover] we know from astronomical evidence that the universe as a whole is expanding, so it cannot be anywhere near a state of thermodynamic equilibrium.[91]

As some of the early critics of the heat death hypothesis, including Grove and Engels, seem to have vaguely suspected in part, it is recognised today that

---

89 Jevons 1900, pp. 751–2; see also Barrow 1994, p. 25.

90 Martinez-Alier 1987, p. 161.

91 Coveney and Highfield 1990, pp. 154–5.

> there is a struggle between gravity, which pulls stars together and provides the energy which heats them inside to the point where nuclear fission begins, and thermodynamics, seeking to smooth out the distribution of energy in accordance with the second law ... The story of the Universe is the story of that struggle between gravity and thermodynamics.[92]

The British cosmologist, theoretical physicist, and mathematician John D. Barrow provides even stronger reasons to doubt the heat death hypothesis in *The Origin of the Universe*:

> It is only recently that cosmologists have realized that the predicted heat death of ever-expanding universes in a future state of maximum entropy will not occur. Although the entropy of the universe will continue to increase, the maximum entropy it can have at any given time increases even faster. Thus the gap between the maximum possible entropy and the true entropy of our universe continually widens ... The universe actually gets farther and farther away from the 'dead' state of complete thermal equilibrium.[93]

Stephen Toulmin has argued on logical grounds that to say that the second law of thermodynamics is a universal law for isolated systems is a different matter from saying that it applies to the 'universe-as-a-whole'. Indeed, it is impossible to know what it would mean to refer to the universe as an isolated system because if it is bounded it has to be bounded by something. Thus, he concludes that 'the conditions necessary for us to apply the Second Law of Thermodynamics to the universe-as-whole are such as *cannot* be satisfied'. 'The

---

92 Gribben 1998, p. 5.

93 Barrow 1994, pp. 26–7; Alex Vilenkin (2006, pp. 170–5), professor of physics and director of cosmology at Tufts University, claims in his *Many Worlds in One* that there are two cosmological theories advanced by physicists today that definitely go against the heat death theory: One is the hypothesis of an inflationary universe (in which 'an eternally inflating universe consists of an expanding "sea" of false vacuum, which is constantly spawning "island universes" like ours'); the other is that of a cyclic universe (in which 'the universe recollapses and immediately bounces back to start a new cycle. Part of the energy generated in the collapse goes to create a hot fireball of matter'). Although such cosmologies of today's physicists are still being strenuously debated, the importance of clearly distinguishing the second law from the heat death theory is obvious.

most it [the second law] could do would be to *imply* something about the universe, and it could do that only if we also knew how far the universe was itself a thermally isolated system'.[94]

Interestingly, although Georgescu-Roegen, the leading figure in twentieth-century ecological economics, frequently referred to the heat death hypothesis of classical physics, he too found it 'intellectually unsatisfactory'. He considered a number of conceivable alternatives presented by physicists, including (a) the Boltzmann-derived hypothesis that 'entropy may decrease in some parts of the universe so that the universe both ages and rejuvenates', and (b) the 'steady state' theory 'in which individual galaxies are born and die continuously'. Both were consistent with the second law, but neither was completely acceptable to him. In the end, he concluded that 'the issue of the true nature of the universe is far from settled'.[95] In considering alternatives to the heat death theory that were consistent with the second law and arguing, in Epicurean-like terms, that the answer must 'await confirmation' with an increase of our knowledge, Georgescu-Roegen's general intellectual position on this issue was not unlike that of Engels a century earlier.

What then is left of the claims of Martinez-Alier, Bell, Kołakowski, Jaki, Bensaïd, Kragh, Stokes, Frenay and others that Engels rejected the second law of thermodynamics? Literally nothing. All of the passages in Engels cited above are criticisms of the heat death of the universe hypothesis extrapolated from the second law of thermodynamics, not of the entropy law itself. All of the arguments that Engels used were similar to, or derived from, those of physicists, astronomers, and scientific commentators in general in his day. If Engels's references in his notes to the 'reconcentration' (or 're-utilisation') of energy are to be taken as proof of the rejection of the second law of thermodynamics, then we would arrive at the absurd conclusion that some of the leading foundational figures in thermodynamics, including Mayer, Helmholtz, Rankine, Grove, and Boltzmann (even Thomson), also rejected the second law. In speculating on this issue (and reaching no definite conclusion), Engels was consistent with the best physics of his day as presented in the top scientific journals, such as *Philosophical Magazine* and *Nature*. Furthermore, given that modern physics has continued to question the heat death of the universe hypothesis – but of course not the second law of thermodynamics – the confusion of the heat death hypothesis with the second law by critics of classical Marxism becomes even more untenable.

94 Toulmin 1982, pp. 40–3.

95 Georgescu-Roegen 1976, p. 8; cf. Georgescu-Roegen 1971, pp. 201–2.

There is no doubt that Engels always adhered to the entropy law with regard to the terrestrial physics of the earth (and the solar system). *The Dialectics of Nature* contains numerous discussions of friction and other entropic processes – passages that verify Engels's deeply held conviction on the correctness of the second law. As Bukharin observed in his *Philosophical Arabesques*,

> Engels ... considered inevitable both the decline of humanity and its extinction, together with the ending of life on the earth as a planet. In other words, human history cannot be divorced in any way from the history of the earth as the base, *locus standi* and source of nourishment of society.[96]

Marxism, the Entropy Law, and Ecology

Unfortunately, what was at first a relatively minor point in a critique of Engels's *Dialectic of Nature* has been transformed into a major criticism (though just as completely devoid of foundations, as we have seen) in a wide-ranging debate on the status of classical Marxism within ecological economics. The criticisms of Engels (and by imputation Marx) levelled by Martinez-Alier, Bensaïd, Stokes, and Frenay in particular are all based on the claim that because Engels allegedly rejected the second law of thermodynamics, he thereby severed at the very start any possible relation between classical Marxism and ecological economics. Usually, this is appended to the wider charge that in failing to take seriously Podolinsky's analysis of energy transfers in agriculture, Marx and Engels closed the door in their day to the development of a Marxist ecological economics, making Podolinsky himself an anomaly in advancing what Martinez-Alier calls his 'ecological Marxism'.[97] (In Bensaïd's case, it should be noted, Engels's critique of Podolinsky's crude energetics is fully accepted, but Engels is still mistakenly chided for 'rejecting' the second law).

What makes this a high-stakes debate is, of course, the history and future of ecological economics and with that ecological analysis as a whole, including its relation to Marxism. For Martinez-Alier, the ecological failure of classical Marxism could not be more straightforward: 'Marx did live after the second law of thermodynamics was established by Sadi Carnot, Clausius, William Thomson, etc. ... he took no account of it in his economic and historical

96 Bukharin 2005, p. 259.

97 Martinez-Alier 1987, p. 62.

doctrines'.[98] Similarly, as he stated in 1995, 'Although Marx and Engels were contemporaries of the physicists who established the laws of thermodynamics in the mid-19th century, Marxian economics and economic history were based on social and economic analysis alone'.[99] Ecological socialist James O'Connor, following Martinez-Alier, asserts that 'Marx did *not* pay sufficient attention to energy economics', including the fact 'that capitalist production (like all production) is based on energy flows and transformations'.[100] As our previous chapters demonstrate, none of these charges stand in the face of criticisms that have been levelled. The allegation that Marx and Engels simply ignored Podolinsky or rejected his important ecological ideas has been thoroughly refuted above.

Martinez-Alier has also argued more broadly that although Podolinsky in his 'ecological Marxism' was a pioneer in trying to quantify energy flows in the economy (and was inspired in this respect by Marx and Engels), there is 'no ecological Marxist history based on quantitative studies of material and energy flows'.[101] This is hardly surprising, of course, if one recognises that attempts to quantify such energy flows on the level of the economy as a whole (or even a sector or industry) are enormously difficult if not practically impossible and prone to all sorts of fallacious conclusions. The shortcomings in Podolinsky's energy accounting are a case in point (see Chapter 2). Here it is worth adding that Georgescu-Roegen actually sided with Engels regarding reservations about Podolinsky's analysis. After briefly relating Martinez-Alier and Naredo's version of the dispute between Marx-Engels and Podolinsky, Georgescu-Roegen comes out in support of Engels's objections regarding 'the fallacy of the energy theory of economic value', writing,

> But thoughts such as Podolinsky's must have been ventilated earlier, for Engels had already protested in an 1875 note: 'Let someone try to convert

---

98 Martinez-Alier and Naredo 1982, p. 209.

99 Martinez-Alier 1995, p. 72.

100 O'Connor 1998, p. 122. Some other ecosocialists are even more critical of Marx in this regard. Thus, Enrique Leff has made the blanket statement that 'Marx's theory of production does not incorporate natural and cultural conditions that participate in the production of value' (1993, p. 48). Subsequent research has definitively overthrown this view in both respects. See especially Burkett 2014.

101 Martinez-Alier 2007, pp. 223, 229; see also Martinez-Alier 1987, p. 62. However, it is worth noting in this context that Soviet community ecologists in the 1920s and early 1930s were pioneers in the development of trophic dynamics, that is, in conducting studies in nutrient cycling related to modern ecosystem analysis (Weiner 1988).

> any skilled labor into kilogram-meters and then to determine wages on this basis!', a thought that ought to kill in the bud any temptation to replace economics by some energetics.[102]

Apart from the Podolinsky episode, a rejection of the second law of thermodynamics would be scarcely conceivable in terms of Marx's (and Engels's) own materialist critique of political economy. Marx's *Capital* is permeated throughout with thermodynamic concepts, the basis for which lay in Marx and Engels's very detailed scientific investigations into physics, chemistry, physiology, agronomy, and so on (see Chapter 3). The very concept of 'labour power', so central to Marx's analysis, arose in part from the new thermodynamics, beginning with Helmholtz. Marx's detailed analysis of steam engines and other forms of machine power (hydraulic and electrical) led him to address thermodynamic conceptions, as did his analysis of the physiological basis of labour. There is no doubt that Marx's *Capital* was the first major economic treatise – and the only one in the nineteenth century – to incorporate within its analysis thermodynamic concepts together with economic value categories. Nor was this an accident. It arose from his dialectical treatment of capitalist commodity production as a contradictory relation of use value and exchange value, and of labour and labour power. It was part of his larger materialist and dialectical conception of history.

Conclusion: The Dialectics of Nature and Society and the Second Law

Marx and Engels's deep concern with thermodynamics and their recognition of its importance for the dialectics of nature and society were appreciated by early Marxists, particularly in the Soviet Union of the 1920s and early 1930s. As a leading early Soviet physicist and sociologist of science Boris Hessen wrote in 1931,

> As soon as the thermal form of motion appeared on the scene ... the problem of energy came to the forefront. The very setting of the problem of the steam engine (to raise water by means of fire) clearly points to its connection with the problem of the conversion of one form of motion into another. It is significant that Carnot's classic work has the title: 'On the

102 Georgescu-Roegen 1986, pp. 8–9; compare Martinez-Alier and Naredo 1982.

> Motive Force of Fire' … [The] treatment of the law of the conservation and conversion of energy given by Engels, raises to the forefront the qualitative aspect of the law of conservation of energy, in contradistinction to the treatment which predominates in modern physics and which reduces this law to a purely quantitative law – the quantity of energy during its transformations. The law of the conservation of energy, the teaching of the indestructibility of motion has to be understood not only in a quantitative but also in a qualitative sense … in the circumstance that matter itself is capable of all the endless variety of forms of motion … in their self-movement and development.[103]

In this context, it is indeed ironic that Martinez-Alier claims with respect to Engels's remarks on the heat death of the universe hypothesis that 'the dialectics of nature failed him there' (a criticism also levelled by Kołakowski).[104] In fact, it was Engels's dialectical conception of nature that allowed him to maintain a healthy scepticism regarding the heat death theory – a scepticism shared by many leading scientists of the day – while still supporting the second law of thermodynamics.

Ironic too – given the repeated claims that Engels rejected the second law of thermodynamics – Marx and Engels's critique of the static character of classical mechanistic physics and of its failure to comprehend the open, dynamic aspects of natural evolution at all levels (including the cosmological) has been lauded by none other than Ilya Prigogine, the 1977 Nobel Prize winner in chemistry and a pioneer in nonequilibrium thermodynamics:

> The idea of a history of nature as an integral part of materialism was asserted by Marx, and, in greater detail, by Engels. Contemporary developments in physics, the discovery of the constructive role played by irre-

103 Hessen 1931, pp. 202–3. The powerful, genuinely materialist and dialectical approach to science reflected by Hessen was severely circumscribed in the Soviet Union in the 1930s, coinciding with Stalin's rise to power. Hessen was arrested in 1936 and executed later that year (Rosenfeld 2012, p. 143; Graham 1993, p. 151). The dominant traditions in 'Western Marxism', meanwhile, largely abandoned natural-scientific concerns. An exception was the work of British Left scientists in the 1930s to 1960s. This included figures such as Haldane, Hyman Levy, J.D. Bernal, Lancelot Hogben, Benjamin Farrington, and Joseph Needham, all of whom adhered to the materialist philosophical outlook dating back to Epicurus and were inspired by the work of Marx and especially Engels's *Dialectics of Nature*. See, for example, Needham 1976.

104 Martinez-Alier 2006, p. 275.

> versibility, have thus raised within the natural sciences a question that has long been asked by materialists. For them, understanding nature meant understanding it as being capable of producing man and his societies.
>
> Moreover, at the time Engels wrote his *Dialectics of Nature*, the physical sciences seemed to have rejected the mechanistic world view and drawn closer to the idea of an historical development of nature. Engels mentions three fundamental discoveries: energy and the laws governing its qualitative transformation, the cell as the basic constituent of life, and Darwin's discovery of the evolution of species. In view of these great discoveries, Engels came to the conclusion that the mechanistic world view was dead.[105]

For a while, the heat death hypothesis seemed to provide a viable mechanistic answer that aligned with Christian theological conceptions. In his own development of this idea, Thomson had quoted the 102nd Psalm: 'They shall perish, but thou shalt endure: yea, all of them shall wax old like a garment; as a vesture shalt thou change them, and they shall be changed'.[106] Marx and Engels resisted this rigid mechanical philosophy and theology. In the end, they developed a materialist-dialectical conception of history that is far more evolutionary in perspective, more in tune with the complexity of the physics of open systems, and thus more in line with an analysis of ecological necessity – a complex, contingent necessity that does not rule out human freedom.

From the days of Newton and Leibniz, attempts have often been made to wed mechanistic models of the universe to both a strong determinism and a religious cosmology. Gottfried von Leibniz saw God as the supreme and all-seeing clock-maker who determined the world and its outcomes down to the minutest details for all time. As he put it,

> 'In the least of substances, eyes as piercing as those of God could read the whole course of things in the universe, *quae sint, quae fuerint, quae mox futura trahantur*' (those which are, which have been, and which shall be in the future).[107]

The second law of thermodynamics was interpreted by such early pioneers as Clausius, Thomson, and Tait as an inexorable and unstoppable tendency

105 Prigogine and Stengers 1984, pp. 252–3.

106 C. Smith 1998, p. 111.

107 Quoted in Prigogine 1997, p. 12.

toward a predetermined final end, eternal death (the heat death of the universe), which fit in with a Christian eschatology.

This, however, ran into a conflict with a more open, dialectical view. This is presented by Prigogine in the first chapter of his *The End of Certainty* as 'Epicurus's Dilemma'. The ancient Greek atomistic philosopher Epicurus had built his physics on mechanical principles in the movement of atoms that he had drawn from Democritus. But Epicurus introduced a subtle change in the theory. In falling toward the earth, atoms on occasion swerved almost imperceptibly from a straight line, creating contingency. A strict determinism was thus impossible. Prigogine argues that only now are we beginning to understand fully the significance of Epicurus's swerve in the development of nonequilibrium physics, both in relation to ecological developments on earth and in phenomena within the cosmos. Epicurus's swerve 'no longer belongs to a philosophical dream that is foreign to physics. It is the very expression of dynamical instability'.[108]

The first modern thinker to focus on and explore in great detail Epicurus's Dilemma, presenting the swerve as an attempt to generate a nonmechanistic materialism rooted in an immanent dialectic, was Karl Marx, who wrote his doctoral dissertation on this problem.[109] Marx saw in Epicurus's nonmechanistic materialism the development of an 'immanent dialectics' in the materialist conception of nature itself – a rejection of determinism along with teleology.[110] This was to define Marx's materialist conception of nature and his open, historical approach to natural phenomena. As Ernst Bloch explained in an eloquent chapter titled 'Epicurus and Karl Marx' in his book *On Karl Marx*, it was Marx who understood the full implications of the materialist dialectic to be found in Epicurus and who adopted 'Epicurus' cuckoo egg, which he alone had laid in the nest of rigid mechanics'. The result was a materialist approach to nature and history that allowed for both subjective and objective factors, freedom and determinacy – 'and, O Epicurus, vise versa, in mutuality'.[111]

Ironically, Marx and Engels anticipated what Karl Popper was to call – in the title of one of his most important works – *The Open Universe*, anticipating as well Popper's rejection of the heat death of the universe hypothesis.[112] In

108 Prigogine 1997, pp. 9–17, 55, 127.

109 Bailey 1928; Bloch 1971, pp. 153–8; Farrington 1967; Foster 2000, pp. 21–65; Marx and Engels 1975a, Vol. 1; Schafer 2006.

110 The quotation comes from Voden's (n.d., pp. 332–3) summary of Engels's comments on Marx's dissertation on Epicurus.

111 Bloch 1971, p. 158.

112 Popper 1982, pp. 172–4.

contrast to Martinez-Alier and Kołakowski, we can then definitively say that the dialectics of nature *did not fail Engels and Marx* here. Rather, it was their conception of nature and the cosmos as a complex, open, dynamic, contingent system, building on Epicurus's Dilemma, that represented the core of their dialectical-ecological view.

CHAPTER 5

The Reproduction of Economy and Society

Introduction

Of all the charges levelled against Marx by ecological economists, perhaps the most specific is that the reproduction schemes in Volume II of *Capital* ignore or downplay the dependence of production on natural conditions as well as the environmental impacts of production. Marx uses these schemes to analyse the basic exchanges required for capitalist reproduction as a material and social class process. The present chapter evaluates the standard ecological critique of the reproduction schemes in light of both Marx's methodology and his response to Quesnay's *tableau économique*.

The next section details the core claim of the ecological critique: that Marx's schemes are self-contained circular flows. As a basis for addressing this claim, we then consider whether Quesnay's *tableau* is subject to the same charge. This is important insofar as ecological economists have often unfavourably compared Marx's work with Physiocratic analysis, despite the well-known connection between Marx's reproduction schemes and the *tableau*. The main result of this section is that the *tableau* definitely does not reduce economic reproduction to self-contained circular flows.

We then show that Marx endorsed the *tableau's* conceptual differentiation of material reproduction and circular monetary flows, and that Marx also paid tribute to Quesnay for theorising the interplay between economic reproduction and its natural environment. We go on to demonstrate that Marx's own schemes maintain and further develop the distinction between material reproduction and circular monetary flows. We also document that Marx's schemes explicitly recognise the dependence of capitalist reproduction on natural conditions.

The reproduction schemes in Volume II of *Capital* do not address the environmental impacts of production, and the dependence of production on the environment is not their dominant theme. However, these two features can be understood in terms of Marx's methodology, especially the strictly delimited role of the reproduction schemes in his overall analysis of capitalism. Marx develops the general dependence of capitalist production on natural conditions in other portions of *Capital*, and this dependence is encapsulated in the categories employed by the reproduction schemes. Marx's analysis of capitalist environmental crises is also developed elsewhere in *Capital*, so that the absence of such crises from the reproduction schemes does not establish any overall ecological weakness in Marx's analysis of capitalism. We then summarise the chapter's argument.

Ecological Economists on Marx's Reproduction Schemes

Ecological economists' criticisms of Marx's reproduction schemes are all rooted in the claim that these schemes treat the economy as a self-reproducing system not dependent on its natural environment. Nicholas Georgescu-Roegen thus argues that 'In Marx's famous diagram of reproduction, ... the economic process is represented as a completely circular and self-sustaining affair'.[1] Marx's schemes evidently mimic 'the standard textbook representation of the economic process by a circular diagram, a pendulum movement between production and consumption within a completely closed system'.[2] Georgescu-Roegen even asserts that for both Marx and 'the standard economist', the 'patent fact that between the economic process and the material environment there exists a continuous mutual influence carries no weight'.[3]

Similarly, Herman Daly claims that 'Marx's models of simple and expanded reproduction are basically isolated circular flows'.[4] In Daly's view, both Marx's schemes and the circular flow diagrams found in mainstream economic principles texts are guilty of 'mixing up abstractions' insofar as they do not distinguish monetary circular flows from the 'linear throughput' of material production.[5] According to Martinez-Alier and Naredo, 'The mechanical analogy common to mainstream economics is shared by Marx, for instance in the schemes of "simple reproduction" where there is no question that the process could be continued indefinitely. No emphasis is given to the question of where the raw materials come from, or what is the motive power of this machine'.[6]

The purported ecological blind-spot in Marx's reproduction schemes is often blamed on Marx's treatment of natural wealth as a 'free gift' of nature, which is in turn blamed on Marx's labour theory of value. Daly, for example, surmises that 'Contacts with the environment are played down' in Marx's schemes 'because resources are held to be free gifts of nature, not a source of value inde-

1 Georgescu-Roegen 1973, p. 50.

2 Georgescu-Roegen 1973, p. 49.

3 Georgescu-Roegen 1973, p. 50.

4 Daly 1992, p. 196.

5 Daly 1992, pp. 196–7.

6 Martinez-Alier and Naredo 1982, p. 208.

pendent of labour'.[7] Georgescu-Roegen inveighs that the reason Marxists give 'no weight' to economy-environment interactions is that they 'swear by Marx's dogma that everything nature offers man is a spontaneous gift'.[8] This assertion is echoed by Charles Perrings, who says that Marx's 'free gifts assumption' explains why he 'assumed that the economy may expand without limit at the expense of the environment'.[9]

The ecological critique of Marx's reproduction schemes is a central element of the broader argument, common among ecological economists and other environmental theorists, that 'Marxian theory' embraces a 'closed-system' view of the economy which 'ignores environment as an interaction field'.[10] In this general interpretation, the environment 'plays only a benign and passive role' in Marx.[11]

Production and Circular Flows in the *tableau économique*

With their proclamation that 'the land is the unique source of wealth', it is not surprising that Quesnay and the Physiocrats have received a sympathetic hearing from ecological economists.[12] In fact, ecological economists have often favourably compared Physiocracy with Marxism. While the Physiocrats' 'steadfast belief that Nature was the source of wealth became a recurring theme throughout biophysical economics', says Cutler Cleveland, 'few of their biophysical principles are evident in ... Marxist theory'.[13] Paul Christensen praises the Physiocrats' 'early attention to the physical side of economic activity', especially their 'reproductive' approach that 'regarded production in terms of the transformation of materials and food taken from the land'.[14] He goes on to exclude Marxism from this reproductive tradition, asserting that Marxian economics shares 'the mechanistic sins of modern [neoclassical] economics' – both having neglected 'the biophysical foundations of economic activity'.[15]

7 Daly 1992, p. 196.
8 Georgescu-Roegen 1973, p. 50.
9 Perrings 1987, pp. 5 and 7.
10 Hawley 1984, p. 912.
11 Perrings 1987, p. 5.
12 Quesnay 1963c, p. 232.
13 Cleveland 1987, p. 50.
14 Christensen 1989, p. 18.
15 Christensen 1989, pp. 17–18.

There is a paradox in the purported ecological superiority of Physiocracy over Marxism, however: how does it square with the close relationship between Marx's schemes of reproduction and Quesnay's *tableau économique*? In this regard, Georgescu-Roegen even asserts that Marx 'borrowed' the reproduction schemes from Quesnay's *tableau*.[16] Yet, as we have seen, he also condemns Marx's reproduction schemes for neglecting economy – environment interactions. One would think that if Marx's schemes are ecologically incorrect, then so is the *tableau*. To the present writers' knowledge, however, neither Georgescu-Roegen nor any other ecological economist has ever drawn this conclusion – let alone reconciled it with Physiocracy's supposed ecological advantages over Marxism.

Is it possible that the *tableau* has the same ecological shortcomings that Marx's schemes purportedly suffer from? Does the *tableau* depict both the material and the monetary dimensions of economic activity as self-sustaining circular flows? If one takes certain statements of Physiocracy scholars out of context, then one may get the impression that the answer is yes, i.e. that the *tableau* may be viewed simply as a precursor of the circular flow diagrams in contemporary mainstream economic principles texts. Joseph Schumpeter says that the Physiocrats 'visualized the (stationary) economic process as a circuit flow that in each period returns upon itself'.[17] Ronald Meek states that in trying 'to illuminate the operation of the basic causes which determined the general level of economic activity', the Physiocrats

> believed that it was useful to conceive economic activity as taking the form of a sort of 'circle', or circular flow as we would call it today. In this circle of economic activity, production and consumption appeared as mutually interdependent variables, whose action and interaction in any economic period, proceeding according to certain socially determined laws, laid the basis for a repetition of the process in the same general form in the next economic period.[18]

Similarly, David McNally argues that 'the major theoretical achievement of the Physiocrats' was 'their general model of economic interdependence organized around the circular flow (or 'reproduction') of economic life'.[19] Such statements do not clearly distinguish the *tableau* from the circular flow diagrams in today's

---

16 Goergescu-Roegen 1971, p. 263.

17 Schumpeter 1954, p. 243.

18 Meek 1963, p. 19.

19 McNally 1988, p. 85.

principles textbooks. The latter typically show firms purchasing productive factors ('resources') from households who use the income so obtained to buy goods and services from the firms – with no apparent input from, or throughput to, the natural environment.[20] Nonetheless, a closer look reveals that the *tableau* does not suffer from the same ecological shortcomings.

In showing how the circulation of (monetary and material) wealth among the productive (cultivator), proprietary (landowning), and sterile (non-agricultural) classes enables annual reproduction, the *tableau* does trace out several monetary circuits in which expenditures are followed by a return flow or 'reflux' of money to the class which originally spent it. The rent paid by farmers to the landowners, for example, ends up flowing back to the farmers through the sale of agricultural products – partly to the landowners, partly to the sterile class.[21] But such monetary refluxes are not to be confused with any circularity in the actual material flows comprising economic reproduction. Quesnay explains this to his hypothetical interlocutor in 'Dialogue on the Work of Artisans':

> Thus there is no circle to be seen here other than that of expenditure followed by reproduction, and of reproduction followed by expenditure, a circle which is run through by means of the circulation of the money which measures the expenditure and the reproduction. You should therefore stop confusing the measure with the thing measured, and the circulation of the one with the apportionment of the other.[22]

As Spencer Banzhaf perceptively puts it, 'Quesnay's analysis of the circulation of wealth ... is, in one sense, not a circulation at all, but a one-way flow of wealth followed by consumption'.[23] The *tableau's* monetary circular flows are merely social vehicles through which the non-circular process of material reproduction takes place. In depicting 'production and consumption' as a 'oneway flow of subsistence and raw materials from nature through the economy', the *tableau* clearly presumes 'a reconstitution and regeneration of [the land's] vital forms and motive potencies'.[24] For Quesnay, this reconstitution and regeneration was to be undertaken through wise land-management practices (including investments in the land and the use of cattle to fertilise the soil) by farmers and

20 See, for example, Hall and Lieberman 2003, p. 156.
21 Meek 1963, pp. 273–5.
22 Quesnay 1963b, p. 226.
23 Banzhaf 2000, p. 546.
24 Christensen 1994, p. 277.

landed proprietors.[25] These productive class functions are inexplicable if one interprets the tableau as a self-reproducing circular flow of material and monetary wealth. Does the latter interpretation apply any more accurately to Marx's reproduction schemes?

Before approaching this issue, however, another antinomy is to be noted. As discussed earlier, ecological economists have linked the purported environmental shortcomings of Marx's reproduction schemes to his treatment of natural wealth as a 'free gift'. Yet, the supposedly more ecologically correct Physiocrats also speak of the land's unique surplus-generating capacity as a 'pure gift' and a 'spontaneous gift' of nature – as did all other classical economists and even modern neoclassical economists.[26] Apart from questionable interpretations of Marx's value theory (see the Introduction to the present work), perhaps part of the problem lies in the false presumption that Marx conflated two distinct claims: (1) that many productive use values are gifts of nature (i.e. wealth that does not result from human labour); and (2) that nature's gifts are limitless and/or substitutable, so that their use does not have any real economic cost. The Physiocrats clearly were not guilty of the latter claim and, as has been detailed elsewhere, neither was Marx.[27]

Marx on the *tableau économique*

Without going so far as to assert that Marx 'borrowed' his reproduction schemes from Quesnay, one can say that the *tableau économique* strongly influenced Marx's own analysis of capitalist reproduction.[28] Marx's praise for the *tableau* is, by his standards, absolutely effusive. He calls it 'an extremely brilliant conception, incontestably the most brilliant for which political economy had up

25 Quesnay 1963c, pp. 232–5, 242–3.

26 Turgot 1898, pp. 9 and 14; Quesnay 1963a, p. 60. On the widespread use of the notion of the 'free gift of nature' to describe the reality of the capitalist economy from classical times to today, see Foster, Clark and York 2010, pp. 61–4.

27 Burkett 2014, Chapter 6.

28 In addition to the discussion in Chapter 6 of *Theories of Surplus Value* (Marx 1963b, pp. 308–44), the *tableau* heads Marx's survey of 'former presentations of the subject' (of reproduction) in Chapter 19, Section 1, of *Capital*, Volume II (Marx 1978, pp. 435–6). See also Marx's 6 July 1863 letter to Engels, outlining an 'Economic Table' of 'the whole process of reproduction' which, Marx tells his friend, he is planning to 'use in place of Quesnay's Table' (Marx and Engels 1975b, pp. 132–3).

to then been responsible'.[29] In the chapter he contributed to Engels's *Anti-Dühring*, Marx describes the *tableau* as 'this both simple and, for its time, inspired representation of the annual process of reproduction through the medium of circulation'.[30]

That Marx speaks of 'reproduction through the medium of circulation' provides a clue as to what he finds most attractive about the *tableau*: it depicts the monetary circulation as an outgrowth of commodity production and exchange. As Marx puts it in *Theories of Surplus Value*, 'The first point to note in this *tableau* ... is the way in which the money circulation is shown as determined purely by the circulation and reproduction of commodities, in fact by the circulation process of capital'.[31] Marx's concept of the circulation of capital includes commodity production and exchange as one of its moments.[32] And for Marx, both the production and the exchange of commodities are material-social processes fully constrained by the laws of nature.[33]

Hence, when Marx applauds the *tableau*'s treatment of monetary circulation as a function of the 'circulation and reproduction of commodities', he is simply endorsing Quesnay's materialist perspective, according to which material production shapes the forms of (commodity and money) circulation. This jibes with Marx's more general praise for the Physiocrats' 'analysis of the various *material components* in which capital exists and into which it resolves itself in the course of the labour-process'.[34] After all, it was the Physiocrats' 'great merit that they conceived these forms as physiological forms of society: as forms arising from the natural necessity of production itself, forms that are independent of anyone's will'.[35]

Just as important, Marx's *tableau* discussion does not presume that material reproduction is self-reproducing apart from its natural environment. Rather, Marx strongly endorses Quesnay's depiction of the interplay between material production and natural conditions. After approving the *tableau's* 'material standpoint', from which 'it is always the previous year's harvest that forms the starting-point of the production period', Marx states the following:

29 Marx 1963b, p. 344.
30 Marx 1939, p. 275.
31 Marx 1963b, p. 308, cf. pp. 343–4.
32 Marx 1978, Part 1.
33 Burkett 2014, Parts 1 and 2.
34 Marx 1963b, p. 44, emphasis in original.
35 Ibid.

> The process of economic reproduction, whatever its specific social character may be, is in this area (agriculture) always intertwined with a process of natural reproduction. The readily apparent conditions of the latter illuminate those of the former, and keep at bay those confusions which are only introduced by the illusions of circulation.[36]

Marx's endorsement of the *tableau's* materialism and naturalism having been established, it remains to be seen if his own schemes fall prey to 'the illusions of circulation' by reducing material reproduction to self-contained circular flows, as his critics allege.[37]

Production, Nature and Monetary Flows in Marx's Schemes

The following analysis of Marx's reproduction schemes represents a dialectical middle ground between the Sraffian physical input-output interpretation and the view, held by some Marxists, that the main purpose of the schemes is to analyse the reproduction of quantities of money capital.[38] In the present view, Marx's schemes are designed to reveal the basic exchanges required by capitalist reproduction as a class-divided unity of production and circulation of wealth.[39] The schemes accordingly pose the problem of reproduction in terms of the contradictory unity of exchange value and use value characterising capitalist production, that is, production of commodities by wage-labour.[40]

Marx divides the aggregate social capital into two great Departments of production: means of production (Department I) and means of consumption (Department II). Monetary circular flows do play an important role in Marx's consideration of the intra- and inter-Departmental exchanges that must

36 Marx 1978, p. 435.

37 This does not imply agreement with Georgescu-Roegen's (1971, p. 263) notion that Marx 'borrowed' his reproduction schemes from Quesnay. Marx's analysis, unlike Quesnay's, considers the tensions between use value and exchange value in the reproduction process. Marx treats both simple and expanded reproduction, whereas Quesnay's analysis is limited to simple reproduction (Rosdolsky 1977, pp. 457–8). There is also the radical difference in the class structures (and corresponding exchanges) depicted by Quesnay's and Marx's respective schemes, as well as their differing presumptions concerning the origins of surplus value (Burkett 2003).

38 Moseley 1999.

39 See Foley 1986, p. 63.

40 Rosdolsky 1977, p. 457.

occur if capitalist reproduction is to take place. Marx shows, for example, that the equality between Department II's purchases of means of production and Department I's purchases of consumption goods corresponds to a circular flow of money between the two Departments.[41] More basically, he emphasises the role of 'money capital … as prime mover, giving the first impulse to the whole process' of reproduction. He thus interprets the aforementioned inter-Departmental circular flow in terms of 'the proposition that' the capitalist class 'must itself cast into circulation the money needed to realize its surplus-value (and also to circulate its capital, constant and variable)', and he calls this 'a necessary condition of the whole mechanism'.[42] Reproduction certainly requires that 'all components of the capital that consist of commodities – labour-power, means of labour and materials of production – must always first be bought with money and later on purchased again'.[43]

Marx immediately qualifies the last statement, however, observing that, 'as we already showed in Volume I, it in no way follows from this that the field of operation of capital, the scale of production, even on the capitalist basis, has its absolute limits determined by the volume of money capital in operation'.[44] The reason for this qualification is obvious: Marx's schemes are not meant just to map out monetary flows, but rather to establish the basic exchanges required by capitalist reproduction as a unity of production and circulation, of use value and value, and, above all, as a class process that is both material and social.

Marx thus insists that capitalist reproduction is 'conditioned not just by the mutual relations of the value components of the social product but equally by their use-values, their material shape'.[45] Hence, 'reproduction has to be considered from the standpoint of the replacement of the individual components of [commodity capital] *both in value and in material*'.[46] The 'natural form of the commodity product' is of the utmost importance in this context.[47] Capital-

41 Marx 1978, pp. 474, 586–597.

42 Marx 1978, p. 430, 497, cf. Ibid. pp. 549–551; de Brunhoff 1976, p. 53.

43 Marx 1978, p. 431; The role of 'prime mover' is only one of money's functions in capitalist reproduction in Marx's view. Money also serves as a liquid store of value, as a means of settling debts, and more basically as a standard of value and unit of account. See de Brunhoff (1976), Nell (1998, pp. 206–209) and Itoh and Lapavitsas (1999) on the interpretative difficulties raised by money's multiple functions in Marx's theory, especially in light of the fact that this theory was bequeathed to us in a preliminary, largely unpublished form.

44 Marx 1978, p. 431.

45 Marx 1978, p. 470.

46 Marx 1978, p. 469; emphasis added.

47 Marx 1978, p. 470.

ist reproduction entails the reproduction of the wage-labour relation and with it 'the capitalist character of the entire production process'.[48] And wage-labour requires 'the transformation of variable capital into labour-power, the payment of wages', the 'incorporation of labour-power into the capitalist production process', the 'sale of commodities [to] the working class', as well as 'the workers' individual consumption' – not to mention the provision of consumer goods to the capitalist class.[49] This whole 'movement' of production and consumption involves 'not only a replacement of values, but a replacement of materials'.[50]

Consistent with these materialist themes, Marx emphasises that 'the constant repetition of the process of production is the condition for the transformation that the capital undergoes again and again in the circulation sphere'.[51] He speaks of the 'fluxes and refluxes of money which take place *on the basis of* capitalist production', again clearly indicating the primacy of material production over its monetary forms.[52] He also insists that 'money in itself is not an element of real reproduction', that 'simple hoard formation ... is not an element of real reproduction', and (yet a third time) that 'surplus-value hoarded up in the money form ... is not additional new social wealth', even though it does 'represent new potential money capital'.[53]

Moreover, the 'constant repetition' of production is not materially self-reproducing in Marx's view. It is not simply driven by labour independent of its natural environment. Contrasting his analysis of reproduction with that of Destutt de Tracy, Marx rejects de Tracy's assertion (adopted from Adam Smith) that 'labour is the source of all wealth'.[54] Accordingly, he emphasises that 'labour ... could not have been transformed into products without means of production, i.e. means of labour and production materials, independent of it'.[55] He points out that 'living labour', as both 'useful, concrete labour' and 'value-forming labour', is dependent on the 'use-value', i.e. the 'concrete, natural form' of the 'means of production and means of consumption'.[56] For example, 'raw materials and ancillaries consumed in the production of commodities have to be replaced in kind so that the reproduction of commodities can begin (and gen-

48 Marx 1978, p. 468.

49 Marx 1978, p. 428.

50 Marx 1978, p. 470.

51 Marx 1978, p. 427.

52 Marx 1978, p. 555, emphasis added.

53 Marx 1978, pp. 566–7.

54 Marx 1978, p. 563.

55 Marx 1978, p. 504.

56 Ibid.

erally so that the process of commodity production can be continuous)'.[57] Marx also describes the equilibrium exchange between the two Departments as one in which 'values that exist in the hands of their producers in the natural form of means of production are exchanged for ... values that exist in the natural form of means of consumption'.[58]

Marx's reproduction analysis also refers to industries in which the pace and forms of labour are dictated by natural processes. The requirement that 'the means of production must always be renewed' clearly becomes more complex 'where labour is seasonal, or different amounts of labour are applied in different periods, as in agriculture'; and Marx pays close attention to the corresponding variations in 'the circulation operation by which the means of production are renewed or replaced'.[59] Marx had laid the foundation for these analyses in Chapter 13 of *Capital*, Volume II, with its detailed treatment of divergences between production time and working time, including their effects on the circulation of capital both material- and value-wise. 'What is involved' in these cases is 'an interruption independent of the length of the labour process, an interruption conditioned by the nature of the product and its production, during which the object of labour is subjected to natural processes, of shorter or longer duration, and has to undergo physical, chemical or physiological changes while the labour process is either completely or partially suspended'.[60]

The Analytical Background for Marx's Schemes

Marx's reproduction schemes encapsulate his prior specification of capitalist production in open-system terms, that is, as a material system of production that draws resources from and emits waste into its natural environment.

When analysing commodities and money in Part 1 of *Capital*, Volume I, Marx makes it clear that value is both a social (people-people) and material (people-nature) relation. A commodity is a useful good or service that is put up for exchange. Recognising that this 'use value ... is conditioned by the physical properties of the commodity', Marx sees commodity use values as 'the material content of wealth' under capitalism.[61] As is well known, Marx insists that both nature and human labour contribute to the production of all use

57 Marx 1978, p. 525.
58 Marx 1978, p. 474.
59 Marx 1978, pp. 525–6, cf. pp. 433–4, 555–6.
60 Marx 1978, p. 316; for details, see Burkett 2014, pp. 41–7.
61 Marx 1976a, p. 126.

values.[62] In analysing commodities and money, therefore, he emphasises that 'the physical bodies of commodities, are combinations of two elements, the material provided by nature and labour'.[63] And he recognises the role of energy ('natural forces') in the processing of natural materials by human labour.[64]

Even when Marx considers commodities as values, he does not separate this value dimension from the use value dimension of commodities with its natural basis.

Hence, 'Value [as abstract labour] is independent of the particular use-value by which it is borne, but a use-value of some kind has to act as its bearer'.[65] And since nature and labour co-create use value, value clearly encompasses the people-nature relation in production. Whether considered in terms of values or of use values, commodity exchange is a material-social dynamic – 'a process of social metabolism'.[66]

In Marx's view (unlike the Physiocrats and the Classical economists), commodity exchange is not a process dictated by natural laws, but is rather an outgrowth of 'the metabolic process of human labour' in its specifically capitalist form: wage-labour. The wage-labour relation is built on the social separation of workers from necessary conditions of production – above all from the land.[67] This separation, and workers' corresponding need to sell their labour power in order to obtain means of subsistence, forms the basis for production to become mainly organised through market relations among competing enterprises employing labour power for a profit, and it is this specific set of production relations that explains why capitalism reduces value to the (homogenous, socially necessary) labour time objectified in commodities.[68]

Unlike Adam Smith and David Ricardo, Marx does not base the reduction of value to labour time on a normative and/or empirical presumption that labour is more important or primary than nature as a production input. Rather, Marx argues that the apparent independence of value (as abstract labour) from natural conditions reflects workers' alienation from these conditions, i.e. the

62 Burkett 2014, p. 26.

63 Marx 1976a, p. 133.

64 Marx 1976a, pp. 133–4.

65 Marx 1976a, p. 295.

66 Marx 1976a, p. 198; cf. Sheasby 2002.

67 Marx 1976a, Part 8.

68 Although markets pre-date capitalism, the dominant position of commodity production and exchange in social production owes itself to the commodification of labour power and its employment by autonomous (private or state) enterprises controlling the (now 'separate') means of production.

conversion of labour power into a buyable commodity. Marx's recognition of the historical specificity of value does not imply a lack of concern with material production and its natural conditions. But unlike the Physiocrats, who saw value as a direct reflection of material (specifically natural) wealth, Marx strives to understand how capitalist value relations shape the production of wealth *and* vice versa.[69]

For example, Chapter 7 of *Capital*, Volume 1, treats useful labour (production of use values) and abstract labour (production of values) not as separate processes, but as two contradictory aspects of a single capitalist labour process.[70] Value production cannot be separated from 'production of use values', i.e. from labour as 'a process between man and nature ... an appropriation of what exists in nature for the requirements of man ... the universal condition for the metabolic interaction between man and nature'.[71] Marx thus emphasises again and again the essential role played in the labour process by 'many means of production which are provided directly by nature and do not represent any combination of natural substances with human labour'.[72] Regardless 'of the level of development attained by social production', Marx insists, 'the productivity of labour remains fettered by natural conditions' including the availability of energy sources 'such as waterfalls, navigable rivers, wood, ... coal, etc.'.[73]

Marx even argues that 'The Physiocrats were ... correct in seeing all production of surplus-value, and thus also every development of capital, as resting on the productivity of agricultural labour as its natural foundation'.[74] He does suggest that the Physiocrats were wrong to confuse this natural basis with the substance of surplus value itself, i.e. to identify value with use value.[75] But Marx insists that 'in no case would ... surplus product arise from some innate, occult quality of human labour'.[76]

Marx's conception of labour power and its exploitation by capital is itself developed in both thermodynamic and biophysical terms.[77] In analysing the limits to the length and intensity of worktime, for example, Marx often employs

69 Saad-Filho 2002, pp. 21–34; Burkett 2003.
70 Marx 1976a, p. 304.
71 Marx 1976a, pp. 283, 290.
72 Marx 1976a, p. 290.
73 Marx 1976a, pp. 647–8; cf. Marx, 1976b p. 34.
74 Marx 1981, p. 921.
75 Marx 1963b, pp. 46–60 and 1976a, p. 672; cf. Burkett 2003, pp. 145–9.
76 Marx 1976a, p. 651.
77 Marx 1976a, pp. 128–35, 270–7, 664; Burkett 2014, Chapter 4.

an analogy between overexploitation of the soil (resulting in loss of fertility) and overexploitation of labour power (resulting in loss of the worker's vital force).[78] And he develops this analogy using a metabolic 'energy income and expenditure' framework.[79]

The creation of surplus value and its appropriation by the capitalist naturally require not just exploitable labour power, but material conditions amenable to labour power's exploitation and to the objectification of the worker's labour in vendible use values. Capital accumulation as a value process is thus extraordinarily dependent on the appropriation of natural wealth in Marx's view. Hence, among the 'circumstances which … determine the extent of accumulation', Marx includes not only 'the soil itself' but also 'objects of labour … provided by nature free of charge, as in the case of metals, minerals, coal, stone, etc.'.[80] After all, 'The mass of labour that capital can command does not depend on its value but rather on the mass of raw and ancillary materials, of machinery and elements of fixed capital, and of means of subsistence, out of which it is composed, whatever their value may be'.[81]

Under capitalism, wealth or use value takes the form of 'an immense collection of commodities', which translates into an immense processing of materials serving as bearers of value.[82] This material throughput accelerates with the rising productivity of labour (use values produced per labour hour) generated by capitalists' competitive efforts to accelerate the extraction of profits from workers. As Marx indicates, 'the increasing productivity of labour is expressed precisely in the proportion in which a greater quantity of raw material absorbs a certain amount of labour, i.e. in the increasing mass of raw material that is transformed into products, worked up into commodities, in an hour, for example'.[83] Rising labour productivity means an increase in the quantity of natural forces and objects that capital must appropriate as materials and instruments of production in order to achieve any given expansion of value. Marx explicitly includes energy sources in capitalism's growing demand for 'auxiliary' or 'ancillary' materials, i.e. materials which, while not forming part of 'the principal substance of a product', are nonetheless required 'as an accessory' of its production.[84] As Marx observes, 'After the capitalist has put a larger capital

78 Burkett 2014, pp. 138–9.
79 Marx 1971, pp. 309–10.
80 Marx 1976a, pp. 747, 751–2.
81 Marx 1981, p. 357.
82 Marx 1976a, p. 125.
83 Marx 1981, p. 203.
84 Marx 1976a, p. 288, cf. p. 311.

into machinery, he is compelled to spend a larger capital on the purchase of raw materials *and the fuels required to drive the machines*'.[85] This energy consumption is greatly boosted by capitalism's development of increasingly large-scale machinery. Marx accordingly assigned a central place to energy consumption and transmission in his analysis of 'Machinery and Large-Scale Industry' in Chapter 15 of *Capital*, Volume I. (This chapter, representing the core of Marx's analysis of capitalist development, takes up nearly one-fifth of the volume).

The Reproduction Schemes and Environmental Crises

The reproduction analysis in Volume II, Part 3, of *Capital* recognises capitalism's reliance on natural conditions, but it does not explicitly treat situations where shortages of natural resources prevent reproduction from occurring. This absence of environmental crises is explained by the fact that Marx's schemes are not designed to analyse breakdowns in reproduction; nor are they meant to address crises of capital accumulation. For Marx, the processes by which capital accumulation leads to crises involve changes in technology and other parameters that alter the material and value structure of production.[86] These dynamics are introduced in Volume I of *Capital* (especially Parts 4, 5 and 7), but they are explicitly excluded from the schemes in Part 3 of Volume II, which – consistent with their focus on the circulatory equilibrium of the aggregate social capital – 'assume not only that products are exchanged at their values, but also that no revolution in values takes place in the components of the productive capital'.[87]

In Marx's view, the formal possibility of capitalist economic crises stems from the temporal separation of commodity sales and purchases in a monetary economy, i.e. from the fact that money obtained from sales may be hoarded (or used to settle debts) rather than spent on other commodities.[88] However, the 'simple circulation of commodities' implies 'the possibility of crises, but

85 Marx 1976c, p. 431, emphasis added.

86 Weeks 1979, pp. 267–9; Foley 1986, p. 64.

87 Marx 1978, p. 469, cf. p. 565. In the present interpretation, the 'accumulation' referred to in the title of Chapter 21 of *Capital*, Volume II ('Accumulation and Reproduction on an Expanded Scale'), is meant in a more limited quantitative sense, excluding changes in commodity and capital values rooted in changing material conditions (both natural and technological). See Zarembka (2000) for alternative views on Marx's usage of the term 'accumulation of capital'.

88 Marx 1968, pp. 50–15, 1976a, p. 209; cf. Kenway 1980.

no more than the possibility'.[89] The actual dynamics of accumulation and crises 'can only be educed from the real movement of capitalist production, competition and credit'.[90] In this connection, Marx makes it quite clear that 'In so far as crises arise from *changes in prices and revolutions in prices*, which do not coincide with *changes in the values* of commodities, they naturally cannot be investigated during the examination of capital in general, in which the prices of commodities are assumed to be *identical* with the *values* of commodities'.[91] Because the reproduction schemes exclude such value-price dynamics, they cannot be used to analyse crises.

Of course, since the schemes are an intersectoral analysis of the monetary exchanges required for capitalist reproduction, they do reveal various crisis *possibilities*. As Marx says, the schemes illustrate 'certain conditions for normal exchange, i.e. conditions for the normal course of reproduction, whether simple or on an expanded scale, which turn into an equal number of ... possibilities of crisis'.[92] Among these 'conditions for an abnormal course', Marx cites several involving production's reliance on natural resources, including 'the replacement of a part of department II's commodity capital by natural elements of constant capital' as well as the need to insure against crop failures by maintaining 'a stock of raw materials etc. ... that surpasses the immediate annual need (this is particularly true of means of subsistence)'.[93]

The actual dynamics of environmental crises of capital accumulation are initially approached elsewhere, in *Capital*, Volume I's analysis of capital's growing appetite for raw materials including energy sources. Here, Marx shows that the rising labour productivity generated by capitalist technology creates a growing divergence between a rising quantity of material throughput and a declining per unit value of this throughput. Partly on this basis, Marx concludes that 'this mode of production ... comes up against no barriers but those presented by the availability of raw materials and the extent of sales outlets'.[94]

But insofar as materials shortages induced by capital accumulation lead to changes in the rate of profit and deviations of market prices from values, their analysis had to wait for Volume III of *Capital* where the relevant phenomena are dealt with for the first time. There, we find a detailed discussion of how 'a rise in the price of raw material can cut back or inhibit the entire reproduc-

89 Marx 1976a, p. 209.
90 Marx 1968, p. 512.
91 Marx 1968, p. 515, emphases in original.
92 Marx 1978, p. 571.
93 Marx 1978, pp. 571, 544.
94 Marx 1976a, p. 579.

tion process' by making it 'impossible to continue the process on a scale that corresponds with its technical basis'.[95] 'Violent fluctuations in price thus lead to interruptions, major upsets and even catastrophes in the reproduction process'.[96] Materials shortages generate 'disturbances in the reproduction process' in both material and value terms, since they not only raise the cost of constant capital (thereby reducing the rate of profit) but also physically disrupt production. As a result, 'A part of *fixed capital* stands idle and a part of the workers is thrown out on the streets'.[97] In this context, Marx stresses the contradiction between capital accumulation and the 'uncontrollable natural conditions' influencing the materials supplies on which accumulation depends.[98]

Marx's analysis of materials-supply disturbances shows that capital accumulation, far from being an automatically self-reproducing process material- or value-wise, is in a permanent state of tension with its own natural (including energy) requirements. But Marx also detects a contradiction between capital accumulation and a sustainable development of production appropriate to human beings socially coevolving with nature. After all, capital's basic requirements (exploitable labour power and conditions under which wage-labour can be objectified in vendible commodities) are, materially speaking, fulfillable under any degradation of natural conditions short of human extinction. This helps explain why the most prominent type of environmental crisis in *Capital* is not materials supply disturbances to accumulation, but rather the crisis in the natural conditions of human development produced by capitalist industrialisation.

Culminating his monumental analysis of machinery and large-scale industry, Marx points out how, having socially separated the producers from the land, capitalism develops a division of labour between agriculture and urban manufacturing that 'disturbs the metabolic interaction between man and the earth'.[99] This metabolic rift 'prevents the return to the soil of its constituent elements', which in turn 'hinders the operation of the eternal natural condition for the fertility of the soil' and 'destroys at the same time the physical health of the urban worker'.[100] In short, capitalism 'only develops the techniques and the degree of combination of the social process of production by simultaneously

95 Marx 1981, p. 204.

96 Marx 1981, p. 213.

97 Marx 1968, p. 516, emphasis in original.

98 Marx 1981, p. 213; cf. Burkett, 2014, pp. 112–19.

99 Marx 1976a, p. 637.

100 Ibid.

undermining the original sources of all wealth – the soil and the worker'.[101] Unlike materials-supply disturbances, this environmental crisis tendency need not involve a crisis of capital accumulation. Nonetheless, it shows that Marx did not see capitalist economy as a self-driven perpetual motion machine, but rather as an open system in which value accumulation is underwritten by the depletion and despoliation of natural wealth, both human and extra-human.[102]

Conclusion

Ecological economists have argued that Marx's reproduction schemes depict self-contained circular flows, and that this reflects a more general tendency on Marx's part to treat the environment as 'simultaneously a horn of plenty and a bottomless sink'.[103] The present response began by posing a paradox: given the close relationship between Marx's schemes and Quesnay's *tableau économique*, how does this standard critique jibe with the view, also common among ecological economists, that Physiocratic theory is ecologically superior to Marx's economics? A solution to this paradox was then developed.

Essentially, this solution is that the standard ecological critique of Marx's reproduction schemes is incorrect. Both Quesnay's *tableau* and Marx's evaluation of it clearly recognise that monetary circular flows are underpinned by a material production process, the repetition of which is dependent on natural conditions. Moreover, a contextual investigation of Marx's own schemes reveals that they encapsulate the environmental dependence of production as Marx develops it in *Capital*. Likewise, the absence of environmental crises from Marx's schemes does not reflect adversely on the ability of Marx's general approach to account for the environmental impacts of production. Marx's analyses of environmental crises are set out at other points in *Capital*, consistent with the strictly delimited purpose of the reproduction schemes in his overall analysis of capitalism.

101 Marx 1976a, p. 638.
102 Foster 2000, pp. 141–77; Burkett 2014, Chapters 9–10.
103 Perrings, 1987, p. 5.

CONCLUSION

Marx and Metabolic Restoration

> But by destroying the circumstances surrounding that metabolism [between humanity and nature] ... it [capitalist production] compels its systematic restoration as a regulative law of social production, and in a form adequate to the full development of the human race.[1]

∴

Scientists during the last few years have sought to designate 'a safe operating space for humanity', constituted by nine planetary boundaries, nearly all of which now have either been crossed or are in the process of being crossed, as seen in: climate change; ocean acidification; destruction of the ozone layer; biosphere integrity; disruption of bigeochemical flows; land-system change, freshwater use, aerosol loading; and the introduction of novel entities (new chemical and biological substances).[2] Climate change is the overriding global environmental concern at present, since it points to the prospect within a generation (under business as usual) of extremely dangerous, even catastrophic global warming, spiralling beyond human control. Nevertheless, the other eight planetary boundaries amount to so many additional swords of Damocles hanging over humanity as a whole.

Compared to today's planetary emergency, viewed in this way, the kinds of ecological issues that Marx was concerned with in the nineteenth century related to soil-nutrient depletion, industrial pollution, deforestation, and desertification may seem of relatively little importance. Why then should we

1 Marx 1976a, pp. 637–8.

2 Steffen, et. al. 2015; Rockström et al. 2009. The nine planetary boundaries designated by Rockström et al. are based on conditions prevailing in the Holocene epoch. The boundaries for climate change, biosphere integrity, biogeochemical flows, and land-system change have already been crossed. Those associated with ocean acidification, land-system change, and freshwater use are currently in danger of being crossed. In contrast, the ozone layer appears to be stabilising at present. Boundaries for aerosol loading and novel entities have not yet been quantified, but these phenomena are nevertheless considered to be of rapidly growing significance.

concern ourselves, as we have in this book, with the question of *Marx and the Earth*? The answer, as one of us wrote more than twenty years ago, is that 'the crisis of the earth is not a crisis of *nature* but a crisis of *society*'.[3] It is our historical social relations and their intersection with the natural environment that are responsible for our current planetary emergency. What is needed, then, is an understanding of the social system and how it interacts with its earth-system environment. This understanding needs to be sufficiently dialectical and revolutionary in its method to grasp the unprecedented changes in which humanity is now caught up, and the means with which to construct a new human praxis in response.

Standard, possessive-individualist social science, and much of contemporary environmental theory – insofar as it is an offshoot of the mainstream liberal tradition – have found themselves incapable of going to the root of the problem in this respect. It is here that Marx's critique of political economy, which was also a critique of the growing rift in the universal metabolism of nature generated by capitalist production, has proven so indispensable. Precisely because he centred his analysis on the 'social metabolism', or the specific way in which society reproduced itself in relation to underlying natural conditions, Marx provided the beginnings of a systems perspective with respect to social ecology, encompassing not only the environment but also humanity itself.[4] In this dialectical, open-system view, the effects of the economy on the earth were not seen as mere 'externalities', as in capitalist economics, which treats the economy as a closed system, but were understood as organically connected to social and environmental reproduction.[5]

Although Marx's approach to the ecological alienation of capitalist society obviously needs to be brought up to date in order to deal with the specific threats that present themselves to us today, the complex, dialectical analysis that he developed in this area provides the foundations for a unified critical approach – one that prefigured many of the later developments in science. This is particularly evident in his theory of metabolic rift. As Del Weston eloquently explained in *The Political Economy of Global Warming: The Terminal Crisis*:

> The metabolic rift ... refers to a rupture in the metabolism of the whole ecological system, including humans' part in the system. The concept is built around how the logic of accumulation severs the basic processes of

3 Foster 1994, p. 12.

4 Marx 1981, p. 949.

5 Tsuru 1994, p. 376; Kapp 1976, pp. 96–8, 101.

> natural reproduction, leading to the deterioration of the environment and ecological sustainability and disrupting the basic operations of nature. It neatly captures the lack of balance between 'expenditure and income' in the Earth's metabolism under the capitalist system. In theorizing this, Marx goes to the very basis of society, how humans interact with the environment, socially and materially, to provide their means of survival. It is from this most fundamental relationship that the ecological contradictions inherent in the very specific mode of production found in the capitalist system, the foundation for the growing disequilibrium in the biosphere and the pending demise of human civil life, are to be found.
>
> In developing the theory of the metabolic rift, Marx maintained that capitalism generated an unhealthy circulation of matter from urban industry and industrial agriculture, which damaged the reproductive capabilities of both human labour power and the land. Whereas Marx saw that humans' pre-industrial interaction with nature enabled harmonious and sustainable production, capitalism was not able to maintain the social relations or the conditions for the recycling of nutrients back to the soil. Thus was born the metabolic rift ... Today, this rift has grown both in dimensions and complexity, to the point where the economic activities of human society are causing an unprecedented change in the Earth's biosphere, its lands, forests, water and air, potentially bringing to an end the Holocene era as a result of anthropogenic global warming.[6]

Marx's Ecology after Marx (and after Engels)

If Marx's ecological analysis was embedded within his political economy in this way, and carried such powerful implications for our contemporary ecological view, why, then, was this so seldom appreciated in later Marxian social theory?[7] Why were Marx's notions of social metabolism and the rift in that metabolism (and thus in the universal metabolism of nature), not better known within the Marxian tradition? We have frequently referred in this volume to the erroneous claim that this failure can be attributed to Marx himself, due to his failure to

6 Weston 2014, pp. 66–7.

7 One crucial factor, which long affected English-language studies of Marx, is that the original translation of Marx's *Capital* into English rendered '*Stoffwechsel*' as 'material exchange' rather than as 'metabolism', thereby obscuring the complex, systemic, interdependent nature of Marx's analysis in this area, and particularly his notion of 'social metabolism'.

embrace Sergei Podolinsky's early attempts at energetics.[8] Far from ignoring Podolinsky, Marx and Engels took him extremely seriously, but were nonetheless wary of the egregious errors in Podolinsky's analysis, such as leaving out at critical points solar energy, fossil fuels, and fertilisers in his calculations of energy input and output, as well as the mechanism that pervaded his analysis. Far from ignoring or rejecting the discoveries in thermodynamics in their day, Marx and Engels wove these discoveries in physics into the basic analysis of historical materialism. Indeed, Marx's *Capital* is distinguished from other economic works of the nineteenth century particularly by its attention to the two sides of labour/production: both its material (use value) and value (exchange value) forms.

Others have charged that Marx's concept of social metabolism and the material-physical bases of his analysis were simply neglected by later Marxian theorists, who thus abrogated any claim to environmental thought in this regard. However, the actual story, we contend, is a good deal more complex than this.[9] The underlying problem can be traced to the split in historical materialism that developed in the late 1930s, between the broad Third International tradition, on the one hand, and what came to be known as 'Western Marxism', on the other. Soviet Marxism, particularly from the late 1930 to the 1960s, tended to promote a kind of mechanism, in which the notion of dialectic was all too often used to legitimate its opposite; while Western Marxism for its part was to be characterised, in philosophical terms, by its extreme anti-positivism and its rejection of the notion of the dialectics of nature, and hence a systematic distancing from issues related to natural science.[10]

Marx's concept of metabolism was taken up by Nikolai Bukharin in his *Historical Materialism*, and by the later Lukács. But Bukharin was to fall victim to the purge under Stalin, while the Western Marxist tradition generally rejected Lukács's post-1920s work given his partial abandonment of earlier views, enunciated in *History and Class Consciousness*.[11] All of this tended to produce a discontinuity in the Marxian tradition on both sides of the great divide. The Frankfurt School and most Marxian theory in the West became associated with a social science that was divided off from natural science – as if history no longer involved the coevolution of society and nature. Soviet theory, for its part,

8 Molina and Toledo 2014, pp. 48–9; Martinez-Alier 1987.

9 Molina and Toledo 2014, p. 49.

10 Colletti 1973, pp. 191–3; Jacoby 1983, pp. 523–6; Merleau-Ponty 1973, p. 32; Sartre 2004, p. 32; Schmidt 1971, pp. 59–61; Vogel 1996, pp. 14–19. On extreme anti-positivism, which denies the possibility of naturalism, see Bhaskar 1998.

11 Bukharin 1925, pp. 108–12; Lukács 1968, p. xvii.

all too often sank into a kind of mechanical dogma in which allusions to dialectics were reduced to mere form – a weakness that was partially alleviated in 'late Soviet ecology'.[12]

Yet, despite the splitting of the Marxian dialectic in this way, knowledge of Marx's material-metabolic analysis persisted in some circles, particularly among critical natural scientists, a relatively small number of Marxian philosophers (especially those conversant with the later Lukács), and in the work of some Marxian political economists.[13] Moreover, continuing explorations in the dialectical method by Marxists virtually guaranteed that as environmental problems became more serious, renewed investigations into the relation between the materialist conception of history and the materialist conception of nature would emerge. Its reliance on a combination of dialectical, historical, and materialist thinking gave Marxian theory an enormous advantage over mainstream natural and social sciences in this respect, generating a natural affinity for ecological complexity.

Nevertheless, there remains a widespread impression, prevalent even among socialists themselves, that Marxism somehow entirely missed out on the resurrection of environmentalism which occurred in the West in the 1960s and '70s. As Martinez-Alier charged as late as 1987 in his influential *Ecological Economics*: 'Proof of the absence of an ecological Marxism could be found in the practice of economic planning in the Soviet Union, where discussion of the inter-generational allocation of exhaustible resources does not exist'. He then proceeded in the same paragraph to extend this 'telltale proof' to the West itself by indicating that the 'great Marxist historians', like Maurice Dobb and E.P. Thompson, in their books written in the late 1940s and early 1960s, had ignored ecological issues.[14]

Neither of these criticisms, however, carried much weight. There is no denying that Soviet ecology was severely damaged almost to the point of annihil-

---

12 Foster 2015, pp. 1–20.

13 Dialectical conceptions in the natural sciences continued to be pursued by Marxian scientists in the West in the 1930s and '40s and after, and played a considerable role in the rise of ecological science. See Sheehan 1985; Foster, Clark and York 2010, pp. 242–7. Mészáros (1970, 1995) was to develop a Marxian ecological critique in line with Lukács's emphasis on Marx's social metabolism concept. Marxian political economists in the United States, such as Scott Nearing, Paul Baran, and Paul Sweezy, continued to present views in which both the materialist conception of history and the materialist conception of nature were evident, and thus crossed the boundaries between natural and social science. See Foster 2011.

14 Martinez-Alier 1987, pp. xi, 227–8; Dobb 1963; Thompson 1963.

ation by the purge beginning in the late 1930s under Stalin, and only slowly recovered over decades. But while the USSR in the 1970s and after would hardly have commended itself to environmentalists by its manner of resource use, to claim that this was a 'proof of the absence of ecological Marxism' throughout the society, as Martinez-Alier did in 1987, meant assuming a monolithic character to the state (and to the state planning apparatus) that was insupportable. One need only mention that it was Soviet climatologists, led in particular by M.I. Budyko, who played a leading role in raising the alarm with respect to climate change in the 1960s and 1970s. Budyko went on to examine the overall question of 'global ecology', inviting comparison in this respect with a figure like Barry Commoner in the West. It was Budyko who, in the 1960s, introduced the analysis of the ice-albedo feedback, which was to make global warming a pressing global concern.[15]

Budyko was closely connected to E.K. Fedorov – the former director of the Hydrometerological Service in the Soviet Union, and a member in the 1970s of the Presidium of the Supreme Soviet. Fedorov played an important role in the First World Conference on Climate held by the World Meteorological Organization in Geneva in 1979. He was the author of 'Climate Change and Human Strategy' (1979) and *Man and Nature: The Ecological Crisis and Social Progress* (1981). Fedorov argued, in Budyko's words, that 'the fundamental social and economic features of the capitalistic system prevent the rational use of natural resources', and that the global ecological crisis could only be overcome through 'a planned socialist economy'.[16] Fedorov's *Man and Nature* was an extraordinarily insightful work, addressing the question of the sustainable use of natural resources, emphasising the dangers of global climate change, and exploring the social bases of the environmental crisis. In this work Fedorov strongly sided with the analysis of Barry Commoner in the United States, and applied Marxian theory, including Marx's own ecological ideas, to the emerging planetary crisis. He wrote the 'Concluding Remarks' to the

---

15 Foster 2015; Budyko 1969, 1974, 1977, 1980, 1986; Budyko, Golitsyn and Izrael 1988; Commoner 1971. It was Soviet scientists too who first demonstrated that the climate would be catastrophically altered by nuclear warfare via nuclear winter. See Budyko, Golitsyn and Izrael 1988, pp. v–vi.

16 Fedorov 1979, 1981; Budyko 1977, p. 235; Weart 2003, pp. 85–8. Both Bodyko and Fedorov floated numerous ideas on how to deal with climate change, including geoengineering, but also encompassing 'restrictions on energy use' (Fedorov 1979, p. 31), along with ecological planning and social change. Fedorov (1981) was based on his earlier works, *The Interaction of Nature and Society* (1972), and *The Ecological Crisis and Social Progress* (1977).

Russian edition of Commoner's *The Closing Circle*.[17] Indeed, a powerful, critical environmentalism with an integrated dialectical perspective was re-emerging among leading Soviet thinkers (in the human sciences as well) in the 1970s and early 1980s in response to the onset of global ecological crisis.[18]

Likewise, to point, as Martinez-Alier does, to Maurice Dobb's failure to address ecological factors in his *Studies of the Development of Capitalism* – a book first written in the late 1940s – or to E.P. Thompson's neglect of such factors in his *The Making of the English Working Class*, published in 1963, can hardly be seen as constituting a convincing 'proof of the absence of ecological Marxism'. Dobbs's and Thompson's books were historical studies centring on the early industrial revolution, and were written prior to the main rebirth of ecological movements in the West – and a quarter-century (or more) prior to Martinez-Alier's book.[19] A more reasonable approach would have examined Marxian works in the 1970s and 1980s contemporaneous with Martinez-Alier's own study.

More to the point, Martinez-Alier declared elsewhere in his book that Marxian political economists, like their conventional counterparts, were constitutionally unable to address the core issues related to ecological economics, due to fundamental flaws in their theoretical frameworks. 'Marxist economists', he insisted, 'could not explain for instance why it is unlikely that the present ratio of automobiles to population of North Atlantic countries (and Japan) [could] be extended to the world at large'.[20] In effect, he was claiming that Marxian economists had completely failed to integrate issues like entropy, thermodynamics, material flows, and environmental limits into their analyses, thus falling short of an ecological worldview.

It is important therefore to ask whether there are any notable cases of Marxian theorists (and Marxian political economists in particular) in the 1970s and '80s who recognised the limits to the growth of automobile-based production, and indeed the limits to growth in general. We can even extend this further by asking: To what degree were Marxian theorists alert to the threat of global warming in the 1970s and early 1980s? Here it should be mentioned, however, that the climate change issue was missing from Martinez-Alier's own

17 See especially Fedorov 1981, pp. 55–61, 68–78, 18–51; Budyko 1986, pp. 371, 406.

18 Weiner 1999, pp. 399–400. See also the extraordinary collection of essays in Ursul 1983 in which Soviet scientists and philosophers raised many of the methodological issues arising out of Marx's ecology, applying them to questions of global ecological crisis.

19 It should be noted that E.P. Thompson later did enter into the ecological argument with respect to the commons in particular. See especially Thompson 1991.

20 Martinez-Alier 1987, p. 16.

Ecological Economics, and thus played no part in the charge that he leveled against Marxian theory. Nevertheless, the recognition of global warming as a concern by Marxian theorists – even where Martinez-Alier himself overlooked it – would obviously do much to dispel his charges.

It is hardly surprising that the sharpest critics of the automobilisation of the society arose out of the Marxian tradition. It was Marxian economists Paul A. Baran and Paul M. Sweezy who first introduced the term 'automobilisation' in 1966 in *Monopoly Capital*, emphasising the automobile's central role in the production and consumption system and the enormous economic and environmental waste with which it was associated.[21] Sweezy extended this analysis in 1973 in his 'Cars and Cities', where he discussed the pollution, waste, and urban congestion associated with the 'automobile-industrial complex', arguing not so much that *the world* could not support the universal extension of the current level of US automobile production and consumption, but rather that the excessive use of automobiles was already overwhelming the urban environment in the *developed countries themselves*. He went on in subsequent articles to question unlimited capital accumulation/economic growth. 'Activities damaging to the environment', he wrote in 1989, 'may be relatively harmless when introduced on a small scale; but when they come into general use and spread from their points of origin to permeate whole economies on the global scale, the problem is radically transformed', leading to 'what has become generally perceived as *the* environmental crisis'. This was marked most clearly, in his view, by the accelerating 'greenhouse effect stemming from the combustion of fossil fuels'. This called for a strenuous social effort to 'reverse' current economic-ecological trends.[22]

As noted in our introduction, leading Marxian philosopher István Mészáros had already argued in 1971, prior to the publication of the Club of Rome's 1972 *The Limits of Growth*, that W.W. Rostow's argument that the US stage of 'high mass consumption' marked by intense automobile usage would simply be transferred as a matter of course to the entire world, was an outright ecological impossibility. 'A decade ago', Mészáros stated, 'the Walt Rostows of this world

21 Baran and Sweezy 1966, pp. 131–8, 241–66.

22 Sweezy 1973; Sweezy 1989, pp. 4–5; Magdoff and Sweezy 1974, pp. 9–10; Sweezy 1977; Sweezy 1980. Sweezy often said that Nicholas Georgescu-Roegen was right in his insistence on the need to incorporate the entropy concept into economics. He always kept Georgescu-Roegen's (1971) *The Entropy Law and the Economic Process* handy on his bookshelf. On one occasion in the early 1990s, in what was a very unusual action on his part, he made copies of the opening chapter in Georgescu-Roegen's (1976) *Energy and Economic Myths* for all the members of *Monthly Review*'s informal editorial committee to read. (Note by JBF).

were still confidently preaching the *universal* adoption of the American pattern of "high mass consumption" within the space of one single century. They could not be bothered with making the elementary, but of course necessary, calculations which would have shown them that in the event of the universalization of this pattern ... the ecological resources of our planet would have been exhausted well before the end of that century several times over'.[23]

Japanese Marxist economist Shigeto Tsuru, one of the pioneering figures in the global environmental movement, exhibited concerns about global warming as early as 1972 in his Columbia lecture on 'North-South Relations on Environment'. Tsuru focused throughout his writings on the open-system character of the economy and the contradiction between use value and exchange value. For Tsuru the failures of received economics had to do with the fact that the whole question of social reproduction – in sharp contrast to Marxian economics – lay outside the frame of analysis. Externalities were only 'external' in the narrow sense that they were excluded from the cost-accounting of capitalist firms – inscribed as the whole of reality by neoclassical economics. From the more realistic perspective of Marxian theory, these effects remained internal to society and the planetary environment as a whole, impacting social and natural reproduction.[24]

Marxist environmentalist, Virginia Brodine, editor of the two periodicals *Science and the Citizen* and *The Environment*, and a close colleague of Barry Commoner, highlighted the issue of global warming, along with the pollution associated with the automobile in particular, in her holistic, prescient book on *Air Pollution* in 1972. In Brodine's view, the world was coming up against global environmental limits arising from accelerated commodity production. It was more than conceivable that carbon emissions associated with the existing system of production could warm the earth sufficiently to melt almost all of the world's ice fields. As she put it:

> We may be affecting global heat balance ... by rapidly releasing the carbon that has been stored in coal, oil, and gas over millions of years. The carbon dioxide content of the air has been increasing quite steadily with our increase in the burning of these fossil fuels, and we are therefore steadily reinforcing the ability of the natural atmosphere to absorb and re-radiate more of the infrared radiation given off by the earth. It now appears that half the carbon dioxide added by combustion remains in the atmosphere,

23 Mészáros 1995, pp. 874–5.

24 Tsuru 1994, pp. 107, 278–80, 375–80.

> while the other half is absorbed by the upper layers of the oceans and by green plants. But will this continue to be so? How much carbon dioxide goes into each of the two reservoirs and what are their limitations? How soon will the carbon newly stored in living organic material be returned to the atmosphere? ... One estimate of the effect of increased carbon dioxide is that an increase of global temperature of 0.9° F will occur by the year 2000 and perhaps of 3.6° F if the carbon dioxide content of the atmosphere is doubled.[25]

The greatest environmental threat to the planet, aside from nuclear radiation from a nuclear war or a series of nuclear accidents, Brodine argued, was 'climatic change' associated with such an alteration in the global heat balance. This required shifting priorities in the advanced capitalist countries away from maximising economic growth. 'An ecologically balanced civilization means, eventually, no further growth of the economy'.[26]

In 1973, during his early Marxian critical-theory phase, Jürgen Habermas argued that 'even on optimistic assumptions' there remained '*one* absolute limit on [the] growth' of capital accumulation, namely global warming stemming from increased energy consumption and carbon dioxide emissions. This then would be the ultimate assertion of the need for 'ecological balance' – a limitation on accumulation that could assert itself decisively, he suggested, by 2048. This threat was all the more serious under capitalism due to its unplanned character and hence its drive to capital accumulation as its sole organising principle.[27]

Similarly, Marxist sociologist and political economist Charles H. Anderson described the negative environmental effects of the automobilisation of capitalist production and consumption, in his 1976 book *The Sociology of Survival: Social Problems of Growth*. For Anderson, one of the chief dangers to the planet as a place of human habitation was climate change:

> The amount of carbon dioxide ... being emitted into the atmosphere by an energy hungry society has raised questions among scientists regarding the possibility of climatic changes ... Carbon dioxide, the atmospheric level of which has increased by 15 percent in this century, produces a 'greenhouse effect' by holding heat in and thus raising the earth's temperature ... [This]

25 Brodine 1972, pp. 62–4, 70, 177–8; Brodine 2007.
26 Brodine 1972, pp. 174, 178.
27 Habermas 1973, pp. 41–2.

> could raise the earth's temperature and substantially alter the agricultural situation; or a mere two degrees centigrade increase could destabilize or melt the polar ice caps, raising the oceans 50 meters and flooding coastal populations and agricultural areas.[28]

Anderson's conclusion in his argument on the 'social problems of growth' was that society would need to shift towards a socialist 'stationary-state economy'. He coupled this with a sophisticated analysis of the buildup of what he called 'ecological debt'.[29]

In the 1970s there was already a rapid growth of ecology within the Marxian tradition. It was in 1972 that Herbert Marcuse wrote his *Counter-Revolution and Revolt* with its powerful chapter on 'Nature and Revolution'. In 1977, Howard Parsons published his important book on *Marx and Engels on Ecology*. In 1980, Allan Schnaiberg introduced his neo-Marxian theory of the 'treadmill of production' in *The Environment: From Surplus to Scarcity*.[30]

Marx, Metabolism, and Open-System Economics

All of this turns the critique raised by Martinez-Alier on its head. Instead of asking why Marxian political economists were unable to explain the environmental limits preventing the extension of 'the present ratio of automobiles to population of North Atlantic countries (and Japan) ... to the world at large', we need to find a way to account for the fact that they not only often succeeded in doing exactly that, but were also ahead of the curve in environmental social science more generally. Why were so many Marxian social scientists able to incorporate not only the limits to growth, but also climate change, into their analyses, while their mainstream counterparts in the social sciences, particularly within economics, saw this as an insurmountable barrier – or chose to temporise?[31]

---

28 Anderson 1976, p. 126. Anderson also noted that the scientific models also said that global cooling was also a theoretical possibility. He followed climatologists in pointing to global warming as the most likely possibility, while emphasising that the central issue was that capital accumulation could engender catastrophic climate change raising issues of human survival.

29 Anderson 1976, pp. 54–61.

30 Marcuse 1972; Parsons 1977; Schnaiberg 1980.

31 A classic example of issue avoidance with respect to environmental problems is the orthodox economist Robert Solow's statement in 1974 that 'If it is very easy to substitute

The answer lies in a fundamental difference in method. Although it is true, as Manuel González de Molina and Victor Toledo point out, that the social metabolism argument in Marx was unknown to most radical social scientists in the West in the 1970s and '80s – given the distancing from natural science that had occurred – there were nevertheless notable exceptions to this to be found among both Marxian philosophers and socialist environmentalists.[32] Moreover, a number of leading Marxian political economists were to approach the root of the problem, providing much of the basis for the ensuing revival of the ecological foundations of Marx's thought. Recognising the contradictions associated with capitalism's externalisation of environmental costs, leading socialist political economists and ecological theorists focused on issues of *social and natural reproduction*, transcending the narrow economic view of neoclassical economics. As Tsuru argued in 1971, in his major address to a world social science conference on the environment, ecological crisis was essentially a 'disruption' (rift) in the human relation to nature (and in the biogeochemical cycles of the earth) engendered by capital accumulation and economic expansion.[33]

Such a conception, it is crucial to understand, is completely foreign to neoclassical economics. Thus, leading neoclassical economists William Nordhaus and James Tobin declared, in 'Is Growth Obsolete?' in 1972: 'As for the danger of global ecological catastrophes, there is probably very little that economics alone can say' – which did not prevent them from accompanying this perspective with a defence of unlimited economic growth. 'Natural resources', they insisted against the limits to growth perspective, '*should* grow in relative scarcity'.[34] In Sweezy's words, 'orthodox economics' adopts 'a framework which on principle excludes the consideration of the effects of economic forces on the social and political complexion of society', and which also fails to integrate the natural-environmental effects. Hence, 'the connections between capitalism as an economic system and the ugliest phenomena of the modern world ... are regarded as no concern of the economist'.[35]

---

other factors for natural resources, then there is in principle no "problem". The world can, in effect, get along without natural resources, so exhaustion is just an event, not a catastrophe' (Solow 1993a, p. 174).

32 With respect to philosophy, see Lukács 1968, p. xvii; Lukács 1974; Lukács 2003, pp. 96, 106, 113–14, 13–31; Mészáros 1970, 1995; Schmidt 1971. With respect to environmentalists, see Kapp 1950, pp. 31–6; Commoner 1971, p. 280.

33 Tsuru 1994, p. 233.

34 Nordhaus and Tobin 1972, pp. 16–17.

35 Sweezy 1953, pp. 26–64.

A concrete illustration of the theoretical blinders of neoclassical economics can be seen in the World Bank's drafting of its 1992 *World Development Report* on the theme of *Development and the Environment*. The few environmental economists associated with the World Bank protested at the refusal of conventional economists to draw a circumference, representing the limits of the planet, around a portrayal of the standard circular-flow diagram of the economy designed for the report. In the neoclassical view, the economy was a closed, self-subsistent entity. To question this was heresy. After a heated back-and-forth, the mainstream World Bank economists dropped the circular-flow diagram of the economy altogether, rather than indicate that the economy existed within planetary limits – or that the planet existed at all![36]

In contrast, Marxian political economy has been distinguished from its inception by its open-system approach of which Marx's metabolic rift theory is the prime example. As the great socialist ecological economist K. William Kapp explained, referring to the outlook common to Marxian and most radical institutional economics:

> Economic systems are intimately and reciprocally related to other systems and are in this sense fundamentally open systems. To view the economy as a closed system may be methodologically convenient ... but this tends to perpetuate a wrong perception of reality which narrows our theoretical horizon ... The environmental crisis forces economists to acknowledge the limitations of their methodological and cognitive approaches and to reconsider the scope of their science. The classical economists – Adam Smith and his successors – could still claim with some justification that economic systems could be understood as semi-closed systems because, in their time, air, water, and so forth were, in a sense, 'free' goods and because they were convinced – wrongly – that rational action – under competitive conditions – had only positive effects. This belief has turned out to be an illusion. To hold on to it in the face of the environmental crisis can only be regarded as a self-deception and a deception to others. Contemporary economists who continue to discuss economic and environmental problems in closed systems have much less of an excuse for this practice than the classical economists ...
>
> In short, we need a new approach which makes it possible to deal with the dynamic interrelations between economic systems and the whole network of physical and social systems ... Systems thinking is inevit-

36 Daly 1996, pp. 5–6.

> ably complex inasmuch as it is concerned with discontinuous non-linear 'feedback' effects which characterized the dynamic interdependencies between the different systems as well as of each subsystem with the composite whole ... It is, by its very nature, multi-dimensional, multi-disciplinary and integrative.[37]

For Kapp, one of the crucial, early examples of such a complex, integrated, dynamic, systems-theory approach with respect to the intersection of economy and ecology was to be found in the analysis of the problem of capitalist industrialisation and the soil-nutrient cycle as developed by Justus von Liebig and Karl Marx. It was undoubtedly Kapp who was to inspire Commoner's recognition of this part of Marx's analysis in *The Closing Circle*.[38]

Marx's dialectical approach to economic and ecological questions led directly to the development of an open-systems approach to social ecology. It was for this reason that Marx was able to see so clearly that the question of the transcendence of the system of capital accumulation was bound up with the issue of 'the inalienable condition for the existence and reproduction of the chain of human generations'.[39] It was this open-system dialectic that allowed him to integrate material (use value) flows, alongside value (i.e. exchange value) flows. Given this background, it is hardly surprising that Marx ended up inspiring nearly all of the early developments in ecological economics.[40]

For Marx, the core contradiction of capitalism is to be found in the alienation of labour, which is intrinsically related to the alienation of nature. Thus the question of the relation of human production to the 'universal metabolism of nature' cannot be separated from that of material-sensuous existence itself, i.e. social ontology, and aesthetics – which is to say that people are simultaneously social and natural beings. Humanity is distinguished, Marx stated, by its ability to form 'objects in accordance with the laws of beauty', and in this lies the potential not only for a radical, materialist aesthetics guided by human sensuousness, but also a developed social ontology of *human praxis* –an ontology that is directly ecological, since based on principles of sustainable human development.[41]

---

37 Kapp 1976, pp. 91–2, 96–8. System theory is not inherently dialectical in nature. On this problem, see Levins 2008.

38 Kapp 1950, pp. 35–6; Commoner 1971, pp. 254–6, 275, 280.

39 Marx 1981, p. 949.

40 Ample evidence for this is provided in Martinez-Alier 1987.

41 Marx and Engels 1975a, Vol. 3, p. 277; Marx and Engels 1975a, Vol. 30, pp. 54–66; Mészáros 1971, pp. 162–214; Marcuse 1972, pp. 59–78; Burkett 2005b.

Against Energeticism

In Marx's critical perspective it is not enough to be a materialist; it is necessary to be dialectical too. It is precisely this, in fact, that is most crucial to the development of a complex, historical, open-systems approach to society and nature. In sharp contrast, an energeticism that fetishises the mere quantitative – thereby mimicking capitalist (exchange-value) relations, while downplaying qualitative (use-value) relations – inevitably leads back to the same old dualistic antinomies, the same timeless form of mechanism and reduction, that characterise the prevailing worldview.

'The idea of nature as an integral part of materialism', Illya Prigogine, winner of the 1977 Nobel Prize in chemistry, declared,

> was asserted by Marx and, in greater detail, by Engels. Contemporary developments in physics, the discovery of the constructive role played by irreversibility, have thus raised within the natural sciences a question that has long been asked by materialists. For them, understanding nature meant understanding it as being capable of producing man and his societies.
>
> Moreover, at the time Engels wrote his *Dialectics of Nature*, the physical sciences seemed to have rejected the mechanistic world view and drawn closer to the idea of an historical development of nature. Engels mentions three fundamental discoveries: energy and the laws governing its qualitative transformations, the cell as the basic constituent of life, and Darwin's discovery of the evolution of species. In view of these great discoveries, Engels came to the conclusion that the mechanistic world view was dead.[42]

Unfortunately, many nineteenth-century and twentieth-century materialists (and socialists) were reluctant to let go of the mechanistic worldview. The so-called 'scientific materialism' (or mechanism), represented by mid-nineteenth-century figures like Ludwig Büchner, Karl Vogt, and Jakob Moleschott, lacked a sufficiently dialectical conception of reality. Hence rather than superseding idealism, such mechanism took on an abstract 'metaphysical' form, constituting itself as idealism's impoverished dialectical twin (identity of oppos-

42 Prigogine and Stengers 1984, pp. 252–3.

ites).[43] The roots of this lay in the philosophical tradition of the Enlightenment. Cartesian dualism, which defined the main tendency of modern Western philosophy, had promoted a rationalist/idealist conception of the mind and human beings, on the one hand, and a mechanistic conception of the body and animals, on the other. Increasingly mind and body, humanity and nature, idealism and mechanism were interpreted as entirely separate entities, operating in discrete spheres, rather than dialectically complex mediations.[44]

It should come as no surprise, then, that among the first reactions to Sadi Carnot's advances in thermodynamics, in which he presented an idealised model of mechanical-engine efficiency in a closed, reversible system, was to see the work of animals and human beings mechanistically. This took the form of direct comparisons of human labour power, horsepower, and steam power – studies with which Marx and Engels were quite familiar.[45]

Podolinsky himself adopted an extreme mechanistic version of this by applying Carnot's model directly, claiming that human labour was the 'perfect machine' – a kind of steam engine able to restart its own firebox. Although drawing out some important relationships, he fell prey to the crude mechanism and energy reductionism that such a view implied. The question of labour power was divorced from its historical and social content, from all qualitative transformations of nature, as well as from humanity's relation to nature, and was viewed from a purely mechanistic and quantitative perspective. Apparently believing that he had unlocked the physical basis of the labour theory of value, Podolinsky failed to perceive the qualitative relations of nature-labour-society that grounded Marx's value theory. Ironically, by applying Carnot's closed-system, perfect-machine thermodynamics to the actual world of human labour, Podolinsky was in effect denying that such labour was tied up with irreversible processes. This meant essentially that entropy was inapplicable to human labour. At the same time, he left out of his quantitative assessment the full complexity of human-nature transformations and even many aspects of more quantitative/energetic relations, such as the solar budget, the use of coal, fertilisers, etc.[46]

---

43 Gregory 1977.

44 Such dualisms and the restriction of materialism to natural science were given a big boost with the rise of neo-Kantianism, following the publication of Frederick Lange's influential book, *The History of Materialism*. See Lange 1950 [1865]; Beiser 2014, pp. 91–4, 166–71.

45 See Morton 1859, pp. 53–68; Marx 1976a, pp. 497–8.

46 See especially Chapter 2 above.

For Podolinsky, the creation and accumulation of value was essentially the same thing as the accumulation of terrestrial energy through the exercise of human labour – the prevention of the dispersion of heat/energy back into space. He did not (and obviously could not be expected to) fully understand what scientists know so well today: 'Earth's temperature is whatever is required to send back to space the same amount of energy that the planet absorbs. If less energy is sent back than is received, the planet warms, "glowing" more brightly and sending more back until a new balance is reached'.[47] This is in fact what is happening in our time with global warming. Through the buildup of carbon dioxide and other greenhouse gas emissions in the atmosphere, humans have finally achieved the goal that Podolinsky sought (and confused with progress) of increasing the energy stored on the earth. But the consequence is to generate a rift in biospheric conditions required for the perpetuation of the human species and innumerable other life forms.

The fact that Marx and Engels were reluctant to embrace Podolinsky's energeticism does not indicate that they rejected thermodynamics or were unsophisticated in their grasp of energy issues. To the contrary, the founders of historical materialism followed the development of the physical sciences very closely, and made sure their analyses were consistent with the latest, most rational developments in thermodynamics and evolutionary theory. Yet their dialectical instincts and emphasis on qualitative forms rather than simply the quantitative aspect of energy transformations (as exhibited in their wider metabolic approach) kept them from capitulating to crude energetics. Attentive to irreversible processes in relation to production, Engels complained of Podolinsky's failure to emphasise the fact that capitalist industrialism squandered limited supplies of coal and other resources. As famed early Soviet physicist and sociologist of science Boris Hessen observed, the 'treatment of the law of the conservation and conversion of energy given by Engels, raises to the forefront the qualitative aspect of the law of conservation of energy, in contradistinction to the treatment which predominates in modern physics and which reduces this law to a purely quantitative law – the quantity of energy during its transformations'.[48]

47 Alley 2000, p. 132.

48 Hessen [1931] 1971, p. 203. In line with these comments by Hessen, systems ecologist Howard Odum was later to turn to Marx in his attempt to bring out the ecological significance of qualitative as well as quantitative transformations in energy (see Odum and Scienceman 2005; Foster and Holleman 2014).

The unity and universality of Marx's worldview, his notion of organic-inorganic relations, his concern with sustainability, were all evident in his treatment of social metabolism [*Stoffwechsel*]. As Peter Dickens has written:

> Marx's early background led him to undertake no less than an analysis of what would now be called 'environmental sustainability'. In particular he developed the idea of a 'rift' in the metabolic relation between humanity and nature, one seen as an emergent feature of capitalist society ... The notion of an ecological rift, one separating humanity and nature, and violating the principles of ecological sustainability, continues to be helpful for understanding today's social and environmental risks. These risks are becoming increasingly global in extent. This is partly because they directly impact on environmental mechanisms operating on a global scale.[49]

Brett Clark and Richard York have extended the metabolic rift analysis to the planetary rift in the carbon metabolism itself, associated with two distinct disruptions of the carbon cycle: (1) the introduction of fossil fuels that accumulated over geological time, serving to break the solar budget; (2) the destruction of carbon sinks, that is, the carbon-absorbing capacity of the ocean and forests, which cease to absorb as much carbon as in the past. Marx's analysis thus helps us to understand the cumulative catastrophic impacts that threaten the earth as a place of human habitation.[50]

Metabolic Restoration: Toward Sustainable Human Development

No matter how serious the planetary ecological emergency that confronts us today, no matter how far we go down the path to catastrophic climate change, the basic answer to the global environmental problem is at all times the same: the struggle to recreate a balance in our relation to the earth – before the earth system (through its equilibrating mechanisms) creates a balance of its own – one outside the contours of what constitutes a safe operating space for humanity. What is required is the creation of a more collective, sustainable, egalitarian mode of production. In order to accomplish this we need to reverse the rift in the metabolism between humanity and nature and begin, as

49 Dickens 2004, pp. 80–1.

50 Foster, Clark and York 2010, pp. 121–50.

Marx himself insisted, the 'restoration' of those very conditions. We will either socially develop in the not-too-distant future the forms of sustainable living that are now required, or this will be imposed on us, under far less favourable environmental conditions, by the earth's 'closing circle'.[51] No one perhaps has better captured the systematic and revolutionary nature of the needed ecological change than Weston, when she wrote:

> The Marxist concepts of the metabolic rift and metabolic restoration are my starting points for establishing the principles for future social organization. Addressing the problems created by the metabolic rift requires radical structural change including the dismantling of capitalist social relations of production that currently dominate humans' relationships to the environment in producing their livelihoods. The metabolic restoration means we have to localise production and consumption and limit consumption to what the local biosphere can support, with a continual cycle of replenishment of the biological basis of production. The capitalist social relations of production need to be transformed into relations of production that are collective, co-operative and non-exploitative of either humans or nature. Of particular importance is the dis-alienation of humans in the relations of production. Thus, just as Keynesian reforms cannot address the very basic faults at the root of the capitalist economy, environmentalism that is not grounded in an understanding of the social relations of production and class analysis cannot solve the contradictions in capitalism that have given rise to global warming and the various other [environmental] crises of capitalism. The need is for human societies to live within metabolic cycles – that is, production, consumption and waste – thereby forming part of a self-sustaining cycle in which the only new inputs are energy from the sun ... Nature, in the new economics, will be recognized as the ultimate source of wealth.[52]

Such a system of social-metabolic restoration, Weston insists, demands: (1) 'restoration of the commons'; (2) 'food security'; (3) 'community of producers'; (4) 'no-growth economies' (as growth is currently understood in terms of GDP accounting); (5) 'equity, ecological justice and redistribution'; and (6) 'real participatory democracy'.[53] Ultimately, such a vision of sustainable human devel-

51 Marx 1976a, pp. 637–8.

52 Weston 2014, pp. 170–1.

53 Weston 2014, pp. 171–80; Burkett 2014, pp. 223–57.

opment corresponds to socialism/communism as defined by Marx: a world in which 'the associated producers govern the human metabolism of nature in a rational way, bringing it under their collective control ... accomplishing it with the least expenditure of energy and in conditions most worthy and appropriate for their human nature'.[54]

54 Marx 1981, p. 959; Burkett 2009, pp. 301–32.

APPENDIX 1

Socialism and the Unity of Physical Forces[1]

Sergei Podolinsky

In accepting the theory of the unity of physical forces or of the constancy of energy, we are also forced to admit that nothing can be *created*, in the strict sense of the word, through labour and that consequently, all the usefulness of labour, the goal for which it strives, can be nothing other than a transposition of a certain quantity of forces. What is the manner in which these transpositions are produced? What are the best ways to apply human labour to nature in order to render a greater fraction of its forces profitable for the satisfaction of human needs? These are the questions to which we shall attempt to respond in the present study.

According to the theory of production formulated by Marx and accepted by socialists, human labour, expressed in the language of physics, accumulates in its products a greater quantity of energy than that which was expended in the production of the labour power of the workers. Why and how is this accumulation brought about?

In order to respond, we must dwell for a moment on the general distribution of energy in the universe.

The total energy, the sum of all of the physical forces of the universe, is a constant quantity, but this is far from the case for the quantities of energy in the different parts of the universe. Certain heavenly bodies send to other bodies, across interstellar space, different kinds of physical forces in considerable quantities, and this permits us to say that the former of these bodies, the *suns*, possess a greater quantity of energy than both interstellar space and the latter celestial bodies, the *planets* and their *satellites*. These last heavenly bodies receive their energy from the nearest suns, in the form of light rays, calorific rays, chemical rays, etc. An exchange such as this, between bodies possessing a greater amount of energy and those with a lesser amount, must inevitably, after some time, lead to a universal equilibrium of energy.

1 This article was originally published in *La Plebe* in 1881 (Vol. XIV). The first instalment appeared in issue number 3, on pages 13 to 16. The second instalment appeared in issue number 4, on pages 5 to 15. Editorial notes were compiled by Paul Burkett and John Bellamy Foster. Translation by Angelo Di Salvo and Mark Hudson. [Editorial note].

It is thought that all of the transformations of physical forces, inevitable in their period of equilibration, are accompanied by a general tendency of certain kinds of physical forces to assume another form than that which they possess, and that there may be a determined form, that of *heat* uniformly distributed across the universe, that all forms of physical force accept, at least partially, in the course of each transformation.

The energy of the universe is thus transformed constantly, leaving behind forms that are not very stable in order to acquire others that are more stable. Consequently, the ease of the subsequent transformations always tends to diminish. After a long series of centuries, the total energy should eventually acquire a form that is incapable of transformation, which would consist of a certain degree of heat uniformly distributed throughout the entire universe. When this occurs, every kind of mechanical movement perceptible to our senses, and consequently, every living phenomenon familiar to our understanding, would not take place because a difference in temperature is *absolutely* necessary to bring about a transformation of heat into any other kind of physical force. This tendency of energy toward a universal equilibrium is called *dispersion of energy*, or according to Clausius's terminology, *Entropy*.[2] This last term expresses the quantity of energy transformed that cannot undergo additional subsequent transformations. These two principles of Clausius derive from it: *The energy of the universe is constant. The entropy of the universe tends towards a maximum*.[3]

Thus, in the strict mechanical sense of the word, the energy of the universe will certainly always and completely be conserved. But this equilibrated energy, that is, the heat uniformly distributed in the universe, will no longer be capable of provoking different presently observed phenomena in the inorganic world and in the living world, because a difference in temperature is absolutely necessary to bring about a transformation of heat into any other kind of physical force.

It is true that we continue to receive on earth from the sun enormous quantities of physical forces still capable of going through all of the transformations of which the physical and biological phenomena on our globe are faithful manifestations. According to Secchi, one square metre of the sun's surface furnishes 5,770,540 kilogrammometers in the form of 76,642 horsepower of work.[4] Several square metres of the sun's surface would be sufficient to set in motion all of the machines that exist on the earth. The total work of the sun is estimated

2 Clausius, *Théorie mécanique de la chaleur*, t. I, p. 411, Paris, 1863 [Footnote in original]. Clausius 1879, pp. 106–7, 195–7 [Editorial note].

3 This paragraph does not appear in the French version (Podolinsky 1880) [Editorial note].

4 Secchi, *Le Soleil*, II, p. 258. Parigi, 1875 [Footnote in original]. Secchi 1875–7 [Editorial note].

at 470 quintillion horsepower. If we accept the widespread theory that explains the source of solar heat by the sun's own condensation, we find that it would take 18,257 years for the visible diameter of the sun to be diminished one single second, and 3,820 years for the sun's temperature to drop one single degree. This figure will not seem exaggerated at all, if one keeps in mind that the substance of the sun is found probably almost in that state of chemical indifference, which is produced by the raising of the temperature, that is known by the name of *dis-association*.[5, 6]

Thus we see that the danger of one day lacking forces capable of being transformed on the earth's surface is still very distant, but at the same time, looking more closely, we would argue that the distribution of these forces is not always the most advantageous for the living world's needs in general and for the existence of the human race in particular. We believe, furthermore, that to a certain extent, it is within the power of humanity to produce certain modifications in this distribution of solar energy, in such a way as to render a greater portion profitable to humans.

In reality the major part of the physical forces that are found on the earth, and which are thus useful for meeting human needs, are by no means found in a form which would be the most advantageous for achieving this goal.

Humanity having above all a need for nutritive, combustible and mechanical forces for work, the most profitable forms of physical forces for it would be: first, the more or less free chemical affinity, represented in the form of nutritive substances of animal or vegetative origin, or in the form of combustible material; second, effective and available mechanical movement that can serve as an engine for the machines that work to the benefit of human beings.

Now, we observe that the earth, in itself, offers us relatively few physical forces that have these advantageous forms for humanity. If it is true that the earth's interior is still in a state of incandescence [heating], which presumes that we would find there many dissociated chemical elements, and that thanks to the elevated temperature one would find a great quantity of virtual movement, we scarcely profit at all from this, and feel, on the contrary, the destructive effects during earthquakes and volcanic eruptions. We are, however, compensated for these disasters by the exceptional fertility of volcanic terrain and by a notable increase in the temperature in the vicinity of volcanoes. On the slopes of Mount Etna, says Eliceo Reclus,

5 H. Sainte-Claire Deville, *Leçons sur la Dissociation*, Paris, 1862 [Footnote in original]. Sainte-Claire Deville 1864 [Editorial Note].

6 This sentence and Note 5 do not appear in the French version [Editorial note].

> the land is so fertile, that its products are able to suffice for a population three or four times more dense than that of the other counties of Sicily and of Italy. More than three hundred thousand inhabitants are clustered on the slopes of this mountain, which from a distance is considered a place of terror and imminent danger, and from time to time this proves to be the case as it is uncovered to flood its countryside with a deluge of fire. At the base of the volcano the cities touch and follow one another like pearls in a necklace.[7]

But generally the surface layers of the earth's crust consist principally of chemical combinations which contain almost no free affinity, nor any noticeable mechanical movement. Such is also the case for the waters and the atmosphere that cover the surface of our planet and with which we are always in contact. All the movements of the air and water, the ebb and flow, the movement of waves produced by the wind, the river currents, the force of falling rain, the wind itself, take their force from the energy radiated from the sun, or are produced by the gravitational pulls of the moon and sun. The chemical affinity, accumulated in the form of coal in the bowels of the earth, is equally an effect of solar heat, a product of the sun's rays over the past centuries. Even free oxygen in the atmosphere, according to certain geological hypotheses, originated in combination with the carbon that now constitutes coal.[8]

All of these examples clearly demonstrate that the energy radiated from the sun is more or less the only source of all the forces profitable to humankind, which are found on the earth's surface.

But the quantity of energy radiated toward us from the sun would be, according to a very well-known physical law, reflected into interstellar space, in the same proportion that it is received, if it did not undergo certain transformations which permit it to prolong its stay on the earth and to constitute there an accumulation of solar energy. This occurs when the rays which arrive from the sun, hot, luminous and capable of producing certain chemical modifications, are collected from material in such fashion as to transform them into free chemical affinity or into a mechanical movement.

---

7 Popalazione chilometrica dell'Italia, 94; della regione dell Etna, 550 (Elisée Reclus, *Nouvelle Géographie Universelle*, I, 538, Paris, 1875) [Footnote in original]. Reclus 1876–94; Reclus 1882–95, Volume I, p. 315 [Editorial note].

8 Sterry Hunt, *Congresso della Società Britannica*, 1878 [Footnote in original]. See Sterry Hunt 1891, pp. ix–xi, 40–7 [Editorial Note].

In this last case, a part of the heat radiated from the sun is no longer reflected into interstellar space, according to the well-known law of Kirchhoff,[9] but is instead captured for a more or less long time on the surface of the earth, since it would take on forms that protect it from immediate dispersal. 'Energy rises by degrees', is how this is expressed by the celebrated philosopher William Thomson. The words of Secchi may serve as the best illustration here:

> The sun's rays that fall on the plants are not reflected by them to the same degree as would be found for the desert or mountainous rock. They are captured by the earth's surface in a greater measure, and the mechanical force of the undulations of the particles is used to produce decompositions of oxygen with carbon and with hydrogen. The dissociation of the stable combinations, their dissolution so to speak, as in the familiar case of water and carbonic acid, is an inevitable consequence of the activity of the sun's rays on the plants.[10, 11]

(End of First Instalment)

Socialism and the Unity of Physical Forces: Continuation and End

What happens in this case? A portion of the solar heat, which seems to dissipate, is in reality captured by the earth's surface *without raising its temperature*, that is, without increasing its losses in space. The losses are equal, but the surface of the earth has received from the sun more energy, or rather, having received the same amount of energy it disperses it less. In whatever way we consider this process, we see under the influence of the plants an accumulation of energy. But it is not dispersed energy anymore, like heat, electricity, and light, but an energy of a higher degree that will still be preservable on the surface of the earth for hundreds of years, and will be capable also of all the other transformations. Thus, the plants in general and cultivation in particular are the most feared enemies of the dispersion of energy into interstellar space.[12]

---

9 Kirchhoff's Law can be expressed as: The quantity of radiated heat is directly related to the difference between the temperature of the heat-source and the environment that surrounds it [Footnote in original]. Kirchhoff 1862–3, p. 17; Kirchhoff 1901, pp. 75–6 [Editorial note].

10 Secchi, *Le Soleil*, t. II, p. 300 [Footnote in original].

11 This paragraph and Notes 9 and 10 do not appear in the French version [Editorial note].

12 This paragraph does not appear in the French version [Editorial note].

But the ways in which the radiating force of the sun effects these transformations are not very numerous. They are above all:

1. The production of the *wind*, that is the impulse to movement given to the air by the modifications of its temperature.
2. The elevation of water through evaporation.
3. The dissociation of stable combinations, for example, of water, of carbonic acid, carried out by the growth of plants.
4. The muscular-nervous work produced by men and animals.
5. The work of machines constructed by men which, in a direct or indirect way, as with Mouchot's solar machine, have as their only motor the sun's heat.

We will see that the quantity of solar forces converted into free chemical affinity and into effective mechanical motion is not always the same and that it can be modified, among other causes, by the efforts of humans. For *man, by certain acts of will, can increase the quantity of solar energy accumulated on the earth and diminish the dispersed energy.*

In cultivating vegetables in places where there were not any, or even where they existed in small quantities; by draining marshes, irrigating dry counties, and introducing a perfected system of cultivation; in applying machines to agriculture, and protecting cultivated plants and vegetables against their natural enemies; man can reach the first goal.

In driving away and exterminating animals that are harmful to the richness of the vegetation, he works to reach the second goal. In the two cases we have an absolute or relative increase of solar power retained on the earth.

Here are some examples, taken from agricultural statistics of France, which tend to prove the truth of our assertions as regards the decisive influence of the work of men, or of animals directed by men, on the quantity of solar energy accumulated by a given terrestrial surface.

France possesses now almost 9,000,000 hectares of forests that produce annually 35,000,000 stères or almost 81,000,000 metric quintals [1 metric quintal = 100 kilograms] of wood. The average production is then yearly 9 metric quintals for each hectare. Accepting the figure of 2550 calories[13] produced by the combustion of one kilogram of wood, we see that the 9 quintals of

13 The *calorie* is a unit of measure of heat which represents the quantity of heat necessary to raise the temperature of a kilogram of water by one degree [Footnote in original].

wood represent a value of 900 × 2,550 = 2,295,000 calories per hectare accumulated in the course of one year.[14]

The 4,200,000 hectares of *natural pastures* in France produce an annual average of 105,000,000 metric quintals of hay, or 25 metric quintals per hectare, which represent, given the same number of calories produced by combustion of hay as for wood, 2500 × 2550 = 6,375,000 calories accumulated per hectare.

Thus, in France, the accumulation of solar heat in the form of chemical affinity produced by the growth of vegetables in their natural state fluctuates between 2,295,000 and 6,375,000 calories per hectare, under conditions where the natural vegetation is richest, that is, in the forests and on the pastures. Let us see now the effect produced by labour.

The *sown pastures* of France occupy a surface of 1,500,000 hectares and produce, on average, with a deduction made for the caloric value of the seeds, 46,500,000 metric quintals of hay or 31 metric quintals per hectare. This production gives 3,100 × 2,550 = 7,905,000 calories. In other words, the surplus compared to the natural pastures is 1,530,000 calories per hectare. Now in order to cultivate one hectare of sown pasture (once every four years) and harvest the hay every year, one must expend approximately 50 hours of a horse's labour and 80 hours of a man's labour, which altogether represents around 37,450 calories. Consequently, each calorie spent in work yields: 1,530,000:37,450 = 41 calories of accumulated solar heat.

The cultivation of *wheat* in France (taking a figure a little less than the actual average) extends over 6,000,000 hectares. The average production under the same conditions, after a deduction for the grain planted, reaches 60,000,000 hectoliters of grain and 120,000,000 metric quintals of straw, that is, 10 hectoliters or roughly 800 kilograms of grain and 20 metric quintals of straw per hectare. The 800 kilograms of grain contain almost 3,000,000 calories, using the figures for the combustion of albumens, starch, etc. The 2,000 kilograms of straw would produce through their combustion 5,100,000 calories, and altogether the harvest of one hectare gives 8,100,000 calories. The surplus over the natural pasture is 1,725,000 calories. This surplus is produced by 100 hours of horse labour and 200 hours of human labour, together representing a total value of 77,500 calories. Therefore, each calorie spent on labour, during the cultivation of a field of wheat, accumulates 1,725,000:77,500 = 22 calories on the earth's surface.[15]

---

14 In the French version (p. 357), a short paragraph follows which does not appear in the Italian. It is restored in the present text [Editorial note].

15 Veggasi Ch. Laboulaye, *Dictionnaire des arts et de l'agricolture*, 4.ª edizione, articoli: 'Agricolture' e 'Carbonification'; *Statistique de la France*, 1874, 1875, 1878; Pelouze et Frémy, *Traitè de Chimie*; Hermann, *Grundzüge der Physiologie*, 5.ª ed. 1877 [Footnote in original]. See Laboulaye 1874; Pelouze and Fremy 1865–6; Hermann 1875 [Editorial note].

The effects produced by irrigation also show the importance of the influence of human work on the quantity of solar energy stored in the earth. The average product of one hectare of wheat on the non-irrigated lands of the Spanish provinces of Valencia and Murcia only yields 6 times the quantity of planted grain, while in the fields traversed by innumerable canals, diverted from Gibraltar, from Jucar, from Segura and from the other rivers of the eastern coast of Spain, the yield is 36 times the weight of the seeds.[16]

What then is the real cause of this increase in the quantity of solar energy, which remains on the earth's surface, in the form of nutritive substances or combustible materials, instead of being immediately reflected, according to the simple law of temperature differences, into frigid interstellar space? It is the *useful work* that we can define in this way: *every expenditure of muscular work of humans or of animals that has as a result an increase in the solar power accumulated on the earth.*

The increase of force can be carried out in two ways: by the immediate conversion of a certain quantity of solar energy into motion or into a nutritive substance, or, just as well, in a mediated way, through the conservation of a quantity of energy existing on the earth, which without the intervention of labour would be inevitably dispersed. In this last category belongs, for example, the useful work of artisans, such as shoemakers, tailors, tool and machine inventors, etc.

It is clear, according to this definition, that useful work can be attributed only to men and certain animals that, like domestic animals, are guided by men, or that, like ants, are busy with cultivation and the raising of domestic animals, driven by their own instincts.

The motion of the air, the wind, can never be categorised as useful work in the true sense of the word, because, left to its own free course, it does not ever produce with the expenditure of its energy a new accumulation of energy on the earth. The same reasoning is applicable to the moving force represented by water currents.

Plants, which in fact accumulate energy in the substance of their own organisms, cannot put it in motion by themselves in the vast majority of cases, neither can they expend it in a useful way, that is, in a way that may increase the quantity of accumulated energy that exists on earth.

Machines constructed by the work of men, left to themselves, even if they could remain for a long time in motion, would not give any useful work, because it is still impossible to imagine an artificial construction that, without any intervention of the work of men's muscles, could produce a continuous increase of solar energy accumulated on the earth.

---

16 This paragraph does not appear in the French version [Editorial note].

Finally, the nervous work of men, also intellectual work, that can contain the *possibility* of an immense accumulation of energy, does not really become useful work for the human race before having been applied to a specific muscular labour, *because we are not acquainted with any means (method) to achieve with intellectual work the goal of all useful work, that is, the absolute or relative increase of the energy that is found at the disposal of humans.*[17]

How animals contribute to increasing the accumulation of energy on the earth's surface is very difficult to establish. Nourishing themselves with vegetables, they make them diminish, but they also make them increase indirectly by accumulating their products in an indirect manner, for example the guano and other products of the digestion of vegetable substances, etc., that serve to make plants grow. The same thing happens with the animals that serve in the nourishment and clothing of humans, animals which, although certainly formed from the vegetable kingdom, supply it at the same time with a surplus of energy. It is known that certain species of animals accidentally or instinctively produce useful work. Thus, the bees that make honey or wax, many insects and birds that fertilise plants and increase in such a way the production of seeds (Darwin, *Origin of Species*), show us clearly that plants and insects adapt to their conditions of mutual coexistence and at times help each other in their struggle for existence.

It would be impossible indeed to establish mathematically the usefulness or harmfulness of different animals, by which we should mean their usefulness for the welfare of the human race.

In sum, strictly speaking, only in the cultivation of the earth is our definition of work best exemplified, since it is evident that a hectare of uncultivated land or of virgin forest, without the influence of humans, produces less nutritive materials, but this quantity can be multiplied by ten or even more. Certainly, man can create neither material nor energy. All the material existed in the earth, in the seeds and in the atmosphere; all of the energy was given by the sun. But thanks to the activity of human work, one hectare of land can accumulate in its vegetation ten times or more energy than without this influence. It is not necessary to believe that this energy is only diffused into space by the work of humans. This would not be correct, since agriculture does not exhaust the soil unless it is carried out in an unscientific and wasteful manner. On the contrary, a perfected agriculture gives better results precisely in the countries where it has flourished for a longer time, in Egypt, China, Japan, Lombardy, France, Belgium and England, etc.

---

17 Compare Marey, *Du mouvement dans les fonctions de la vie*, pagina 205, Paris, 1868 [Footnote in original]. See Marey 1868 [Editorial note].

We believe we are right then in affirming that scientific agriculture is the best example of useful work, that is, of work that increases the quantity of solar energy on the earth's surface.[18]

We now attempt to apply this theory to the satisfaction of human needs. We assume as demonstrated that the satisfaction of any need is accompanied by an exchange of physical forces between the organism and the external environment. A certain quantity of energy indispensable to the satisfaction of our needs is offered to us without efforts on our part and is in a matter of speaking a gift of nature, for example, oxygen in the air, etc. We have not worried ourselves about this part of the energy necessary for our existence. But all the rest must be acquired through work. Now we have seen that the only useful work, the only kind that really increases the quantity of energy that is available to humans, is the work done with the muscles. All intellectual work, even that of the man of genius, *does not in fact* increase the accumulation of energy on the earth if it does not increase the productivity of the muscular labour of the worker who adapts himself to the forces of new inventions, to the perfected tool, to the machine, to a new system of agriculture and industry. Without this essential expenditure of muscular force, the most splendid invention would be useless. There exists only one way for work to increase the quantity of energy on earth, and that is to make the muscular labour of the workers more productive with the help of machines and of perfected processes, etc. Sensing this truth, Adam Smith specified the necessary work in production, in the wider sense, in the historical sense of the word, as the only measure of the value of things.[19]

From the experiments of Hirn and Helmholtz, we now know that a relationship exists between the quantity of oxygen inhaled during work and the quantity of work supplied. It turns out that estimating the quantity of work accounted for by the combination of breathed-in oxygen with the elements of our bodies, and comparing this quantity with the work supplied by the muscles, we have an almost constant relationship which does not stray far from the ratio of 5:1. For this reason, the fraction 1:5 is considered to be the *economic coefficient* of the human machine, whether measured in terms of the quantity of oxygen inhaled, or, almost equivalently, by the quantity of food ingested.[20]

18 The preceding four paragraphs do not appear in the French version [Editorial note].

19 Smith, *Ricerche sulla natura e sulle cause della riccessa delle nazioni*, Collezione dei principali economisti, v. 1 [Footnote in original]. See Smith 1937, Book I [Editorial note].

20 Verdet, *Théorie mécanique de la chaleur*, II, pag. 216 [Footnote in original]. See Verdet 1868–72 [Editorial note].

But in reality one should reduce considerably the fraction which represents the economic coefficient of the human machine, because nutrition combined with respiration, is not the organism's only indispensable need. As an approximate calculation, the value of the food may be considered to represent half of the work demanded for the satisfaction of our needs. For the whole, then, it is necessary to allow double the amount of work and thus to reduce the economic coefficient of humans by two times, that is to put it at 1:10, rather than 1:5. This means that the satisfaction of all of our needs, presently considered as indispensable, represents a quantity of work almost ten times greater than the human muscular labour. This surplus must be accounted for by the greater productivity of human muscular labour, guided by intelligence, by the muscular power of domestic animals, or finally, by inanimate forces both natural and artificial.

It is easy to see that the economic coefficient of the human machine is in no way a constant quantity. It varies considerably through the centuries and from place to place. The savage who satisfies his own needs in large part from the free bounty which nature offers him, without much productive work on his part, and whose needs are limited almost to nourishment, apparently possesses an economic coefficient that is higher than civilised man. But, on the other hand, if the work produced by the muscular system of the civilised man is a smaller fraction of his expenses than that of the savage, then the usefulness of his work is much greater, because his work satisfies a greater quantity of needs and satisfies them on average much better than with the savage primitive or during the intermediate stages of civilisation.

Production, which is equivalent to the possibility of satisfying our needs, increases in the civilised countries by reason of their progressive movement, and this increase proceeds, in the majority of cases, more rapidly than the growth of population. Let us take the example of France.

France in:[21]

1820	had	29,700,000	inhabitants;	wheat production was	44,000,000	hectoliters
1830	"	31,500,000	"	"	52,000,000	"
1850	"	35,000,000	"	"	87,900,000	"
1860	"	36,100,000	"	"	101,000,000	"
1868	"	37,300,000	"	"	106,000,000	"

21 Gustave Heuzé, *La France agricole*, atlante N. 18, 1875; e *Annuaire du Bureau* [Footnote in original]. Possibly referring to the following: Heuzé 1875 [Editorial note].

According to other calculations,[22] the consumption of wheat per inhabitant in France was:[23]

In	1821	1.53	hectoliters
"	1835	1.59	"
"	1852	1.85	"
"	1872	2.11	"

The yields of the hectares cultivated in cereals have greatly increased in average numbers of hectoliters harvested. In 1840,[24] the average yield of:

Wheat was ...	12.28	hectoliters per hectare;	in 1866 it was	15.70	hectoliters
Rye was ...	10.79	"	"	13.83	"
Barley was ...	14.00	"	"	18.91	"
Oats was ...	16.30	"	"	24.50	"
Corn was ...	14.27	"	"	17.61	"

All of these figures, even if they are not perfectly exact in their particulars, still reveal that in France, human work has shown a clear tendency to become more productive over the course of the present century, and that its usefulness has grown in an indisputable manner.

The statistics furnished by Switzerland provide more detailed evidence. Comparing the increase in the population with the export of cereals, because Switzerland exports grain in larger quantities than it imports, we see that in spite of the continuous increase in the well-being, and consequently in the consumption in this country, the exportation of grains increased much faster than the population:[25]

In 1840,	the population was	3,100,000;	cereal exports were	1,500,000	metric feet
" 1850	"	3,500,000	"	4,500,000	"
" 1860	"	–	"	8,500,000	"
" 1875	"	4,400,000	"	17,500,000	"

22 Maurice Block, *Statistique de la Francé*, t. II, pag. 389 [Footnote in original]. See Block 1875 [Editorial note].

23 This sentence and the immediately following figures do not appear in the French version [Editorial note].

24 Gustave Heuzé, loc. cit., pag. 27 [Footnote in original].

25 Elis Sidenbladh, *Royaume de Suède – Exposé statistique* 1878, pag. 40 [Footnote in original]. See Sidenbladh 1878; Sidenbladh 1876 [Editorial note].

Sweden in:

1830	possessed	1,857	factories with production valued at	13,000,000	kronen
1850	"	2,513	"	37,000,000	"
1870	"	2,183	"	92,000,000	"
1875	"	2,719	"	173,000,000	"

Finally, Spain, which is usually not considered to be a country following a path of rapid progress, shows us examples according to which, under good conditions for the increase in population, the increase in agricultural production took place at a still more rapid pace. It has now been one hundred years since in the Basque Country the annual harvest was 200,000 bushels (a Spanish measure a little larger than a hectoliter) of wheat and 400,000 of corn, and at that time there were resources sufficient for a population of scarcely 100,000 souls. Since then the population has doubled, but the crops harvested have increased in a greater proportion: The Basque Country today produces annually 600,000 fanegas of wheat [one fanega equals 1.58 bushels], more than one million fanegas of corn of which a portion is exported to England and to Germany, and 80,000 fanegas of dry vegetables.[26]

We thus see that useful work can accumulate energy in greater proportion than the population increase, but certainly, this is not the general case. In the majority of countries, especially where unbridled luxury reigns from developed capitalist production, a great part of the work is, contrary to its purpose, almost exclusively directed to the production of luxury objects, that is, to the gratuitous dispersion of energy rather than to its accumulation.[27]

What is the cause of this apparent contradiction? Since the economic coefficient of primitive man is greater, it is necessary to consider his body as a better organised machine than the body of civilised man; still, the latter produces more with his work. To find the solution to this problem it is necessary to return to the noted considerations of Sadi Carnot on *thermal machines*, that is, machines that transform heat or other forces into mechanical motion.[28] Man is also a thermal machine.

---

26 Louis-Lande, *Basques et Navarrais*, pag. 205, Paris, 1878 [Footnote in original]. See Louis-Lande 1878 [Editorial note].

27 The preceding four paragraphs and Notes 25 and 26 do not appear in the French version [Editorial note].

28 *Réflexions sur la puissance motrice du feu*, pag. 20, Paris, 1824 [Footnote in original]. Carnot, *Réflexions sur la Puissance Motrice du Feu et sur les Machines Propres a Développer cette Puissance*. Paris: Chez Bachelier, 1824. See Carnot 1977, pp. 11–13 [Editorial note].

According to Sadi Carnot, in order to be able to judge the degree of perfection of a thermal machine, one needs to know not only its economic coefficient, *but also its capacity to recycle the heat spent at work*. A machine having the capacity to reheat itself, making the heat spent at work rise toward its firebox, would be a *perfect machine*, and only such a machine could provide a true conception of the transformation of heat and *vice versa*. Now, no machine constructed by the hands of men possesses this faculty. No machine heats its own firebox with its own work alone, and no machine works on a *reverse cycle*, that is, the transformation of work into heat is unknown. As a consequence, the true laws of these transformations cannot be found with the aid of inanimate machines. The plant world, producing almost no effective mechanical motion, also cannot even be remotely considered as an example of a *perfect thermal machine*.

But, observing the work of humans, we see in front of us just exactly what Sadi Carnot calls a *perfect machine*. From this perspective, humanity would be a machine that would not only transform heat and other physical forces in work, but that would also produce the *complete reverse cycle*, which converts its work into heat and other forces essential for the satisfaction of its needs, that is to say it would recycle to its firebox the heat produced by its own labour. In reality, a steam engine, even admitting that it will run an entire year without the intervention of muscular human labour, could never produce all the elements necessary to sustain its work in the following year. The human machine, by contrast, will have created a new crop, will have raised young domestic animals, will have constructed new machines, and will still be able to continue with success its new work in the following year. The reason is evident: the human machine is a perfect machine, whereas an inanimate machine never achieves the conditions of perfection that Sadi Carnot requires.

The degree of perfection of the human machine is determined not only by its economic coefficient, but above all by its capacity to carry out the reverse cycle, that is, to convert its work into an accumulation of the physical forces necessary for the satisfaction of human needs. This is how we should explain that primitive man, with his coefficient of almost 1/6, is less perfect as a machine than civilised man with his coefficient of only 1/10. It is that primitive man profits only from the free gifts of nature, while the civilised man satisfies almost all of his needs with the help of his work and produces in this way an accumulation of solar energy on earth, whose quantity surpasses ten times the force of his muscles.

The essential conditions for the continuation of work by an imperfect inanimate machine do not depend on its work, nor on its qualities, because we know that a steam engine, for example, cannot by itself renew the heat of its firebox.

All of these machines depend on human labour to furnish them with the substances that produce this heat. By contrast, the conditions of labour, i.e. of the existence of the human machine, can be rigorously established.

As long as muscular labour supplied by the human machine is converted into an accumulation of energy necessary for the satisfaction of human needs, which represents a quantity in excess of the sum of the muscular work of the human machine, by as many times as the denominator of the economic coefficient exceeds the numerator – the existence and the possibility of the labour of the human machine are guaranteed.

Every time that the productivity of human labour falls below the inverse coefficient of the human machine, there will be misery and perhaps a decrease in the population. Every time, instead, that the usefulness of labour surpasses this number, there will be an increase in well-being and probably an increase in the population.

The considerations that we have set forth lead us to the following conclusions:

1. The total quantity of energy which the earth's surface receives from its interior and from the sun tends to diminish. The amount of energy accumulated on the earth's surface tends to increase.
2. This increase of accumulated energy (beyond the limited quantity produced by the increase of non-cultivated vegetation) has as its only cause the muscular work of man and of certain animals. Every expenditure of mechanical work by humans and by any other organised beings, which is accompanied by an increase in the overall quantity of energy that exists on the earth, should be qualified with the name of useful work.
3. The economic coefficient of humans tends to diminish as their needs increase.
4. The usefulness of muscular work tends instead to increase, because, at the present time, a certain expenditure of muscular labour makes the accumulation of energy on the earth increase more than it did during the primitive era of civilisation.
5. As long as humans have on average a quantity of free chemical affinity and of mechanical work at their disposal, in the form of nutritive substances, of mechanical power of the animals and of mechanical motors, which together surpass the mechanical force characteristic of humans so many times as the denominator of his economic coefficient exceeds the numerator – the existence of humans is materially assured, because in this case the entire human race presents an example of a perfect thermal machine, according to the conception of Sadi Carnot.

6. The principal goal of work should be, consequently, the absolute increase of the quantity of solar energy accumulated on the earth, not simply the transformation into work of a greater quantity of solar and other forms of energy already accumulated on the earth. Yet in the latter kind of transformation, the increased amount of energy – for example, the work done by means of the burning of coal – is accompanied so much more by inevitable losses through dispersion [of energy] into space, that it reaches a higher percentage of heat and of the other physical force used in the work.[29]

We have reached the second question posed at the beginning of our study: what are the forms of reproduction that are most advantageous for the satisfaction of human needs? We may respond: *those that produce the greatest accumulation of solar energy on the earth.*

Clearly, this is not primitive cultivation. This cultivation, in truth, is not yet cultivation, because it is not based on the accumulation of energy, on useful work, but only on the utilisation of forces previously accumulated from the vital processes of nature. Primitive man, nourishing himself with fruit and with roots, hunting wildlife, fishing, only disperses into interstellar space the solar energy accumulated on the surface of the earth.

Neither is it production under slavery. This social form, based on perpetual warfare, excludes a considerable proportion of workers from all participation in the labour which accumulates energy on the earth, i.e. from the work which is really useful for the satisfaction of social needs. Without speaking of the number of workers killed and injured during the continuous wars, we mention only *the standing armies, the owners of slaves* and their *multitudes of guards*, as completely unproductive elements of humanity, during the entire era of the reign of slavery.

Feudalism already contains several elements of progress. The serf owns a plot of land that he can cultivate without fearing the eye of the master, without feeling the whip of the overseer.

But how restricted this element of progress is! The small plots of the serf are but parcels in comparison to the fields of the seigneur, which extend as far as the eye can see, and the time allowed for free work is but a brief recreation after the hard day's labour done for the lord. It is not necessary then to marvel at the fact that the productivity of labour during serfdom never reached even the average of its present-day usefulness.

29 This entire set of six conclusions does not appear in the French version [Editorial note].

We now arrive at the capitalist production of today's society. This form of production knew how to utilise the division of labour, and, as this division no longer sufficed, it applied the machines of industry and agriculture on a large scale. It obtained unhoped for results and it takes pride in this. But let us look more closely. All of these results were obtained not from capitalism, but from the *accumulation of the labour of generations of workers in the past*, or of the *present-day workers' associations* (*co-operatives*). Capitalism, on the contrary, does nothing more than throw, in times of crisis generated by it, thousands of workers onto the street, and it operates this way in a manner analogous to wars, slavery, to epidemics, that is, it disperses a portion of the energy at the disposal of humanity, instead of increasing its accumulation on the earth's surface.

The relation between labour and capital is analogous to that which exists between the human machine and an imperfect machine. An imperfect machine, a steam engine for example, is *absolutely incapable* of transforming work into heat, of producing a *combustible, nutritive substance*, without the intervention of human muscular and intellectual labour. And even the intellectual work of humans does not become productive until it induces a certain amount of muscular work. In the same way, capital, the product of the accumulated labour of ancient times, left to itself, would produce *absolutely nothing*, would not carry out *any new accumulation of energy* on the earth, without the indispensable co-operation of the expenditure of a certain amount of presently existing muscular labour. With the progress of industry this quantity of absolutely indispensable muscular work will probably fall to a minimum that is difficult to define, but we strongly doubt that this quantity could ever be completely eliminated. This is why we think that *only work, and indeed muscular work, should ultimately serve as the basis for the definition of the value of production and that as a consequence it will enter as the preponderant element in every socialist theory of the correct* (or, which is the same, a *precisely* egalitarian) *distribution of products*. This deduction is a death sentence for all systems of production other than socialism. It remains only for us to demonstrate that socialism is precisely the mode of production capable of carrying out the greatest amount of accumulation of energy on the earth, the mode which, in other words, serves to satisfy most easily and agreeably all the needs of humanity. It is also that which gives us the best chance for the continuous progress that fosters more and more along its way a harmonious and peaceful path.

We can limit ourselves to several examples.

It goes without saying that labour and production in common, through the association of the work forces, are more advantageous, regarding the accumulation of energy, than is individual work. Apart from the fact that the egalitarian association of the workers is the best means to profit in a reasonable manner from all the advantages of the division of labour, avoiding its deadly influence on the health of the workers and on their intellectual development, it is also the only system through which the machines become real organs for social organisms, instead of being, as occurs frequently under capitalism, destructive weapons in the hands of the privileged few that direct them against the proletarian majority. With the current system, every new improvement in the big industries deprives a certain number of workers of work, and as such causes a certain number of workers to die of hunger. Instead of increasing the accumulation of energy on the earth, it accelerates the gratuitous dispersion of the force of the workers excluded from production. Under the socialist system, by contrast, every mechanical or other improvement will have as an immediate consequence a decrease in the number of hours of work of all the workers and it will furnish them with time for new production, or indeed for education, for art, etc.

If the influence of the future socialist order is already very clear in that which concerns production, it is still more decisive in reference to exchange. The entire class of idle and rapacious businessmen will simply be eliminated.

A higher level and a more equal distribution of the quantity and quality of food will inevitably lead to an increase in the muscular and mental power of the immense majority of individuals comprising the human race. Hence, a new growth in production, a new accumulation of solar energy on the surface of the earth.

A precise and conscientious system of accounting, in which the figures are neither hidden nor distorted, will certainly have a considerable role to play in a more egalitarian, which is to say more just, distribution of the efforts of production and of the enjoyment accompanying the satisfaction of our needs.

Rational public health, and the possibility for everyone to adapt their private hygiene to the demands of science, will quickly raise the life expectancy and, in parallel fashion, the productivity of the human organism to a level that is currently found only in exceptional cases.

We can no longer deprive ourselves of the advantages in the accumulation of energy which the socialist system offers to us, for the sake of the security of all existing persons, also for the sake of the assistance that could be offered to the elderly, the sick, and the infirm. But we must also indicate the immense advantage that socialist education will offer, insofar as muscular labour, and the habit of undertaking it without great toil, will be taught to all, no one excepted. In our opinion, such a pursuit of integrated instruction will tend not only to raise the productivity of the social organism, but will serve above all to prevent any foolish ambition on the part of a selfish minority for the re-establishment of the oligarchical order under which we now live.

Such are, in the form of a brief and very general outline, the relations that exist between the theory of the accumulation of energy and the different forms of production. We hope to be able to revisit this issue soon in a more extensive and better-illustrated work.

S. Podolinsky
May 1881

APPENDIX 2

Human Labour and the Unity of Force[1]

Sergei Podolinsky

1 The Doctrine of Energy

If we acknowledge the correctness of the theory of the unity of force, of the constancy of energy,[2] we are also obliged to accept that nothing can be created by labour and that its goal and its utility consists only in a conversion of certain quantities of forces. In what way do these conversions come about? What are the best means of employing human labour in order to draw upon a greater fraction of natural forces for the satisfaction of human needs? We want to try to give an answer to this question in the present essay.

We know that human labour can accumulate greater quantities of energy in its results than was necessary to produce the labour power of the worker. Why and in what way does this accumulation of energy arise?

In order to answer this question we have to pay closer attention to the general diffusion of energy in space.

The total energy, the whole sum of physical forces of the universe, is a constant quantity. It is entirely otherwise, however, with the quantities of energy in the different parts of the universe. Some celestial bodies send significant quantities of different physical forces through the universe to other celestial bodies. This fact allows us to say that the first bodies, the suns, have energy in a greater quantity than the second, the planets and their satellites. The latter celestial bodies receive their energy from their closest suns in the form of illuminating, heating and chemically potent rays. Such an exchange of force between the bodies that have more energy and those that are endowed with less must lead, after a more or less long time, to a universal equilibrium of energy.

This equilibrium, however, cannot be accomplished other than by means of a whole series of transformations of physical forces. Observation teaches us that all such transformations of physical forces are accompanied by a tendency of those physical forces to assume a determinate form, namely that of the

1 Sections 1–3 of this article were first published in *Die Neue Zeit*, no. 1(9), pp. 413–24, 1883. Sections 4–6 were first published in *Die Neue Zeit*, no. 1(10), pp. 449–57, 1883. This translation is by Peter Thomas, with editing and annotations by Paul Burkett and John Bellamy Foster.

2 The capacity of the development of force is called energy [Footnote in original].

heat uniformly distributed throughout space. This last form of energy is the enduring form which is transformed with the most difficulty, while all other forms of energy – light, electricity, chemical affinity, etc. – are transformed frequently into heat, at least partially, in the course of their transformations.

In this way, a conversion of the energy of the universe constantly occurs through energy losing its less enduring forms and other more immutable forms taking their place. Thus further transformations of energy gradually become more difficult. After a long series of millions of years, therefore, all energy has to take on an enduring form, namely, that of heat uniformly diffused throughout the universe. If the universe endures for long enough, every type of mechanical movement perceptible to our senses and thus also every type of the phenomena of life will be completely absent, for a difference in temperature is absolutely necessary in order to transform heat into any other force. This tendency of energy towards a general equilibrium is called dispersion [*Zerstreuung*] of energy or, following the terminology of Clausius, entropy.[3] This latter expression signifies the quantity of transformed energy that is no longer capable of any reverse cycle transformations. From this follow the two laws of Clausius: *the energy of the universe is constant. The entropy of the universe has a tendency to reach a maximum.*

Thus, in the strict mechanical sense of the word, the energy of the universe is an always and absolutely constant quantity. However, this energy, brought completely into equilibrium, would be incapable of generating all those phenomena in the inorganic and organic world that we now observe, and which represent, fundamentally, nothing more than an expression of the different transformations of energy. That part of physical force that has now already been transformed into uniformly diffused heat constitutes, in a manner of speaking, a leftover of the world's activity, a leftover that gradually grows more and more.

Presently, however, we still receive on our earth enormous quantities of physical forces that are capable of experiencing the most varied transformations, as whose expression all the physical and biological phenomena appear.[4] According to Secchi, each square metre of the sun's surface delivers 5,770,540 kilogram-metres or 79,642 horsepower of labour.[5] A few square metres of the sun's surface would suffice to set all the machines on the earth into motion. The total labour power of the sun is estimated at 470 quintillion horsepower. If we

3 Clausius, *Théorie mécanique de la chaleur*. T.I., p. 411. Paris, 1868 [Footnote in original]. See Clausius 1879, pp. 106–7, 195–7 [Editorial note].

4 Biology is the doctrine of living animals [Footnote in original].

5 Secchi, *Le Soleil*, II. p. 258. Paris, 1875 [Footnote in original]. See Secchi 1875–7, two volumes [Editorial note].

accept the widespread theory that establishes the source of the sun's heat as its own condensation [*Verdichtung*], we find that 18,257 years would be necessary for the reduction of the apparent diameter of the sun by a single second and 3820 years would need to pass before the temperature of the sun would fall by a single degree. The last figure will in no way appear to be exaggerated if we consider that the sun's substance is probably almost constantly in that state of chemical indifference, caused by the high temperature, which is known by the name of *dissociation*.[6]

We thus see that the danger of one day suffering a lack of transformable forces on the surface of the earth is still a long way off; at the same time, however, we note upon closer inspection that the distribution of these forces is not always the most advantageous for the satisfaction of the needs of the organic world in general and of the human species in particular. We believe, however, that it is in the power of humanity, to a certain extent, to effect changes in this distribution that will enable us to use a greater part of the world's energy for the benefit of humanity.

In reality, the greater part of the physical forces on the earth's surface is far from being in the most advantageous condition for the satisfaction of human needs.

Humans above all need significant quantities of food, combustible material and mechanical work forces; the most advantageous forms of physical forces would thus be: (1) the more or less free chemical affinity in the form of nutritious substances deriving from plants and animals or in the form of combustible material; and (2) any mechanical movement which could serve as a driving force for the machines working for the benefit of humanity.

We see, though, that our globe in itself provides very few physical forces shaped into such advantageous forms for humanity. If the interior of the earth is really still in a state of incandescence and thereat are found large quantities of dissociated elements which, thanks to the high temperature, contain significant quantities of potential work, we nevertheless do not use these. Rather, we experience only their destructive effects at the time of earthquakes and volcanic explosions. Incidentally, we are nevertheless partially recompensed by the exceptional fertility of volcanic earth and by the increase in temperature in the vicinity of volcanoes. 'On the slopes of Etna', E. Reclus says, 'the earth is so fertile that its products are able to suffice for a population three or four times more dense than that of the other counties of Sicily and of Italy. More

6 H. Saint-Claire Deville, *Leçons sur la dissociation*. Paris, 1862 [Footnote in original]. See Sainte-Claire Deveille 1864 [Editorial note].

than three hundred thousand inhabitants are clustered on the slopes of this mountain, which from a distance is considered a place of terror and imminent danger, and from time to time this proves to be the case as it is uncovered to flood its countryside with a deluge of fire. At the base of the volcano the cities touch and follow one another like pearls in a necklace'.[7]

In general, however, the surface strata of the earth's crust are made up of chemical compounds, which contain almost no free chemical affinity and consequently have very little potential (possible) force of movement. We find the same thing in relation to the bodies of water and atmosphere that surround the surface of our globe, and with which we continually come in contact. All movements of air and water, ebb and flow, the movement of the waves caused by the wind, the currents of the rivers, the force of falling rain, even the wind, borrow their forces from the sun's energy or are caused by the gravity of the moon and the sun. The chemical affinity that is accumulated in the form of coal inside the earth is likewise an effect of solar heat, a product of the sun's rays over the course of many thousands of years. Even the free oxygen of the atmosphere, according to new geological hypotheses, was previously combined with the carbon that now constitutes coal – and was freed from it only through the influence of the sun's rays by means of a very rich growth of plants.[8]

All these examples show us very clearly that the radiant energy of the sun is almost the only source of all the forces on the earth's surface useful to humans.

We know, however, that the quantity of the energy that is radiated by the sun towards the earth would be thrown back into space in the same amount if this energy did not undergo certain transformations, which allow it a longer stay and even to become an accumulation of solar energy on the earth's surface. This actually occurs whenever the sun's rays that arrive to us, warm, illuminating and chemically effective, are so received by matter that they are transformed into free chemical affinity or into mechanical movement. In this last case, a part of the radiating solar energy is no longer, according to the well-known Kirchhoff's Law,[9] simply thrown back into space. Rather, it can then be accumulated for a longer time on the earth's surface, taking on forms that temporarily guard

7 E. Reclus, *Géographie universelle*. I. 538. Paris, 1875. Kilometric population of Italy 94, of the Etna region 550 [Footnote in original]. See Reclus 1876–94; Reclus 1882–95, Volume I, p. 315 [Editorial note].

8 Sterry Hunt, *Congress of the British Society*, 1878 [Footnote in original]. See Sterry Hunt 1891, pp. ix–xi, 40–7 [Editorial note].

9 Kirchhoff's Law can be expressed as: the quantity of radiated heat is directly related to the difference between the temperature of the heat-source and the environment that surrounds it [Footnote in original]. See Kirchhoff 1862–3, p. 17; Kirchhoff 1901, pp. 75–6 [Editorial note].

it against dispersion. 'Energy rises by degrees', says the famous English physicist William Thomson about this process.[10] The following words of Secchi illustrate the point well: 'The sun's rays that fall on the plants are not reflected by them to the same degree as would be found for the desert or mountainous rock. They are held back in a greater measure and the mechanical force of their vibration is used for the decomposition of compounds of oxygen with carbon and with hydrogen, of saturated and enduring compounds, which are known by the names of carbon dioxide and water'.[11]

What occurs during this process? A part of the sun's heat perishes as such. It is held by the earth's surface without raising its temperature, that is, without increasing its losses [into space]. With the same loss [into space], the earth's surface obtains more energy, or receiving the same [quantity of energy], it loses less of it. However we may look at this process, we obtain under the influence of plants an accumulation of energy – and not radiated energy, like, for example, heat, electricity and light, but energy of a higher degree – which can be kept for hundreds of years and still retains the capacity for further transformations. Thus the plants on the earth's surface are the worst enemy of the dispersion of energy into space.

2 The Transformable Energy on the Earth's Surface

We thus see that the radiating energy of the sun has not yet completely lost the ability of taking on further higher forms on the earth's surface. Nevertheless, the way in which this process happens ranges within relatively narrow limits. More specifically, this transformation happens in the following ways:

1. The generation of winds, that is, through the impetus which the air gains from changes in temperature.
2. The elevation of water by means of evaporation.
3. The dissociation of enduring compounds, for example, of water, of carbon dioxide, of ammonia during the growth of plants.

---

10 William Thomson (Lord Kelvin, 1824–1907), Irish-Scottish mathematical physicist and engineer, one of the founders of the science of thermodynamics, who also supported the controversial interpretation of the entropy law as implying the eventual 'heat death' of the universe. The quotation in the present text may be a reference to Thomson 1848, pp. 66–71 [Editorial note].

11 Secchi, *Le Soleil.* T. II, p. 300 [Footnote in original]. See Secchi 1875–7, Vol. II, p. 300 [Editorial note].

4. The muscular and nervous labour of animals and humans.
5. The work of the machines made by humans, which have the sun's heat as their only driving force, in either a mediated or an immediate way, as in the case of the now widely known solar machine of Mouchot.[12]

Of course, there are also enormous quantities of transformable energy on our earth, outside of this list of processes we have compiled. These, however, have been left unused by humanity up until now. First place, according to its size, is taken by the energy of the movement of the earth around the sun and around its own axis. Both movements are forms of energy that are still very transformable or, according to Thomson, high-grade energy, as are in fact all mechanical movements. There is a well-known calculation according to which the immediate stop of the earth in its cycle around the sun would be expressed in the development of a quantity of heat, for whose generation it would be necessary to burn a quantity of coal exceeding the mass of the earth fourteen times. The energy of rotation [*Umdrehung*] around the earth's axis is likewise of a very significant amount. However, the influence of both movements on the distribution of energy on the earth's surface has not been precisely determined. Concerning the energy of rotation around the axis, however, this conclusion is perhaps not completely correct because it is known that a part of this energy is transformed into heat under the influence of friction against the mass of water remaining behind in the change from low to high tide. This increases the temperature of the water, while the movement of the earth, even if very insignificantly, is slowed down.[13] By using the tide as a moving force for machines, for example, mills, we hold water up during its highest point at the time of the high tide and use the receding water during the low tide. On the whole, however, the tides are still relatively rarely used as motors.

We have already seen that the inner heat of the globe likewise does not play a very significant role in the economy of energy on the earth's surface. If we view magnetism as an expression of the energy found in the earth's interior, it

12 A reference to the work of the French mathematics teacher and engineer/inventor Augustin Mouchot (1825–1912). Mouchot was awarded a Gold Medal at the Worlds Fair of 1878 for his research relating to the use of solar heat. In 1861 he had patented the first machine capable of producing electricity by exposure to the sun. His device used glass-enclosed water to evaporate water in an iron bucket, with the resulting steam providing a motive force for a simple engine. See Kryza 2003, Chapter 6 [Editorial note].

13 The credit for the first thought of such an influence of the tide goes to Kant. See his *Theory of the Heavens*. Königsberg, 1785 [Footnote in Original]. The correct reference is Kant 1900, pp. 1–11. See also the discussion in Engels 1964, p. 106 [Editorial note].

of course represents a relatively significant quantity of force that is not to be scorned because it is used during navigation and for the fabrication of many scientific apparatuses. At any rate, the absolute size of the earth's magnetic force is not very noticeable in comparison to the solar energy effective on the earth's surface.

Thermal springs furnish us with a not large but nevertheless advantageously applicable quantity of transformable energy. Their heat can be used for various technical ends, for example, for the heating of houses, for the preparation of mortar, etc. We still do not know how to apply the heat of the thermal springs as motor power; to a small degree, such an application is of course entirely conceivable.

There is very little free chemical affinity on the earth's surface, except for that (already mentioned) of the oxygen in the atmosphere. Inside the earth there are certainly significant masses of metals and sulphur in a free state, but we feel little of the efficacy of their chemical energy on the earth's surface.

Turning now to the forms of transformable energy already enumerated at the beginning of this section, we see that the movement of the air or the wind is a very high-grade and, in the human sense of the word, useful form of energy that can furnish a large quantity of mechanical work. Nonetheless, it is not very difficult for us to show that the movement of air is nothing other than a part of solar energy, comprehended as in retrogressive transformation. In order to generate the active force of the wind, the sun must deliver a many times greater amount of energy, of which a significant part is dispersed into space. It cannot happen otherwise, however, because the sun's heat, a low-grade energy, according to the general laws of dispersion, cannot ever be completely transformed into the mechanical movement of air, a higher-grade energy. Even that part of energy that is transformed into movement passes over into dispersion, for the wind is nothing other than a result of the tendency towards equalisation of temperatures.

What has been said about the force of movement of the wind is likewise applicable to the forces of the water currents and in general of falling water. By falling on the millwheels, water gives a higher fraction of useful work than either the steam engine or electromagnetic machine or the more advantageously equipped organisms of pack animals or of humans can deliver. We should not forget here, however, the enormous mass of solar energy that has served to raise the water by means of evaporation.

We see from all this that regardless of the significant quantity of solar energy retained by the earth's surface, it is nevertheless in no way rich in transformable energy such as, for example, mechanical movement or free chemical affinity. Even heat is not in abundance. We find free chemical affinity accumulated in large quantities only in combustible materials of organic derivation. This mass as such is of course significant. According to approximate calculations, the English coalfields amount to 190,000,000,000 tonnes of coal and the North-American even 4,000,000,000,000.[14] This whole quantity, however, just as with all the other organic combustible materials, e.g. peat, petrol etc., is formed by the influence of solar energy, i.e. from the plants on the earth's surface from different epochs. We believe, that is, that plants, with the help of solar rays, have in the course of centuries transformed a saturated substance deprived of free chemical affinity, carbon dioxide, into coal, which contains a large quantity of such energy. At the same time, the oxygen of the atmosphere was freed from the carbon dioxide to which it was previously bound, and its energy of chemical affinity thereby freed up to nourish the life of the higher organisms, of animals and humans.

3 Energy Accumulation

We can begin our investigation from the moment when the earth's land surface was formed to such an extent that the earth's crust frustrated a significant influence of the earth's inner heat on its surface temperature. As this filling up was already so advanced that the temporarily dissociated water could be transformed into steam and a large part of the steam could be transformed into fluid water (which, dissolving the salt that had been condensed up until then, formed the oceans in the depressions of the earth's crust), most of the chemical processes in the inorganic substance of the earth's crust were already finished. Chemical affinity was already saturated to approximately the same degree as today, if we leave out of consideration the processes of plant life. We even believe that thanks to its influence the saturation of chemical affinity is not as extensive now, for, according to the above-mentioned hypothesis, the whole quantity of coal now found in the earth's layers was then in compound with the oxygen of the atmosphere. We know, that is, that the plants draw their carbon from the carbon dioxide of the atmosphere. We have no reason to suppose that they would have done differently during the coal period. Therefore, we have every right to believe that at the beginning of organic life the quantity of unsaturated chemical energy on the earth's surface was insignificant. The influence of the transformable energy inside the earth was constantly dimin-

14 *Edinburgh Review* 1860. Coal Fields of North America and Great Britain, pp. 88–9 [Footnote in original]. See Edinburgh Review 1860, p. 88 [Editorial note].

ished by the gradual swelling of the earth's crust. Of course, back then the earth received somewhat more energy from the sun than it does now. However, the dispersion of the same was also much more significant, for the earth was then hotter than it is today and radiated more energy into icy space. The large quantities of energy obtained from the sun increased only insignificantly the energy of the earth, because the chemical solar rays then found no such substances upon which they would have been able to exercise an influence, as now occurs, for example, with the help of plants, that is, through the dispersion of unsaturated compounds. The same thing occurred with heat and light rays. Heat rays were merely absorbed in the same way as its dispersion and did not increase the amount of transformable energy on the earth's surface. With the exception of the movement of heated air and the water, solar energy was not transformed into any other form on the earth's surface, as still now occurs on the plantless sand area of the Sahara desert or on the ice sheets of the polar regions. If one does not consider the heat contained inside the globe, the quantity of transformable energy contained by the sun back then and the preservable solar energy on the earth's surface appear to have been less significant than at the present time. For if we reckon the coal beds within the earth's surface (to which we are completely entitled, given the organic derivation of the coal deposits), we find ourselves today in possession of very significant quantities of transformable energy. This supply consists, on the one hand, in the unsaturated affinity of enormous quantities of carbon, and on the other, in the free affinity of the oxygen of the atmosphere.

If we examine the course of development of this process, we find that energy contained inside the earth plays an ever-smaller role in the course of time in the formation of the energy budget of the earth's surface. The quantity of energy obtained from the sun decreases, slowly, but regularly. In order for an accumulation of energy to be formed on the earth's surface despite the diminished supply of it, it is essential that a process come about that works against the dispersion. This process must be such that a part of the heat obtained from the sun is transformed into other forms of energy, into chemical affinity, mechanical labour, etc., and, indeed, into ever greater masses.

At the moment, the earth's surface has in a higher degree than formerly the quality of converting lower forms of solar energy (heat) into higher forms (chemical affinity, movement). One must have a correct idea of such a conversion working against the process of dispersion, in order to recognise its significant complexity. This is especially the case with regard to the transformation of heat into mechanical activity. The ways and means in which solar energy is transformed into mechanical movement are also certainly not numerous.

It is easy to prove that the quantity of solar energy that is transformed into free chemical affinity or into mechanical work is not always the same and that, among other causes, it can also be influenced by the activity of humans.

One can, that is to say, assume as undoubted that the existence of plants has the quality of effecting an accumulation of solar energy on the earth's surface to a higher degree than that of animals. The coal deposits are a smoking gun in this regard. One should even recognise that despite the new theories (Bernard et al.) about the unity of life in both kingdoms,[15] animals lose a large quantity of their heat through respiration and movement, that is, they disperse much solar energy into space that had been accumulated by plants. It is of course very difficult to ascertain the precise relation of the two quantities; it is certain, however, that *humans, though certain activities dependent upon their wills, can increase the quantity of accumulated energy of plant life and reduce the quantity of energy dispersed by animals*.

By cultivating plants in places where they either do not yet exist, or exist only in a small amount, by draining marshes, irrigating the deserts, applying perfected cultivation systems, using machines for agriculture and, finally, by protecting the cultivated plants against their natural enemies, we reach the first of the two indicated goals.

Through the displacement or extermination of animals that are damaging to the plant kingdom, we work at the same time for the second goal. In both cases, we obtain as a result an absolute or relative enlargement of the solar energy retained on the earth's surface.

We are thus presented with two parallel processes which, taken together, form the so-called life cycle. Plants have the quality of accumulating solar energy; animals, however, by nourishing themselves from plant stuffs, transform a part of this saved energy into mechanical labour and disperse it afterwards into space. If the amount of energy accumulated by plants remains larger than that of the energy dispersed by animals, there arises a build up of energy stores, e.g. in the period of the formation of coal during which it seems plant life had a preponderance over animal life. If, on the other hand, animal life obtained the upper hand, the accumulated energy store would soon be dis-

15 Podolinsky refers parenthetically to the work of the French physiologist Claude Bernard (1813–78), who in 1870 gave a series of lectures at the Paris Museum of Natural History that were later published as *Leçons sur les phénomènes de la vie, communs aux animaux et au végétaux*. Paris: J.-B. Ballière, 1878–9. For the English edition see Bernard 1974 [Editorial note].

persed and animal life would have to return to the mass determined by the plant kingdom. In this way a certain state of equilibrium between the accumulation and the dispersion of energy would develop. The energy budget of the earth's surface would then be of a more or less stable size; the accumulation of energy, however, would fall to nothing or at any rate much lower than at the time of the preponderance of plant life.

Factually, however, we see no such stagnation of the energy budget on the earth's surface. The quantity of accumulated energy is even now generally understood to be growing. The quantity of plants, of animals and of humans is now undoubtedly more significant than in previous times. Many previously infertile strips of land are now cropped and covered with luxurious plant growth. In almost all civilised lands the harvests have increased. The number of domestic animals and especially of humans has substantially increased. If some countries have lost their earlier fertility and number of inhabitants, that depends on far too gross and self-evident business mistakes; otherwise, however, the opposite is the rule, and on the whole a general increase of the amount of nutritious material and of transformable energy on the earth's surface can no longer be denied.

The most important cause of this general increase is the labour performed by humans and the domesticated animals used by them.

Some examples from the agricultural statistics of France will illustrate for us the correctness of this proposition:

At the moment France possesses nine million hectares of forest, which deliver a yearly yield of 35,000,000 cubic metres, or nearly 81 million metric quintals, of dry wood. Thus, each hectare delivers a yearly yield of nine metric quintals or 900 kilograms. Each kilogram of dry cellulose contains 2,550 calories, so consequently the yearly accumulation of energy on each hectare of forest constitutes the quantity of 900 × 2,550 = 2,295,000 calories.

The natural pastures in France cover an area of 4,200,000 hectares and produce each year on average 105,000,000 metric quintals of hay, that is, 2,500 kilograms on each hectare. The accumulation of solar energy thus represents 2,500 × 2,550 = 6,375,000 calories per hectare.

We therefore see that without the contribution of labour, plant growth yields an accumulation of solar energy that does not exceed the amount of 2,295,000 to 6,375,000 calories per hectare, even in the most favourable conditions (as they are encountered in the forest or on the pastures).

Where, however, labour is applied, we immediately see a significant increase. France currently possesses 1,500,000 hectares of artificial pastures that, after deducting the value of the sown seeds, yield in an average year 46,500,000 metric quintals of hay, that is, 3,100 kilograms for each hectare. Consequently the yearly energy accumulation is 3,100 × 2,550 = 7,905,000 calories per hectare. The excess in comparison with the natural pastures thus equals 1,530,000 calories per hectare, and this surplus is due only to the labour used in the creation of the artificial pastures. The quantity of this labour for a hectare of artificial pasture is approximately the following: 50 hours of labour of a horse and 80 hours of labour of a human. The whole labour expressed in terms of thermal units is 37,450 calories. We thus see that each calorie of labour applied in the creation of artificial pastures effects a net energy accumulation of 1,530,000 : 37,450 = 41 calories.

We observe the same thing also in the cultivation of grain. France grows something over 6,000,000 hectares of wheat, which, deducting the seed, gives 60,000,000 hectoliters of grain and a further 120,000,000 metric quintals of straw. Each hectare thus gives 10 hectoliters or 800 kilograms of grain and 2,000 kilograms of straw. The 800 kilograms of grain contain – according to a special calculation of the composition of starch, bran, etc. – approximately 3,000,000 calories, which, together with the 2,000 × 2,550 = 5,100,000 calories found in the straw, make up the sum of 8,100,000 calories.

The surplus in comparison with the natural pastures is 8,100,000 - 6,375,000 = 1,725,000 calories. In order to obtain this, approximately one hundred hours of horse labour and 200 hours of human labour are used, which together have the value of 77,500 calories. Consequently each calorie in the form of labour for cultivation of the pastures generates a terrestrial accumulation of solar energy equivalent to 1,725,000: 77,500 = 22 calories.

Where does this surplus of energy come from, which is indispensable for the elaboration of this mass of nutritious and combustible materials? We can give only one answer: *from the labour of humans and domesticated animals*. What, then, in this connection, is labour? *Labour is such a use of the mechanical and intellectual energy accumulated in the organism, which has as a consequence an increase of the general energy budget of the earth's surface.*[16]

This increase can come about either directly, through the transformation of new quantities of solar energy into more transformable forms, or also in a mediated way, through protection against that dispersion into space, which

16 See 1. *Statistique de la France* 1874, 1875 and 1878. 2. *Dictionnaire des arts et de l'agriculture* de Ch. Laboulaye, 4. édition 1877. *Articles agriculture* par Hervé Mangon, et *Carbonisation*. 3. Pelonze et Frémy, *Traité de Chimie*. 4. Hermann, *Grundzüge der Physiologie*. 5. Auflage, 1877 [Footnote in original]. See Laboulaye 1874; Pelouze and Fremy 1865–6; Hermann 1874; Hermann 1875 [Editorial note].

would have occurred inevitably without the involvement of labour. To this last category belongs, for example, the labour of the tailor, the shoemaker, construction workers and such.

It is clear, from this perspective, that useful labour can only be ascribed to humans and some animals, which are either managed by humans, as with domesticated animals, or which, like ants, partially work on their own, and partially devote themselves to the breeding and raising of domesticated animals, driven by their own instincts.

The movement of air, i.e. the wind, cannot ever be regarded in and for itself as useful labour, for, left to itself, the wind, through the dispersion of its energy, generates no new accumulation of energy on the earth's surface. The same is also the case for water currents as a moving force.

Although plants accumulate energy in the substance of their own bodies, they cannot, in the majority of cases, set such energy into movement independently; they cannot usefully employ it in the sense of a general increase of the quantity of force on the earth's surface.

Man-made machines may, if left to themselves, remain in operation for a long time; but they would nevertheless not yield any useful work, for we still cannot imagine an artificial mechanism that would have the ability to progressively augment the solar energy accumulated on the earth without the participation of the muscle-power of humans.

Finally, even the nervous labour of humans only becomes really useful labour for humanity when it leads to some type of muscular effort. For we do not know any other way of achieving through nervous labour an immediately useful goal, i.e. an absolute or relative increase of the energy available in the human kingdom.[17]

In passing over to the muscular labour of animals and humans, it is similarly difficult to determine with certainty the boundaries of useful labour. If we subject a lowly member of the animal kingdom to observation, we will find out only with great difficulty which of its functions should have the name of labour attached to it. Often labour is confused with mechanical movement; hence, the question becomes: are the fluttering of a butterfly and the crawling of a snail also labour?

From our point of view, we can confidently answer: no. The crawling of a snail and the fluttering of a butterfly are not labour, for they are accompanied merely by a dispersion of energy, but not by an accumulation of energy.

17 Cf. M. Marey, *Du mouvement dans les fonctions de la vie*. p. 205. Paris, 1868 [Footnote in original]. See Marey 1868 [Editorial note].

But, one could reply, the snail crawls around in order to find food, the butterfly flutters about in order to find a good place for the development of its larvae. We, however, reply in turn: nature knows no goals and reckons its account merely from the results. The entire life of the snail, all of its crawling, seeking for food, digestion of the found means of existence and the ability gained from this for new movements, do not transform the slightest quantity of solar energy into such higher forms which by their further deployment could increase the store of energy on the earth's surface. A snail is incapable of dedicating itself to agriculture, thus it also cannot increase the accumulation of solar energy through plants. One might perhaps respond to us that the snail, even if not through its life, then at least through its death, could advance the growth of plants. For a snail can, given good conditions and rich nutrition, destroy a large mass of plant material. If, on the contrary, it is forced to suffer hunger and die in the case of a failed crop of the types of plants most beneficial to it, it thereby gives the plants the possibility of developing in greater numbers, thus increasing the accumulation of energy. This is certainly a curious objection, the answer to which is not difficult. If the luxuriousness of the growth of plants of any particular locality really increases through the loss of the snail, it is then very probable that also the number of enemies of this plant growth will increase. After its death, the snail is no longer in a condition to keep the plants it formerly exploited from their new enemies and therefore the energy conversion remains presumably the same as it was before.

For we should keep in mind that by the word 'labour' must be understood a 'positive act' of the organism, which has as a necessary consequence an accumulation of energy. Therefore the 'passive fact' of death in the struggle for life can never belong to the category of labour.

We have introduced this example – which admittedly may seem peculiar to many – in order to assign the question of the conservation of energy its correct place from the beginning. It could, for example, appear that the death of the snail or the caterpillar actually encourages plant growth simply due to the fact that they no longer destroy any plant material. After all, one says that a capitalist saves when he does not consume all of his income. We have just sought to show, however, that a snail can never perform useful labour because it never increases the accumulation of energy through its activity. The same is the case regarding those conscientiously saving humans [the capitalists].

We hope that we have thus managed to bury the doctrine of saving or, as it were, of negative labour. For labour is always a positive concept, which consists in such an expenditure of mechanical or physical labour that has as its end result an increase of energy accumulation.

Viewed from this perspective, we can conclude that the different movements of animals that are self-evidently goal-less or that have as a goal merely the seeking out of means of nutrition, etc., cannot be counted as labour, precisely because they leave behind no increase of energy accumulation. Thus, for example, the activity of the spider that goes to great pains spinning its web and that of the doodlebug, despite all of the engineering knowledge involved, are still not by a long way useful labour.

In the strict sense of the word, it is only with the agriculture of humans that the correctness of our definition of labour becomes clear. For it is evident that a hectare on a wild steppe or in a virgin forest, without the involvement of humans, produces each year merely a determinant quantity of nutritious material, but the application of human labour can raise this amount ten or twenty fold. Of course, the human creates neither material nor energy. The material was already contained in its totality in the ground, in the seed and in the atmosphere; all of the energy was furnished by the sun. Thanks to the involvement of humans, however, a hectare of land covered with cultivated plants can accumulate perhaps ten times the quantity of energy it would have without their involvement. One should not believe that all of this energy was already aggregated in the soil and merely dispersed in a greater amount by human labour. That would not be correct, for agriculture exhausts the soil only if it is conducted irrationally, that is, wastefully. On the contrary, a perfected agricultural science gives the best harvests precisely in the lands where agriculture has flourished already for a longer time, e.g. in England, France, Belgium, in Lombardy, in Egypt, China, Japan, etc. Therefore we believe we are correct to say that scientifically organised agriculture can be counted as one of the best examples of really useful labour, that is, such labour which increases the amount of solar energy upon the earth's surface.

4 The Labour of the Human Organism

Beginning with the distribution of energy in space, we have arrived at human labour, an important factor in the distribution of energy upon the earth's surface. We have not said anything until now, however, about the emergence of that capacity for labour in the human organism, without which the accumulation of energy on the earth's surface under the influence of labour would be difficult to explain. From where in the organism does the energy necessary for labour derive? Which mechanisms does this activity use? What phenomena accompany it?

We can answer the first question by saying that the whole mechanical labour of animal organisms has its source in nutrition. The free chemical affinity of nutritious material is saturated within the organism by the inhaled oxygen, and thereby converted into heat. A part of the latter passes over into mechanical labour.

Hirn conducted one of the first and most important experiments on the conversion of the heat of the human organism into labour.[18]

He used a large wooden hermetically (airtight) sealed box, but which was furnished with glass openings in order to be able to observe its interior. In the box a human who served as the object of the experiment could find enough free space in order not to touch its walls. The air necessary for breathing was admitted through a pipe and the exhaled gases were removed in the same way. At the beginning of the experiment, the human remained in a state of rest. In the further course of the experiment, however, he performed a determinant sum of labour in the box, climbing up or down a ladder. The mechanism for this was arranged in the following way:

In the lower part of the box was mounted a wheel that turned around an axis, being set into movement by a belt outside the box. During the movement of the wheel, the human who served as the object of the experiment had to imitate the movement with his feet while holding himself up on a handrail mounted in the upper part of the box, just as if he were climbing stairs. Accordingly, rungs were also mounted on the wheel at certain intervals. When the wheel was moved in the opposite direction, the human had to descend onto the wheel and, after an hour, for example, his centre of gravity had covered the same distance as the circumference of the wheel in the opposite direction.

The quantity of heat energy generated by the worker is different in these three cases, according to whether the man was at rest or descended onto or dismounted the wheel. These differences agree completely with the postulations of the mechanical theory of heat. It was the case, namely, that during the pause each gram of oxygen inhaled delivered 5.18 to 5.80 calories, while during labour it only delivered 2.17 to 3.45 calories. This experiment yields very important results. For it gives us the possibility, even if only approximately, of determining the size of the economic coefficient of the human machine, that is, the percentage yield of the heat transformed during labour.[19] Helmholtz managed on

18 Gustave Adolphe Hirn (1815–90), French industrialist and thermodynamic theorist/engineer. He tried to apply to human muscular labour the concepts and measurement methods developed in his experiments involving steam engines [Editorial note].

19 The economic coefficient of a machine is that number which gives the relation of its efficiency to the heat used by it [Footnote in original].

the basis of Hirn's experiment and with the help of some hypotheses commonly acknowledged in physiology to quantify this coefficient.[20]

At complete rest, an adult human delivers a quantity of heat in the course of an hour, which carried over into labour, could raise the body of this human to a height of 540 metres. This height is precisely that at which one arrives when mountain climbing without particular effort in the course of an hour, that is, under the same conditions as in Hirn's experiment. However, during this experiment the respiratory activity of the worker was intensified fivefold. It follows immediately that the economic coefficient of the human machine represents 20 percent or 1/5 of the total heat generated by the organism or, what is the same thing, that the human possesses the ability to transform 1/5 of the total energy added by nutrition into muscular labour. As is generally known, even the most advanced steam engines do not reach this quantity. This extraordinary capacity to convert lower forms of energy into mechanical labour is found to an even higher degree in some of the inner organs of the human body, e.g. in the heart. Helmholtz has found that the heart, by means of its own force, could raise itself up to a height of 6,670 metres in the course of an hour. The strongest locomotives, which e.g. are used on the Tyrol railways, could not raise their own weight up over 825 metres in an hour. Consequently, these locomotives, considered as machines, are eight times weaker than a muscular apparatus similar to the heart.[21]

The causes of this disproportionately significant strength of the muscular apparatus have been partially explained by the latest researches in the field of muscle physiology. In part, however, they still remain shrouded in darkness. Here is not the place to enter into further discussion of this matter. In general, however, we can apply most of the laws of the steam machine or any other thermal machine (set into movement by heat) also to the labouring human.

In this comparison we should not forget that the human organism is much more complicated than any other thermal machine. All artificial machines obtain their sources of movement in one or a few ways, e.g. through the burning of combustible material, through chemical processes in galvanic elements, etc. Similarly, the work of machines proceeds only in one or a few directions. We observe something completely different when it comes to humans. Even though nutrition together with the inhaled gases are likewise almost its only

20 Hermann von Helmholtz (1821–94), German physicist and physician, and one of the co-discoverers of the first law of thermodynamics, which he termed the 'Law of Conservation of Force' [Editorial note].

21 Verdet, *Theorie mécanique de la chaleur*. II., 246 [Footnote in original]. See Verdet 1868–72 [Editorial note].

sources of force, the human organism possesses, on the other hand, certain abilities to prevent the energy from dispersing. These are partially applied instinctively, as satisfaction of needs, and also partially deliberately, in the form of education, learning, and improvement. For instance, houses and walls, which merely satisfy our immediate needs and protect us from the excessive lack of warmth, also lead to a saving and advantageous distribution of energy in the human body just as much as does, for example, instruction in a useful employment of energy during labour.

A second and even more significant difference between the human organism and any other thermal machine consists in the diversity of human labour. Without taking the intellectual activity of humans into account, the mechanical achievements of humans are already so rich and diverse that they are overtaken by a mechanical apparatus only with difficulty. It is precisely this diversity of movements that gives human labour the ability to cause simultaneously all those transformations in the environment, which in their end results make possible an accumulation of energy. Such is the case, for example, with the long series of various kinds of cultivation. This diversity of movements of the human machine is the most important cause of the higher productivity of the labour of humans.

On the other hand, we must also mention those causes that apparently result in a significant decrease in the high economic coefficient of the human machine. Foremost is the necessity of satisfying some purely intellectual needs, which meanwhile cause a great addition to the general energy budget of humanity. Naturally, the higher the development of humanity rises, the greater the role these intellectual needs play in its life.

However, there are not a few purely material needs in addition to the need for nutrition and for air to breath, and it is not easy to determine the quantity of necessary labour for these. Since we still do not have a close measure of this, we hold ourselves to the following calculation, which is certainly inexact but nevertheless is provisionally adequate for our purposes.

In most civilised lands, food expenditure represents approximately half of the budget of the middle classes. Housing, clothing and the satisfaction of intellectual needs claim the second half. We should conclude from this that if the economic coefficient, calculated according to the quantity of nutrition and the inhaled oxygen, equals the fraction of 1/5, and if the whole quantity of energy that is claimed by humanity for the satisfaction of its material and intellectual needs is properly brought into consideration, this coefficient must be decreased to the fraction of 1/10, and then even more so in light of the fact that a human passes a significant part of its life, during childhood, old age and sickness, as unproductive.

Thus, if we consider the human organism as a thermal machine with an economic coefficient of 1/10, it becomes possible to define a little more closely the preconditions of human life on earth. In earlier times of its presence on this planet, humanity did not yet have the means to increase the earth's energy store. We should thus believe that humanity lived exclusively from materials drawn from already existing stores. Actually, humanity did nothing more than hunt wild game, catch fish, gather fruits and consume all of these foodstuffs, without furnishing any type of useful labour; that is, humanity simply dispersed energy into space. If humanity had reached no higher development than the wild animals, it would probably have been made extinct by other animals, or at any rate its number would have been one corresponding merely to the general conditions of the struggle for life. But under the influence of very special conditions, particularly of an advantageous organisation of the brain and the upper extremities, humanity began to employ its mechanical energy in a direction that enabled a general accumulation of energy on the earth's surface. With that, the existence, increase and development of humanity were also made possible. Humanity is no longer bound by the quantity of the energy store; on the contrary, it can independently increase this store. Whether or not it really did this from the beginning, whether or not it currently does this in all cases, is an altogether different question. The possibility, however, is already at hand. Of course, at the beginning of civilisation, the dispersion of energy, due to destruction of forests, unregulated hunting etc., exceeded by a long way the accumulation of energy through agriculture and animal husbandry. With time, however, both influences came into equilibrium and finally the accumulation of energy by means of agriculture began to gain the upper hand over the dispersion of energy. Actually, of 1,300–1,400 million humans, barely 100 million are fed with the products of hunting, fishing or solely of animal husbandry, i.e. with foodstuffs that are not a product of human labour. All the remaining humans, 1,200–1,300 million in number, are obliged to feed themselves at the cost of agriculture, that is, at the cost of an energy accumulation that is the immediate result of human labour. If all present cultivation together with the more than 1,000 million tillers of the soil should ever disappear, the remaining humans would have great difficulties in feeding themselves with natural products, and would certainly not manage without also resorting themselves to tilling the soil. It immediately follows from this that no fewer than 1,000 million humans must now regularly be occupied in working on the accumulation of solar energy on the earth's surface in order to satisfy the needs of the entire population.

As we have seen, the economic coefficient of this labouring human machine, that is, of the entirety of humanity, equals approximately the fraction of 1/10.

Although humanity can transform only 1/10 of its energy into mechanical labour, this quantity already suffices for it to support a more or less steady growth of the human population. Even though humanity's intellectual needs grow with its development and the economic coefficient thereby naturally becomes smaller, the total labour of humanity in general is nevertheless progressing. What are the causes of this apparent contradiction?

Since the development of the mechanical theory of heat, any process that leads to the production of mechanical movement can be compared to the activity of a thermal machine, i.e. a machine that transforms heat into labour. Incidentally, such views were enunciated in the past by Sadi-Carnot in his famous work that appeared in 1824. 'In order to consider in the most general way the principle of the production of motion by heat, it must be considered independently of any mechanism of any particular agent. It is necessary to establish principles applicable not only to steam-engines but to all imaginable heat-engines, whatever the working substance and whatever the method by which it is operated'. Sadi-Carnot says further: 'Whenever there exists a difference of temperature ... it is possible to have also the production of impelling power'.[22]

We know, however, that the entirety of the heat can never be transformed into work, and that in the most advantageous of cases hardly 20 percent of useful work is obtained. All remaining heat is for the most part dispersed. In order to come to an accurate conception of the quantity of work obtained, we must move on to consider the machine's opposed transformation of work into heat, so that we can determine the quantity of heat contained in our work. Following Sadi-Carnot, this would be a reverse cycle or circular process. In his opinion, we can speak of a relation between the contained work and the employed heat only when the cycle is completed. Sadi-Carnot names a machine that carries out this circular process of the transformation of heat into work, and work again into heat (existing only in the imagination, for it has not yet been constructed), the *perfect machine*. Such a machine cannot yet be mechanically made, for it would have to apply the heat itself, by means of its own labour, to its own steam boilers.

When we observe the labour of humanity, however, we have before our very eyes an example of what Sadi-Carnot called a perfect machine. For from this perspective, the human organism would be a machine that not only transforms heat and other physical forces into labour, but which also brings about the

22 Sadi-Carnot, *Réflexions sur la puissance motrice du feu*. Paris, 1824. See p. 8 et sqq [Footnote in original]. Sadi Carnot, *Réflexions sur la puissance motrice du feu et sur les machines propres à développer cette puissance*. Paris: Chez Bachelier, 1824. See Carnot 1977, pp. 6, 8 [Editorial note].

operational reverse cycle, i.e. it transforms labour into heat and into the other physical forces which are necessary for the satisfaction of our needs, heating with its own labour converted into heat, its own steam boilers, so to speak. A steam engine, for example, even if it could function for a longer time without the involvement of human muscular power, does not possess the ability to produce the elements necessary to undertake its own work the following year. The human machine, on the other hand, creates new harvests, raises the young generations of domesticated animals, invents and builds new machines, etc. In a word: humanity regularly creates the material and the elements for the future continuation of its labour. Thus, humanity fulfils Sadi-Carnot's requirement of perfection much better than any artificial machine.

The degree of perfection of the human machine, however, is not always the same and changes not only depending upon its economic coefficient but also particularly with respect to its ability to bring about the operational reverse cycle, i.e. to convert its labour into an accumulation of physical forces necessary for the satisfaction of our needs. Of course, the needs of savages are much easier to satisfy than those of civilised people, and therefore its economic coefficient is significantly greater, perhaps 1/6 instead of 1/10. However, the labour of the savage is much less productive in its end results than that of the civilised human, because the savage lives for the most part from natural produce that he finds already at hand, while the civilised human satisfies his needs with the products of his labour and in this way creates an accumulation of energy on the earth's surface, whose quantity exceeds the force of his muscles by at least ten times.

The necessary conditions for the continuation of the work of an inanimate machine are not immediately dependent upon the work of this machine, upon its qualities. All artificial machines on the contrary are immediately dependent upon the muscular labour of the human who governs it, and supplies it with the heat engendering substance. The conditions of labour or, if we will, of the existence of the human machine, can on the other hand be rigorously established:

> *So long as the labour of the human machine can be transformed into such an accumulation of energy, capable of satisfying our needs, which exceeds the entire force of humanity by so many times as the denominator of the economic coefficient is greater than its numerator, the existence and the possibility of working is guaranteed for the human machine.*

Every time the productivity of human labour falls below the size of the inverse economic coefficient, poverty and often a decrease of population arises. Conversely, when the utility of labour exceeds this size, we have to expect an increase of prosperity and an increase of population.

5 Labour as Means for the Satisfaction of our Needs

The degree in which our needs can be satisfied by the accumulation of an energy supply is dependent on a whole series of factors that we will now subject to our attention. The most important of these are: the energy supply on the earth's surface, the number of humans, the extent of their needs, and the productivity of their labour, i.e. their ability to increase the energy accumulation.

The availability of a sizeable store of energy in the plant kingdom alleviated significantly the struggle of prehistoric man against the wild animals, despite the latters' greater force and ability to procure food for themselves. The use of fire, i.e. the solar energy accumulated by plants, was a powerful ally of humanity during its earliest and most difficult victories.

If humanity achieved all these victories while it was still in a lower stage of development, this occurred mostly because even then the energy store which it knew how to use was greater than that available to all of the stronger animals. The wildest predators could only set the force of their own body against humans, but humans, naturally much more weak, met them with a whole arsenal of offensive and defensive weapons, whose comparatively colossal store of energy only they knew how to use. In the beginning they used their victory in the most wasteful way without thinking about a renewal of the dispersed energy accumulation. Naturally, the energy store in the hands of humanity in such an inefficient economy remained a very insignificant one. Further, since the numerical population is dependent on the size of this store, it will not surprise us if we only rarely encounter a dense population during the hunting and animal husbandry periods. This situation changes only with the general spread of agriculture, which, through the application of the mechanical labour of humanity on energy accumulation, enables a more rapid increase of population.

In order to understand fully the influence of useful labour on the accumulation of energy and consequently also on the increase of population, we must deal a bit more closely with the special character of labour as a means for the satisfaction of our needs.

We can see from the following passages on labour of three famous economists how difficult it is to come to a correct understanding of it without using the methods of contemporary science. Quesnay said: 'Labour is unproductive'. Adam Smith: 'Only labour is productive'. Say: 'Labour is productive, natural forces are productive and capital is productive'.[23]

Is it possible to reconcile such contradictions? Apparently, this is only a semantical dispute. Adam Smith said, for example: 'The yearly labour of a nation is the base fund [*Urfond*] which produces all objects that are necessary or comfortable for life; all of these objects are either the immediate product of labour or they are bought for the value of this product'. Sismondi added: 'We believe with Adam Smith that labour is the sole source of wealth, ... however, we add that utility is the only goal of the accumulation (of products) and that the national wealth only grows with national usage'.[24]

For his part, Quesnay says the following: 'We are not concerned with the formal side of production, how, for example, the hand workers who work any type of material perform their labours, but rather with the real production of wealth. I say real production because I will not deny that the labour of the worker gives the raw material an allowance of value, but one should not confuse a simple addition of commodities with their real production'.[25]

Today we can ascribe this contradiction to the fact that labour of course creates no material, so that the productivity of labour can only consist in adding something to the object that was not created by labour. This 'something' is in our opinion energy. On the other hand, we know that the only means through which humanity is in a position to increase in any situation the quantity of energy is the use of its labour power. Therefore, Quesnay was correct when he said that labour does not create any real commodity precisely because labour cannot create any material. However, Smith was equally correct, because that which we need in any commodity, that which satisfies our needs, can only be attained with the help of labour.

Of course, one should not forget that the earth's surface has the ability, apart from the influence of human labour, to accumulate a certain quantity of energy that can be used by humans. But the older economists already knew that these stores were insufficient in comparison to those furnished by labour. Thus, for example, James Steuart said: 'The natural products of the earth which are presented independently from the will of the humans and always in a merely

23 *Dict. Encycl. du XIX s.* Article Travail [Footnote in original]. See Larousse 1865–90, Vol. 15, pp. 435–6. For Quesnay's view, see Meek 1963, pp. 72–4, 207, 227–9; for Adam Smith, see Smith 1937, pp. lvii–ix, 314–32; for Say, see Say 1832, pp. 26–32 [Editorial note].

24 *Collection des principaux économistes*. T.V., p. 1 [Footnote in original]. See Smith 1937, p. lvii; Sismondi 1991, p. 53 [Editorial note].

25 Quesnay. *Collection des principaux économistes. Physiocrates II*. pp. 187–8 [Footnote in original]. See Meek 1963, pp. 205, 207 [Editorial note].

inadequate quantity resemble the small sum of money that one gives to a young man in order to give him the possibility of beginning his career and of establishing a business venture, with whose help he is supposed to seize his luck himself'.[26]

From all sides, therefore, we obtain evidence that the natural products of the earth are in no position to satisfy all of our needs and that we are obliged to increase the quantity of products artificially. Useful labour serves as a means to this end.

According to everything that has been said, we can arrive at the following conclusions as an answer to the question posed at the beginning of our work:

1. The total quantity of energy that the earth's surface receives from its interior and from the sun is gradually being reduced. Despite this, the accumulation of energy on the earth's surface is growing.
2. This increase takes place under the influence of the labour of humans and domesticated animals. By the word 'labour' we understand any use of mechanical or physical force of humans or animals that leads to an increase of the energy budget on the earth's surface.
3. The human, considered as a thermal machine, possesses a certain economic coefficient that becomes ever smaller with the growth of human needs.
4. At the same time, however, the productivity of labour rises as the economic coefficient sinks, and in this way needs are satisfied more easily and in a greater number.
5. So long as the average human has at his disposal a quantity of chemical affinity and available mechanical labour which exceeds his own force as many times as the denominator of the economic coefficient is larger than its numerator, the existence of humanity is materially assured.

6 Unity of Force and Political Economy

Here we have arrived at the point where we should give an answer to the second question we posed: 'What are the best means of employing human labour in order to draw upon a larger fraction of natural forces for the satisfaction of human needs?'

---

26 James Steuart. *Principles of Political Economy*. Dublin. I. p. 116 [Footnote in original]. See Steuart 1966, p. 118 [Editorial note].

In general terms, we have already given this answer: *the best means are those that cause the largest accumulation of energy on the earth*. Primitive cultivation – which is not yet actually a cultivation, because it is not based upon useful labour, upon an accumulation of energy, but merely on the use of force amassed already through the earlier life processes – cannot be reckoned among these means. The savage, by nourishing himself with fruits or roots, hunting game or catching fish, merely disperses the previously accumulated energy into space.

The slave economy is already an advance; but even it is still very imperfect, for this form of society, which has its foundation in perpetual wars, excludes a large part of the workers from participation in the accumulation of energy, in the labour that is really useful for the satisfaction of human needs. Without speaking of the immense number of workers killed or wounded in the continual wars, we mention only *the standing regular armies, the owners of slaves and their cohorts of overseers* in order to show how many unuseful and unproductive elements are contained in the society founded upon slavery.

Feudalism already contains more elements of progress. At least the serf possesses a parcel of land that he is allowed to work without being overseen by the eyes of the lord and without feeling the whip of the overseer.

But how evanescently small is this progress still! How tiny are the parcels of the serf in comparison to the incalculable goods of the lord. For the serf, free labour-time is merely a short repose after the long days of compulsory labour for the lord. One should therefore not wonder that the productivity of labour under feudalism did not reach even the median of today's productivity.

Thus we come to the capitalist mode of production. This form of production knows how to use the division of labour and, as this no longer sufficed for it, it began to employ machines for industry and for agriculture on a large scale. It achieved magnificent results that exceeded its own expectations. But capitalism also has its dark side.

Instead of increasing the accumulation of energy on the earth, the machines often intensify the useless dispersion of the already available labour powers. They do this by excluding a part of the proletariat from production following upon inevitable overproduction. Under socialism, by contrast, any mechanical or any other improvement would directly reduce the labour time of all workers, giving them the leisure for new production, for intellectual and artistic culture, etc.

A higher level and a more equitable division of the quality and quantity of foodstuffs would inevitably bring about an increase in the muscular and nervous force of humanity. From that would spring a new growth of production and a greater accumulation of energy on the earth's surface.

An exact and precise system of accounting, which neither hides nor falsifies the numbers, would conserve much superfluous labour that is lost in the current anarchy.

Rational public healthcare and the possibility of accommodating all of the demands of science in one's personal hygiene would necessarily raise the life-expectancy of humanity and simultaneously also the productivity of the human organism to such a height which today is only found in exceptional cases.

Such are, in our opinion, in the form of a very short and perhaps overly general sketch, the relations between the accumulation of energy and the different forms of production. We hope to return to this question in a more extensive work in the near future.[27]

---

27 This hope of the gifted writer could unfortunately not be fulfilled. It was not granted to him to explicate further his fruitful idea of applying the results of the physical sciences to political economy, for soon after completing the sketch published here he fell victim to an incurable neuropathy. The Editor. [This footnote was inserted by the editor of *Die Neue Zeit*, Karl Kautsky].

Bibliography

Adorno, Theodor 1997, *Aesthetic Theory*, Minneapolis, MN: University of Minnesota Press.

Alley, Richard B. 2000, *The Two-Mile Time Machine*, Princeton, NJ: Princeton University Press.

Altvater, Elmar 1990, 'The Foundations of Life (Nature) and the Maintenance of Life (Work)', *International Journal of Political Economy*, 20(1): 10–34.

——— 1993, *The Future of the Market*, London: Verso.

——— 1994, 'Ecological and Economic Modalities of Time and Space', in *Is Capitalism Sustainable?*, edited by Martin O'Connor, New York: Guilford.

Anderson, Charles H. 1976, *The Sociology of Survival*, Homewood, IL: Dorsey.

Angelo Secchi 1913, In *Catholic Encyclopedia*, New York: Robert Appleton.

Angus, Ian and Simon Butler 2011, *Too Many People?*, Chicago, IL: Haymarket.

Anker, Paul 2001, *Imperial Ecology*, Cambridge, MA: Harvard University Press.

Arendt, Hannah 1958, *The Human Condition*, Chicago, IL: University of Chicago Press.

Aristotle 1882, *On the Parts of Animals*, Boston, MA: Routledge Kegan Paul.

Asmis, Elizabeth 1984, *Epicurus' Scientific Method*, Ithaca, NY: Cornell University Press.

Ayers, Michael 1991, *Locke: Epistemology and Ontology*, Boston, MA: Routledge Kegan Paul.

——— 1999, *Locke*, Boston, MA: Routledge Kegan Paul.

Bailey, Cyril 1928, 'Karl Marx on Greek Atomism', *Classical Quarterly*, 22(3–4): 205–6.

Baker, Adam 2011, 'Socialism or Ecosocialism', *Alliance Voices*, December, retrieved from: http://alliancevoices.blogspot.com/2011/12/socialism-or-ecosocialism.html.

Baksi, Pradip 1996, 'Karl Marx's Study of Science and Technology', *Nature, Society and Thought*, 9(3): 261–96.

——— 2001, 'MEGA IV/31: Natural Science Notes of Marx and Engels, 1877–1883', *Nature, Society and Thought*, 14(4): 377–90.

Banzhaf, H. Spencer 2000, 'Productive Nature and the Net Product', *History of Political Economy*, 32: 517–51.

Baran, Paul A. and Paul M. Sweezy 1966, *Monopoly Capital*, New York: Monthly Review Press.

Barrow, John D. 1994, *The Origin of the Universe*, New York: Basic Books.

Beete Jukes, Joseph 1872, *The Student's Manual of Geology*, Edinburgh: Adam and Charles Black.

Beiser, Frederick C. 2014, *After Hegel*, Princeton, NJ: Princeton University Press.

Bell, Daniel 1966, 'The "End of Ideology" in the Soviet Union', in *Marxist Ideology in the Contemporary World*, edited by Milorad M. Drachkovitch, New York: Praeger.

Bensaïd, Daniel 2002, *Marx For Our Times*, London: Verso.

Benton, Ted 1989, 'Marxism and Natural Limits', *New Left Review*, 178: 51–86.

——— (ed.) 1996, *The Greening of Marxism*, New York: Guilford Press.

——— 2001, 'Marx, Malthus and the Greens: A Reply to Paul Burkett', *Historical Materialism*, 8(1): 309–32.

Bernard, Claude 1974, *Lessons on the Phenomena of Life Common to Animals and Plants*, translated by Hebbel E. Hoff, Roger Guillemin and Lucienne Guillemin, Springfield, IL: Thomas.

——— 2000, 'Introduction to the Study of Experimental Medicine', in *Nineteenth Century Science: A Selection of Original Texts*, edited by A.S. Weber, Peterborough, Ontario, Broadview Press.

Bhaskar, Roy 1983, 'Materialism', in *The Dictionary of Marxist Thought*, edited by Thomas Bottomore, Cambridge: Blackwell.

——— 1998, *The Possibility of Naturalism: A Philosophical Critique of the Contemporary Human Sciences*, 3rd edition, London: Routledge.

Biancardi, C., A. Donati and S. Ulgiati 1993 'On the Relationship Between the Economic Process, the Carnot Cycle and the Entropy Law', *Ecological Economics*, 8(1): 7–10.

Bloch, Ernst 1971, *On Karl Marx*, New York: Herder and Herder.

Block, Maurice 1875, *Statistique de la Francé*, Paris: Guillaumin et cie.

Bookchin, Murray 1990, *The Philosophy of Social Ecology: Essays on Dialectical Naturalism*, New York: Black Rose.

Bramwell, Anna 1989, *Ecology in the Twentieth Century*, New Haven, CT: Yale University Press.

Brennan, Andrew 1988, *Thinking about Nature: An Investigation of Nature, Value and Ecology*, Athens, GA: University of Georgia Press.

Brodine, Virginia 1972, *Air Pollution*, New York: Harcourt Brace Jovanovich.

——— 2007, *Red Roots, Green Shoots*, New York: International Publishers.

Brush, Stephen G. 1978, *The Temperature of History*, New York: Burt Franklin.

Budyko, M.I. 1969, 'The Effect of Solar Radiation Variations on the Climate of the Earth', *Tellus* 21(5), (October): 611–19.

——— 1974, *Climate and Life*, New York: Academic Press.

——— 1977, *Climatic Changes*, Washington, D.C.: American Geophysical Union.

——— 1980, *Global Ecology*, Moscow: Progress Publishers.

——— 1986, *The Evolution of the Biosphere*, Boston, MA: D. Reidal.

Budyko, M.I., G.S. Golitsyn and Y.A. Izrael 1988, *Global Climatic Catastrophes*, New York: Springer-Verlag.

Bukharin, Nikolai 1925, *Historical Materialism*, New York: International Publishers.

——— 1972, *Imperialism and the Accumulation of Capital*, New York: Monthly Review Press.

——— 2005, *Philosophical Arabesques*, New York: Monthly Review Press.

Burkett, Paul 1998, 'Labor, Eco-Regulation, and Value', *Historical Materialism*, 3(1): 125–33.

——— 2001a, 'Marxism and Natural Limits: A Rejoinder', *Historical Materialism*, 8(1): 333–54.

——— 2001b, 'Ecology and Historical Materialism', *Historical Materialism*, 8(1): 443–59.

——— 2001c, 'Marx's Ecology and the Limits of Contemporary Ecosocialism', *Capitalism Nature* Socialism, 12(3): 126–33.

——— 2002, 'Analytical Marxism: A Rejoinder', *Historical Materialism*, 10(1): 177–92.

——— 2003, 'The Value Problem in Ecological Economics', *Organization & Environment*, 16(2): 137–67.

——— 2004, 'Marx's Reproduction Schemes and the Environment', *Ecological Economics*, 49(4): 457–67.

——— 2005a, 'Entropy in Ecological Economics: A Marxist Intervention', *Historical Materialism*, 13(1): 117–52.

——— 2005b, 'Marx's Vision of Sustainable Human Development', *Monthly Review*, 57(5): 34–62.

——— 2006, 'Two Stages of Ecosocialism?', *International Journal of Political* Economy, 35(3): 23–45.

——— 2009, *Marxism and Ecological Economics: Toward a Red and Green Political Economy*, Chicago, IL: Haymarket Books.

——— 2014, *Marx and Nature: A Red and Green Perspective*, Chicago, IL: Haymarket Books.

Burkett, Paul and John Bellamy Foster 2006, 'Metabolism, Energy, and Entropy in Marx's Critique of Political Economy: Beyond the Podolinsky Myth', *Theory and Society*, 35(1): 109–56.

Capra, Fritjof 1982, *The Turning Point: Science, Society, and the Rising Culture*, New York: Simon & Schuster.

Cardwell, Donald S.L. 1971, *From Watt to Clausius: The Rise of Thermodynamics in the Early Industrial Age*, Ithaca, NY: Cornell University Press.

Carnot, Sadi 1977, *Reflections on the Motive Power of Fire*, Gloucester, MA: Peter Smith.

Caudwell, Christopher 1937, *Illusion and Reality: A Study of the Sources of Poetry*, London: Lawrence and Wishart.

Challey, James F. 1971, 'Carnot, Nicolas Léonard Sadi', in *Dictionary of Scientific Biography*, edited by C.C. Gillispie and F.L. Holmes, New York: Scribner.

Chicherin, Boris N. 1998, *Liberty, Equality and the Market*, New Haven, CT: Yale University Press.

Christensen, Paul P. 1994, 'Fire, Motion, and Productivity: the Protoenergetics of Nature and Economy in François Quesnay', in *Natural Images in Economic Thought*, edited by Philip Mirowski, Cambridge: Cambridge University Press.

Churchich, Nicholas 1990, *Marxism and Alienation*, Rutherford, NJ: Fairleigh Dickinson Press.

Clark, Brett and Richard York 2005a, 'Carbon Metabolism: Global Capitalism, Climate Change, and the Biospheric Rift', *Theory and Society*, 34(4): 391–428.

——— 2005b, 'Dialectical Nature: Reflections in Honor of the 20th Anniversary of Levins and Lewontin's *The Dialectical Biologist*', *Monthly Review*, 57(1): 13–21.

Clark, John P. 1989, 'Marx's Inorganic Body', *Environmental Ethics*, 11: 243–58.

Clausen, Rebecca 2007, 'Healing the Rift: Metabolic Restoration in Cuban Agriculture', *Monthly Review*, 59(1): 40–52.

Clausen, Rebecca and Brett Clark 2005, 'The Metabolic Rift and Marine Ecology', *Organization & Environment*, 18(4): 422–44.

Clausius, Rudolf 1879, *The Mechanical Theory of Heat*, London: Macmillan.

Cleveland, Cutler J. 1987, 'Biophysical Economics', in *The Development of Ecological Economics*, edited by R. Costanza, C. Perrings, C.J. Cleveland, Cheltenham: Edward Elgar.

——— 1999, 'Biophysical Economics', in *Bioeconomics and Sustainability*, edited by John Gowdy and Kozo Mayumi, Northampton, MA: Edward Elgar.

Coles, Peter 2001, *Cosmology: A Very Short Introduction*, New York: Oxford University Press.

Commoner, Barry 1971, *The Closing Circle*, New York: Knopf.

——— 1976, *The Poverty of Power*, New York: Knopf.

Comyn, Marian 1922, 'My Recollections of Karl Marx', *The Nineteenth Century and After*, Volume 91, retrieved from: https://www.marxists.org/subject/women/authors/comyn/marx.htm.

Colletti, Lucio 1973, *Marxism and Hegel*, London: Verso.

Cornu, Auguste 1957, *The Origins of Marxist Thought*, Springfield, IL: Charles C. Thomas.

Coveney, Peter and Roger Highfield 1990, *The Arrow of Time*, London: W.H. Allen.

Daly, Herman E. 1968, 'On Economics as a Life Science', *Journal of Political Economy*, 76(2): 396–8.

——— 1981, 'Postscript: Unresolved Problems and Issues for Further Research', in *Energy, Economics, and the Environment*, edited by Herman E. Daly and Alfred F. Umaña, Boulder, CO: Westview.

——— 1992, *Steady-state Economics*, Second Edition, London: Earthscan.

——— 1996, *Beyond Growth*, Boston, MA: Beacon Press.

Darwin, Charles 1964 [1859], *On the Origin of Species*, Cambridge, MA: Harvard University Press.

——— 1968 [1859], *On the Origin of Species*, New York: Penguin.

de Brunhoff, Suzanne 1976, *Marx on Money*, New York: Urizen.

de Kadt, Maarten and Salvatore Engel-Di Mauro 2001, 'Failed Promise', *Capitalism Nature Socialism*, 12(2): 50–6.

de Lavergne, Léonce 1855, *The Rural Economy of England, Scotland, and Ireland*, London: William Blackwood and Sons.

Deléage, Jean-Paul 1994, 'Eco-Marxist Critique of Political Economy', in *Is capitalism Sustainable?*, edited by Martin O'Connor, New York: Guilford.

Dickens, Peter 2004, *Society and Nature: Changing Our Environment, Changing Ourselves*, Cambridge: Polity.

Dobb, Maurice 1963, *Studies in the Development of Capitalism*, New York: International Publishers.

Dobrovolski, Ricardo 2012, 'Marx's Ecology and the Understanding of Land Cover Change', *Monthly Review*, 64(1): 31–9.

Dobson, Andrew 1990, *Green Political Thought*, London: Routledge.

Draper, Hal 1986, *The Marx-Engels Glossary*, New York: Schocken.

——— (ed.) 1998, *The Adventures of the Communist Manifesto*, Berkeley, CA: Center for Socialist History.

Eckersley, Robyn 1992, *Environmentalism and Political Theory*, New York: SUNY Press.

Edinburgh Review 1860, 'Review of Henry Darwin Rogers, Essays on the Coal Formation and its Fossils, and a Description of the Coal Fields of North America and Great Britain, annexed to the Government Survey of the Geology of Pennsylvania, Edinburgh and Philadelphia, 1858', *Edinburgh Review*, CXI: 68–95.

Einstein, Albert 2006, *Relativity*, London: Penguin.

Eiseley, Loren 1958, *Darwin's Century*, New York: Doubleday.

Empson, Martin 2009, *Marxism and Ecology*, London: Socialist Workers Party.

Engels, Frederick 1892, *The Condition of the Working Class in England*, Chicago, IL: Academy Publishers.

——— 1937, *Engels on Capital*, New York: International.

——— 1940, *The Dialectics of Nature*, New York: International Universitics Press.

——— 1941 [1888], *Ludwig Feuerbach and the Outcome of Classical German Philosophy*, New York: International Universities Press.

——— 1964, *Dialectics of Nature*, Moscow: Progress Publishers.

——— 1978a, *Socialism: Utopian and Scientific; With the Essay on 'The Mark'*, New York: International.

——— 1978b, 'The Mark', in *Collected Works*, Vol. 24, by Karl Marx and Frederick Engels, New York: International Publishers.

——— 1983, 'The Funeral of Karl Marx', in *Karl Marx Remembered*, edited by Philip S. Foner, San Francisco, CA: Synthesis Publications.

Faber, Daniel and Allison Grossman 2000, 'The Political Ecology of Marxism', *Capitalism Nature Socialism*, 11(2): 71–7.

Farrington, Benjamin 1944, *Greek Science I: Thales to Aristotle*, New York: Penguin.

——— 1967, *The Faith of Epicurus*, New York: Basic Books.

Fedorov, E.K. 1979, 'Climate Change and Human Strategy,' *Environment*, 21(4): 25–31.

——— 1981, *Man and Nature: Ecological Crisis and Social Progress*, New York: International Publishers.

Fischer-Kowalski, Marina 1997, 'Society's Metabolism', in *International Handbook of Environmental Sociology*, edited by Michael Redclift and Graham Woodgate, Northampton, MA: Edward Elgar.

Fleming, James Rodger 1998, *Historical Perspectives on Climate Change*, New York: Oxford University Press.

Foley, Duncan K. 1986, *Understanding Capital*, Cambridge, MA: Harvard University Press.

Foster, John Bellamy 1994, *The Vulnerable Planet*, New York: Monthly Review.

——— 1999, 'Marx's Theory of Metabolic Rift: Classical Foundations for Environmental Sociology', *American Journal of Sociology*, 105(2): 366–405.

——— 2000, *Marx's Ecology: Materialism and Nature*, New York: Monthly Review Press.

——— 2002, *Ecology Against Capitalism*, New York: Monthly Review Press.

——— 2005, 'The Treadmill of Accumulation', *Organization & Environment*, 18(1): 7–18.

——— 2009, *The Ecological Revolution*, New York: Monthly Review.

——— 2010, 'Marx's Ecology and Its Historical Significance', in *The International Handbook of Environmental Sociology*, Second Edition, edited by Michael R. Redclift and Graham Woodgate, Cheltenham: Edward Elgar.

——— 2011a, 'Capitalism and the Accumulation of Catastrophe', *Monthly Review*, 63(7).

——— 2011b, 'The Ecology of Marxian Political Economy', *Monthly Review*, 63(4) (September): 1–16.

——— 2014, 'Foreword', in *The Necessity of Social Control*, by István Mészáros, New York: Monthly Review.

——— 2015, 'Late Soviet Ecology', *Monthly Review*, 66(1): 1–20.

Foster, John Bellamy and Paul Burkett 2000, 'The Dialectic of Organic/Inorganic Relations: Marx and the Hegelian Philosophy of Nature', *Organization & Environment*, 13(4): 403–25.

——— 2001, 'Marx and the Dialectic of Organic/Inorganic Relations', *Organization & Environment*, 14(4): 451–62.

——— 2004, 'Ecological Economics and Classical Marxism: The "Podolinsky Business" Reconsidered', *Organization & Environment*, 17(1): 32–60.

Foster, John Bellamy and Brett Clark 2004, 'Ecological Imperialism: The Curse of Capitalism', in *Socialist Register 2004: The New Imperial Challenge*, edited by Leo Panitch and Colin Leys, New York: Monthly Review Press.

Foster, John Bellamy and Hannah Holleman 2012, 'Weber and the Environment: Classical Foundations for a Post-Exemptionalist Sociology', *American Journal of Sociology*, 117(6): 1625–73.

——— 2014, 'The Theory of Unequal Ecological Exchange,' *Journal of Peasant Studies*, 41(2): 199–233.

Foster, John Bellamy and Harry Magdoff 1999, 'Liebig, Marx, and the Depletion of Soil Fertility', in *Hungry for Profit*, edited by F. Magdoff, J.B. Foster and F.H. Buttel, New York: Monthly Review Press.

Foster, John Bellamy, Brett Clark and Richard York 2008, *The Critique of Intelligent Design*, New York: Monthly Review.

——— 2010, *The Ecological Rift*, New York: Monthly Review Press.

Fourier, Charles 1996, *The Theory of Four Movements*, Cambridge: Cambridge University Press.

Fraser, Craig G. 2006, *The Cosmos: A Historical Perspective*, Westport, CT: Greenwood.

Frenay, Robert 2006, *Pulse: The Coming Age of Systems and Machines Inspired by Living Things*, New York: Farrar, Straus and Giroux.

Garber, Elizabeth, Stephen G. Brush, and C.W.F. Everitt 1995, 'Introduction', in *Maxwell on Heat and Statistical Mechanics*, Bethlehem, PA: Lehigh University Press.

George, Susan 1998, 'Preface', in *Privatizing Nature*, edited by Michael Goldman, New Brunswick, NJ: Rutgers University Press.

Georgescu-Roegen, Nicholas 1971, *The Entropy Law and the Economic Process*, Cambridge, MA: Harvard University Press.

——— 1973, 'The Entropy Law and the Economic Problem', in *Economics, Ecology, Ethics*, edited by Herman E. Daly, San Francisco, CA: W.H. Freeman.

——— 1975, 'Energy and Economic Myths', *Southern Economic Journal*, 41(3): 347–81.

——— 1976, *Energy and Economic Myths*, New York: Pergamon.

——— 1979, 'Energy Analysis and Economic Valuation', *Southern Economic Journal*, 45(4): 1023–58.

——— 1986, 'The Entropy Law and the Economic Process in Retrospect', *Eastern Economic Journal*, 12(1): 3–25.

Giampietro, Mario and David Pimentel 1991, 'Energy Efficiency: Assessing the Interaction Between Humans and Their Environment', *Ecological Economics*, 4(2): 117–44.

Gleyse, Jacques, et. al. 2002. 'Physical Education as a Subject in France (School Curriculum, Policies and Discourse): The Body and the Metaphors of the Engine', *Sport Education and Society* 7(1): 5–23.

Goethe, Johann Wolfgang von 1952, *Goethe's Botanical Writings*, Honolulu, HI: University of Hawaii Press.

Gorz, André 1994, *Capitalism, Socialism, Ecology*, London: Verso.

Gould, Stephen J. 1987, *An Urchin in the Storm*, New York: Norton.

Graham, Loren R. 1993, *Science in Russia and the Soviet Union*, Cambridge: Cambridge University Press.

Gregory, Frederick 1977, *Scientific Materialism in Nineteenth Century Germany*, Boston, MA: D. Reidel Publishing.

Gribben, John 1998, *In Search of the Big Bang*, London: Penguin.

Grove, William Robert 1864, 'On the Correlation of Physical Forces', in *The Correlation and Conservation of Forces*, edited by Edward L. Youmans, New York: D. Appleton & Co.

——— 1873, 'The Correlation of Physical Forces', in *The Correlation and Conservation of Forces*, edited by Edward L. Youmans, New York: Appleton.

Guha, Ramachandra and Juan Martinez-Alier 1997, *Varieties of Environmentalism*, London: Earthscan.

Gunderson, Ryan 2011, 'The Metabolic Rifts of Livestock Agribusiness', *Organization and Environment*, 24(4): 404–22.

Habermas, Jürgen 1973, *Legitimation Crisis*, Boston, MA: Beacon Press.

Haeckel, Ernst 1929, *The Riddle of the Universe*, London: Watts.

Hall, Robert E. and Marc Lieberman 2003, *Macroeconomics: Principles and Applications*, 2nd edition, Mason, OH: Thomson-Southwestern.

Harman, Peter M. 1982, *Energy, Force, and Matter*, Cambridge: Cambridge University Press.

Hawley, Amos H. 1984, 'Human Ecological and Marxian Theories', *American Journal of Sociology*, 89(4): 904–17.

Hayward, Tim 1994, *Ecological Thought*, Cambridge: Polity.

Hegel, Georg Wilhelm Friedrich 1959, *Encyclopedia of Philosophy*, New York: Philosophical Library.

——— 1970 [1830], *The Philosophy of Nature*, Vols. 1–3, Atlantic Highlands, NJ: Humanities Press.

——— 1971 [1830], *The Philosophy of Mind*, Oxford: Oxford University Press.

——— 1975a [1830], *Hegel's Logic*, Oxford: Oxford University Press.

——— 1975b, *Hegel's Aesthetics*, Vol. 1, Oxford: Oxford University Press.

——— 1975c, *Hegel's Aesthetics*, Vol. 2, Oxford: Oxford University Press.

——— 1977 [1807], *The Phenomenology of Spirit*, New York: Oxford University Press.

——— 1993 [1835], *Introductory Lectures on Aesthetics*, London: Penguin.

——— 1995, *Lectures on the History of Philosophy*, Vol. 3, Lincoln, NE: University of Nebraska Press.

Helmholtz, Hermann 1873, 'On the Interaction of Natural Forces', in *The Correlation and Conservation of Forces*, edited by Edward L. Youmans, New York: Appleton.

——— 1876, *Populäre Wissenschaftliche Vorträge* [*Popular Scientific Lectures*], Braunschweig: Durck und Verlag von Friedrich Vieweg und Sohn.

——— 1908, *Popular Scientific Lectures*, Vol. 2, New York: Longmans Green.

Henderson, William Otto 1976, *The Life of Friedrich Engels*, London: Frank Cass.

Herber, Lewis 1965, *Crisis in Our Cities*, Englewood Cliffs, NJ: Prentice Hall.

Hermann, D. Ludimar 1874, *Grundriss der Physiologie des Menschen*, Berlin: Hirschwald.

——— 1875, *Elements of Human Physiology*, 5th edition, London: Smith, Elder.

Hessen, Boris 1931, 'The Social and Economic Roots of Newton's Principia', in *Science at the Crossroads*, edited by Nikolai Bukharin, London: Frank Cass.

——— 1971, 'The Social and Economic Roots of Newton's "Principia"', in *Science at the Cross Roads*, edited by Nikolai Bukharin, London: Frank Cass and Co.

Heuzé, Gustave 1875, *Les Jardins de Versailles et l'École d'Horticulture*. Paris: Société Centrale d'Agriculture de France.

Himka, Jean-Paul 1993, 'Podolynsky, Serhii', in *Encyclopedia of the Ukraine*, edited by C.C. Gillispie, F.L. Holmes and V. Kubijovyc, Toronto: University of Toronto Press.

Hobbes, Thomas 1929, *The English Works*, Vol. 1, London: John Bohn.

Hobsbawm, Eric 1969, *Industry and Empire*, London: Penguin.

Holubnychy, Vsevolod 1971, 'History of Ukrainian Economic Thought', in *Ukraine: A Concise Encyclopedia*, Toronto: University of Toronto Press.

——— 1993, 'Political Economy', in *Encyclopedia of the Ukraine*, edited by C.C. Gillispie, F.L. Holmes and V. Kubijovyc, Toronto: University of Toronto Press.

Hornburg, Alf 1998, 'Towards an Ecological Theory of Unequal Exchange', *Ecological Economics*, 25(1): 127–36.

Horton, Stephen, 1997, 'Value, Waste and the Built Environment: A Marxian Analysis', *Capitalism Nature Socialism*, 8(1): 127–39.

Hughes, Jonathan 2000, *Ecology and Historical Materialism*, New York: Cambridge University Press.

——— 2001, 'Analytical Marxism and Ecology: A Reply to Paul Burkett', *Historical Materialism*, 9: 153–67.

Hulme, Michael 2009, 'On the Origin of "The Greenhouse Effect": John Tyndall's 1859 Interogation of Nature', *Weather*, 64(5): 121–3.

Hume, George 1914, *Thirty-Five Years in Russia*, London: Simpkin, Marshall, Hamilton, Kent.

Huws, Ursula 1999, 'Material World: The Myth of the Weightless Economy', in *Socialist Register 1999: Global Capitalism Versus Democracy*, edited by Leo Panitch and Colin Leys, New York: Monthly Review Press.

Imbrie, John, and Katherine Palmer Imbrie 1979, *Ice Ages: Solving the Mystery*, Short Hills, NJ: Enslow Publishers.

Inwood, Michael 1992, *A Hegel Dictionary*, London: Blackwell.

Itoh, Makoto and Costas Lapavitsas 1999, *Political Economy of Money and Finance*, New York: St. Martin's Press.

Jacoby, Russell 1983, 'Western Marxism,' in *A Dictionary of Marxist Thought*, edited by Tom Bottomore, Oxford: Blackwell.

Jaki, Stanley L. 1974, *Science and Creation*, New York: Neale Watson.

Jevons, William Stanley 1900, *The Principles of Science*, London: Macmillan.

——— 1906, *The Coal Question*, London: Macmillan.

Kant, Immanuel 1900, 'Examination of the Question Whether the Earth Has Undergone an Alteration of its Axial Rotation', in *Kant's Cosmogony*, translated and edited by W. Hastie, Glasgow: James Maclehose and Sons.

——— 1952, *The Critique of Judgment*, Oxford: Oxford University Press.

Kapp, K. William 1950, *The Social Costs of Private Enterprise*, Cambridge, MA: Harvard University Press.

——— 1976, 'The Open-System Character of the Economy and Its Implications', in *Economics of the Future*, edited by Kurt Dopfer, Boulder, CO: Westview Press.

Kautsky, Karl n.d., *Ethics and the Materialist Conception of History*, Chicago, IL: Charles H. Kerr.

Kaufman, Robert 1987, 'Biophysical and Marxist Economics', *Ecological Modeling*, 38: 91–105.

Kenway, Peter 1980, 'Marx, Keynes and the Possibility of Crisis', *Cambridge Journal of Economics*, 4: 23–36.

Keynes, John Maynard 1951, *Essays and Sketches in Biography*, New York: Meridan Books.

Khalil, Elias L. 1990, 'Entropy Law and Exhaustion of Natural Resources: Is Nicholas Georgescu-Roegen's Paradigm Defensible?', *Ecological Economics*, 2(2): 163–78.

Kirchhoff, Gustav 1861–3, *Researches on the Solar Spectrum, and the Spectra of the Chemical Elements*, Part 1, Cambridge and London: Macmillan.

——— 1901, 'On the Relation Between the Emissive and the Absorptive Power of Bodies for Heat and Light', in *The Laws of Radiation and Absorption: Memoirs by Prévost, Stewart, Kirchhoff, and Kirchhoff and Bunsen*, edited by D.B. Brace, New York: American Book Company.

Klein, Naomi 2011, 'Capitalism vs. The Climate', *The Nation*, 28 November edition.

——— 2014, *This Changes Everything*, New York: Simon and Schuster.

Koropeckyj, Iwan S. 1984, *Selected Contributions of Ukrainian Scholars to Economics*, Cambridge, MA: Harvard University Press.

——— 1990, *Development in the Shadow: Studies in Ukrainian Economics*, Edmonton: University of Alberta Press.

Kołakowski, Leszek 1978, *Main Currents of Marxism*, New York: Oxford University Press.

Kovel, Joel 2001, 'A Materialism Worthy of Nature', *Capitalism Nature Socialism*, 12(2): 73–84.

——— 2002, *The Enemy of Nature*, London: Zed.

——— 2005, 'The Ecofeminist Ground of Ecosocialism', *Capitalism Nature Socialism*, 16(2): 45–60.

——— 2011a, 'Five Theses on Ecosocialism', *Ecosocialist Horizons*, 25 November, retrieved from: http://ecosocialisthorizons.com/2011/11/five-theses-on-ecosocialism/.

——— 2011b, 'The Future Will Be Ecosocialist', *Ecosocialist Horizons*, 27 November, retrieved from: http://ecosocialisthorizons.com/2011/11/the-future-is-ecosocialist/.

——— 2011c, 'Ecology', *Ecosocialist Horizons*, 25 November, retrieved from: http://ecosocialisthorizons.com/2011/11/marx-and-ecology/.

——— 2011d, 'On Marx and Ecology', *Capitalism Nature Socialism*, 22(1).

——— 2014, 'Ecosocialism as a Human Phenomenon', *Capitalism Nature Socialism*, 25(1): 18.

Kragh, Helge 2004, *Matter and Spirit in the Universe*, London: Imperial College Press.

Krohn, Wolfgang and Wolf Schäfer 1983, 'Agricultural Chemistry: The Origin and Structure of a Finalized Science', in *Finalization in Science*, edited by Wolf Schäfer, Boston, MA: D. Reidel.

Kryza, Frank 2003, *The Power of Light: The Epic Story of Man's Quest to Harness the Sun*, New York: McGraw-Hill.

Kuhn, Thomas S. 1977, *The Essential Tension*, Chicago, IL: University of Chicago Press.

Laboulaye, Charles 1874, *Dictionnaire des Arts et de l'Agricolture*, Paris: Librairie du Dictionnaire des Arts et Manufactures.

Lakatos, Imre 1970, 'Falsification and the Methodology of Scientific Research Programs', in *Criticism and the Growth of Knowledge*, edited by Imre Lakatos and Alan Musgrave, Cambridge: Cambridge University Press.

Lange, Frederick 1950 [1865], *A History of Materialism*, New York: Humanities Press.

Larousse, Pierre (ed.) 1865–90, *Le Grand Dictionnaire Universel du XIX Siècle*, Paris: Administration du Grand Dictionnaire Universel.

Lee, Donald C. 1980, 'On the Marxian View of the Relationship between Man and Nature', *Environmental Ethics*, 2(1): 3–16.

Leff, Enrique 1993, 'Marxism and the Environmental Question', *Capitalism Nature Socialism*, 4(1): 44–66.

Lenin, Vladimir I. 1961, *Collected Works*, Vol. 38, Moscow: Progress.

Lessner, Friedrich 1957, 'Before 1848 and After', in *Reminiscences of Marx and Engels*, edited by Institute for Marxism Leninism, Moscow: Foreign Languages Publishing House.

Levins, Richard 2008, 'Dialectics and Systems Theory', in *Dialectics for the New Century*, edited by Bertell Ollman and Tony Smith, New York: Palgrave Macmillan.

Levins, Richard and Richard Lewontin 1985, *The Dialectical Biologist*, Cambridge, MA: Harvard University Press.

Liebig, Justus von 1864, 'On the Connection and Equivalence of Forces', in *The Correlation and Conservation of Forces*, edited by Edward L. Youmans, New York: D. Appleton & Co.

——— 1865, *Die Chemie in ihrer Anwendung auf Agricultur und Physiologie* [*Chemistry *in its Application to Agriculture and Physiology*], Seventh Edition, Braunschweig: F. Vieweg und Sohn.

Lifshitz, Mikhail 1938, *The Philosophy of Art of Karl Marx*, New York: Critics Group.

Light, A. 1997, 'Deep Socialism? An interview with Arne Naess', *Capitalism Nature Socialism*, 8(1): 69–85.

Lindley, David 2001, *Boltzmann's Atom*, New York: Free Press.

——— 2004, *Degrees Kelvin: A Tale of Genius, Invention, and Tragedy*, Washington, DC: Joseph Henry Press.

Lipietz, Alain 2000, 'Political Ecology and the Future of Marxism', *Capitalism Nature Socialism*, 11: 69–85.

Locke, John 1959 [1690], *An Essay Concerning Human Understanding*, New York: Dover.

Longo, Stefano 2012, 'Mediterranean Rift', *Critical Sociology*, 38(3): 417–36.

Louis-Lande, Lucien 1878, *Basques et Navarrais*, Paris: Disier et cie.

Löwy, Michael 1997, 'For a Critical Marxism', *Against the Current*, 12(5): 31–5.

Lozada, Gabriel A. 1991, 'A Defense of Nicholas Georgescu-Roegen's Paradigm', *Ecological Economics*, 3(2): 157–60.

Lukács, Georg 1968 [1923], *History and Class Consciousness*, Cambridge, MA: MIT Press.

——— 1971, *History and Class Consciousness*, London: Merlin Press.

——— 1974, *Conversations with Lukács*, edited by Theo Pinkus, Cambridge, MA: MIT Press.

——— 1980, *The Destruction of Reason*, London: Merlin.

——— 2000, *A Defense of History and Class Consciousness: Tailism and the Dialectic*, London: Verso.

Luke, Timothy W. 1999, *Capitalism, Democracy and Ecology: Departing from Marx*, Urbana, IL: University of Illinois Press.

Luxemburg, Rosa 1970, *Rosa Luxemburg Speaks*, New York: Pathfinder.

——— 1972, *The Accumulation of Capital: An Anti-Critique*, New York: Monthly Review.

——— 2013, *The Letters of Rosa Luxemburg*, London: Verso.

Magdoff, Fred 2007, 'Ecological Agriculture', *Renewable Agriculture and Food Systems*, 22(2): 109–17.

Magdoff, Fred, John Bellamy Foster and Frederick H. Buttel 2000, 'Introduction', in *Hungry for Profit: The Agribusiness Threat to Farmers, Food, and the Environment*, New York: Monthly Review.

Magdoff, Fred and John Bellamy Foster 2011, *What Every Environmentalist Needs to Know About Capitalism*, New York: Monthly Review Press.

Magdoff, Harry and Paul M. Sweezy 1974, 'Watergate One Year Later', *Monthly Review*, 26(1): 1–11.

Malm, Andreas 2013, 'The Origins of Fossil Capital: From Water to Steam in the British Cotton Industry', *Historical Materialism*, 21(1): 15–68.

Mancus, Philip 2007, 'Nitrogen Fertilizer Dependency and its Contradictions', *Rural Sociology*, 72(2): 269–88.

Mandelbaum, M. 1971, *History, Man and Reason: A Study in Nineteenth Century Thought*, Baltimore, MD: Johns Hopkins University Press.

Marcuse, Herbert 1972, *Counter-Revolution and Revolt*, Boston, MA: Beacon.

——— 1978, *The Aesthetic Dimension*, Boston, MA: Beacon.

Marey, Étienne Jules 1868, *Du mouvement dans les fonctions de la vie*, Paris: G. Baillière.

——— 1874, *Animal Mechanism*, New York: D. Appleton and Company.

Martinez-Alier, Juan 1987, *Ecological Economics*, Oxford: Basil Blackwell.

——— 1995, 'Political Ecology, Distributional Conflicts and Economic Incommensurability', *New Left Review*, 211: 70–88.

——— 1997, 'Some Issues in Agrarian and Ecological Economics, in Memory of Nicholas Georgescu-Roegen', *Ecological Economics*, 22(3): 225–38.

——— 2003, 'Marxism, Social Metabolism, and Ecologically Unequal Exchange', Paper presented at World System Theory and the Environment Conference, Lund University, Sweden.

——— 2005, 'Social Metabolism and Ecological Distribution Conflicts', Paper presented to the Australian New Zealand Society for Ecological Economics, Massey University, Palmerston North, New Zealand.

——— 2006, 'Social Metabolism and Environmental Conflicts', in *The Socialist Register, 2007: Coming to Terms with Nature*, edited by Leo Panitch and Colin Leys, New York: Monthly Review Press.

——— 2007, 'Marxism, Social Metabolism, and International Trade', in *Rethinking Environmental History*, edited by J.R. McNeill and J. Martinez-Alier, New York: Altamira.

Martinez-Alier, Juan and J.M. Naredo 1982, 'A Marxist Precursor of Energy Economics: Podolinsky', *Journal of Peasant Studies*, 9(2): 207–24.

Marx, Karl 1934, *Letters to Dr. Kugelman*, New York: International Universities Press.

——— 1939, 'From the Critical History', in *Anti-Dühring*, by Frederick Engels, New York: International Publishers.

——— 1963a, *The Poverty of Philosophy*, New York: International Publishers.

——— 1963b, *Theories of Surplus Value*, Part 1, Moscow: Progress Publishers.

——— 1966, *Critique of the Gotha Programme*, New York: International.

——— 1967, *Capital*, Vol. 1, New York: International Publishers.

——— 1968, *Theories of Surplus Value*, Part 2, Moscow: Progress Publishers.

——— 1970, *A Contribution to the Critique of Political Economy*, Moscow: Progress Publishers.

——— 1971, *Theories of Surplus Value*, Part III, Moscow: Progress Publishers.

——— 1973, *Grundrisse*, New York: Vintage.

——— 1974, *Early Writings*, New York: Vintage.

——— 1975 *Texts on Method*, Oxford: Basil Blackwell.

——— 1976a, *Capital*, Vol. 1, New York: Vintage.

——— 1976b, *Value, Price and Profit*, New York: International Publishers.

——— 1976c, 'Wages', in *Collected Works*, Vol. 6, by Karl Marx and Frederick Engels, New York: International Publishers.

——— 1978, *Capital*, Vol. 2, London: Penguin.

——— 1981, *Capital*, Vol. 3, London: Penguin.

——— 1991, 'Economic Manuscript of 1861–63, Continuation', in *Collected Works*, Vol. 33, by Karl Marx and Frederick Engels, New York: International Publishers.

Marx, Karl and Frederick Engels 1964, *The Communist Manifesto*, New York: Monthly Review.

——— 1975a, *Collected Works*, New York: International Publishers.

——— 1975b, *Selected Correspondence*, Moscow: Progress Publishers.

——— 2005, *The Communist Manifesto*, edited by Phil Gasper, Chicago, IL: Haymarket.

——— 2011, *MEGA* IV/26, Berlin: Akademie Verlag.

Mayer, Julius R. 1870, 'The Mechanical Theory of Heat', *Nature*, 1: 566–7.

Mayumi, Kozo 1991, 'Temporary Emancipation from the Land', *Ecological Economics*, 4(1): 35–56.

——— 2001, *The Origins of Ecological Economics*, New York: Routledge.

McNally, David 1988, *Political Economy and the Rise of Capitalism*, Berkeley, CA: University of California Press.

Mead, George 1936, *Movements of Thought in the Nineteenth Century*, Chicago, IL: University of Chicago Press.

Meek, Ronald L. 1963, *The Economics of Physiocracy: Essays and Translations*, Cambridge, MA: Harvard University Press.

Merchant, Carolyn 1980, *The Death of Nature*, New York: Harper and Row.

Merleau-Ponty, Mourice 1973, *Adventures of the Dialectic*, Evanston, IL: Northwestern University Press.

Mészáros, István 1970, *Marx's Theory of Alienation*, New York: Monthly Review.

——— 1971, *Marx's Theory of Alienation*, London: Merlin Press.

——— 1995, *Beyond Capital*, New York: Monthly Review Press.

——— 2008, *The Challenge and Burden of Historical Time*, New York: Monthly Review.

——— 2010, *Social Structure and Forms of Consciousness*, Vol. 1, New York: Monthly Review Press.

Mirowski, Philip 1988, 'Energy and Energetics in Economic Theory', *Journal of Economic Issues*, 22(3): 811–30.

Molina, Manuel González de and Victor M. Toledo 2014, *The Social Metabolism*, New York: Springer International Publishing.

Moore, Jason W. 2000, 'Environmental Crises and the Metabolic Rift in World-Historical Perspective', *Organization & Environment*, 13(2): 123–57.

——— 2003, 'The Modern World-System as Environmental History?: Ecology and the Rise of Capitalism', *Theory and Society*, 32(3): 307–77.

——— 2011, 'Transcending the Metabolic Rift', *Journal of Peasant Studies* 38(1): 1–46.

Morton, John Chalmers 1859, 'On the Forces Used in Agriculture', *Journal of the Society of the Arts*, 9 December.

Moseley, Fred 1999, 'Marx's Reproduction Schemes and Smith's Dogma', in *The Circulation of Capital*, edited by G. Reuten, and C. Arthur, London: Macmillan.

Mumford, Lewis 1926, *The Golden Day*, Boston, MA: Beacon Press.

Müntzer, Thomas 1988, *Collected Works*, Edinburgh: T & T Clark.

Naess, Arne 1973, 'The Shallow and the Deep, Long-Range Ecology Movement: A Summary', *Inquiry*, 16: 95–100.

——— 2008, *The Ecology of Wisdom*, Berkeley, CA: Counterpoint.

NASA/Goddard Space Flight Center 2003, 'NASA Satellite Measures Earth's Carbon Metabolism', *Science Daily*, 24 April, retrieved from: http://www.sciencedaily.com/releases/2003/04/030424082215.htm.

Nearing, Scott 1952, *Economics for the Power Age*, East Paltaka, FL: World Events Committee.

——— 1962, 'World Events', *Monthly Review*, 14(7): 389–94.

Nearing, Scott and Helen Nearing 1970, *Living the Good Life*, New York: Schocken Books.

Needham, Joseph 1976, *Moulds of Understanding*, London: George Allen and Unwin.

Nell, Edward J. 1998, *The General Theory of Transformational Growth*, New York: Cambridge University Press.

Nordhaus, William D. and James Tobin 1972, 'Is Growth Obsolete?', in *Economic Research: Retrospect and Prospect*, Vol. 5, *Economic Growth*, Cambridge, MA: National Bureau of Economic Research.

O'Connor, James 1998, *Natural Causes*, New York: Guilford.

O'Connor, J.J. and E.F. Robertson 1997, 'Alexander Aleksandrovich Friedmann', *MacTutor History of Mathematics*, Retrieved from http://www-history.mcs.st-andrews.ac.uk/Biographies/Friedmann.html.

Odum, Howard T. and Scienceman, David M. 2005, 'An Energy Systems View Of Karl Marx's Concepts of Production and Labor Value', in *Emergy Synthesis 3: Theory and Applications of the Emergy Methodology* (Proceedings from the Third Biennial Emergy Conference, Gainesville, Florida, January 2004), edited by Center for Environmental Policy.

Ollman, Bertell 1976, *Alienation: Marx's Conception of Man in Capitalist Society*, Cambridge: Cambridge University Press.

Olmstead, J.M.D. and E. Harris Olmstead 1952, *Claude Bernard and the Experimental Method in Medicine*, New York: Henry Schuman.

O'Neill, John 1993, *Ecology, Policy and Politics*, London: Routledge.

——— 1994, 'Humanism and Nature', *Radical Philosophy*, 66: 21–9.

Oxford English Dictionary (OED) 1971, Oxford: Oxford University Press.

Pannekoek, Anton 1912, *Marxism and Darwinism*, Chicago, IL: Charles H. Kerr.

Papanelopoulou, Faidra 2003, 'Paris-Province: Energy Physics in Mid-Nineteenth-Century France', *Revue de la Maison Francaise d'Oxford*, 1(2): 1–19, retrieved from: http://www.mfo.ac.uk/Publications/Revue%20Fox/papanelopoulou.htm.

Parsons, Howard 1977, *Marx and Engels on Ecology*, Westport, CT: Greenwood.

Parkinson, Eric 1999, 'Talking Technology', *Journal of Technology Education*, 11(1), retrieved from: http://scholar.lib.vt.edu/ejournals/JTEv11n1/parkinson.html.

Pelouze, Théophile Jules and Edmond Fremy 1865–6, *Traité de Chimie Générale*, Paris: V. Masson.

Pepper, David 1996, *Modern Environmentalism*, London: Routledge.

Perrings, Charles 1987, *Economy and Environment*, New York: Cambridge University Press.

Peterson, Keith R. 2010, 'From Ecological Politics to Intrinsic Value: An Examination of Kovel's Value Theory', *Capitalism Nature Socialism*, 21(3): 81–101.

Pimentel, David and Marcia Pimentel 1996, *Food, Energy, and Society*, Revised edition, Boulder, CO: University Press of Colorado.

Podolinsky, Sergei 1880, 'Le socialisme et l'unité des forces physiques', *La Revue Socialiste*, 8: 353–65.

——— 1881, 'Il Socialismo e l'Unita delle Forze Fisiche', *La Plebe*, 14(3): 13–16; 14(4): 5–15.

——— 1883, 'Menschliche arbeit und einheit der kraft', *Die Neue Zeit*, 1(9): 413–24; 1(10): 449–57.

——— 1991, *Human Labor and Its Relation to the Distribution of Energy*, Moscow: Noosfera.

——— 1995, 'El trabajo del ser humano y su relación con la distribución de la Energia' [*Human labor and its relation to the distribution of energy*], in *Los principios de la economía ecolólogica: Textos de P. Geddes, S.A. Podolinsky y F. Soddy*, edited by Juan Martínez Alier, Madrid: Fundación Argentaria.

——— 2004 [1881], 'Socialism and the Unity of Physical Forces', *Organization & Environment*, 17(1): 61–75.

——— 2008 [1883], 'Human Labour and Unity of Force', *Historical Materialism*, 16(1): 163–83.

Popper, Karl 1982, *The Open Universe: An Argument for Indeterminism*, Totowa, NJ: Rowman and Littlefield.

Prigogine, Ilya 1997, *The End of Certainty*, New York: Free Press.

Prigogine, Ilya and Isabelle Stengers 1984, *Order Out of Chaos*, New York: Bantam.

Quesnay, François 1963a, 'Rural Philosophy' (extract), in *The Economics of Physiocracy: Essays and Translations*, edited by Ronald L. Meek, Cambridge, MA: Harvard University Press.

——— 1963b, 'Dialogue on the Work of Artisans', in *The Economics of Physiocracy: Essays and Translations* edited by Ronald L. Meek, Cambridge, MA: Harvard University Press.

——— 1963c, 'General Maxims for the Economic Government of an Agricultural Kingdom', in *The Economics of Physiocracy: Essays and Translations* edited by Ronald L. Meek, Cambridge, MA: Harvard University Press.

Rabinbach, Anson 1990, *The Human Motor: Energy, Fatigue and the Origins of Modernity*, New York: Basic Books.

Rankine, William 1852, 'On the Reconcentration of the Mechanical Energy of the Universe', *Philosophical Magazine*, 4(4): 358–60.

Rappaport, Roy A. 1984, *Pigs for the Ancestors*, New Haven, CT: Yale University Press.

Ravaioli, Carla 1995, *Economists and the Environment*, London: Zed.

Reclus, Elisée 1876–94, *Nouvelle Géographie Universelle: La Terre et les Hommes*, 19 Volumes, Paris: Hachette.

——— 1882–95, *The Earth and Its Inhabitants*, 19 Volumes, edited by E.G. Ravenstein and A.H. Keane, New York: D. Appleton & Co.

Rockström, Johan et al. 2009, 'A Safe Operating Space for Humanity', *Nature*, 461(24 September): 472–5.

Rolston III, Holmes 1988, *Environmental Ethics*, Philadelphia, PA: Temple University.

Rosdolsky, Roman 1977, *The Making of Marx's 'Capital'*, London: Pluto.

Rose, Margaret A. 1984, *Marx's Lost Aesthetic: Karl Marx and the Visual Arts*, Cambridge: Cambridge University Press.

Rosen, George 1959, 'The Conservation of Energy and the Study of Metabolism', in *The Historical Development of Physiological Thought*, edited by C.M. Brooks and P.F. Cranefield, New York: Hafner.

Rosenfeld, Léon 2012, *Physics, Philsophy, and Politics in the Twentieth Century*, Hackensack, NJ: World Scientific Publishing.

Routley, Val 1981, 'On Karl Marx as an Environmental Hero', *Environmental Ethics*, 3: 237–44.

Royal Astronomical Society 1875, 'Johann Heinrich von Mädler', *Monthly Notices of the Royal Astronomical Society*, 35: 171–5.

Rudnytsky, Ivan L. (ed.) 1952, 'Mykhaylo Drahomanov: A Symposium and Selected Writings', *Annals of the Ukrainian Academy of Arts and Sciences in the U.S.*, 1(3): 1–225.

——— 1987, *Essays in Modern Ukrainian History*, Cambridge, MA: Harvard University Press.

Saad-Filho, Alfredo 2002, *The Value of Marx*, London: Routledge.

Sainte-Claire Deville, Henri 1864, *Leçons sur la Dissociation: Professées Devant la Société Chimique le 18 Mars et le 1 er Avril 1864*. Paris: Lahure.

Saito, Kohei 2014, 'The Emergence of Marx's Critique of Modern Agriculture: Ecological Insights from His Excerpt Notebooks', *Monthly Review*, 66(5): 25–46.

Salleh, Ariel 1997, *Ecofeminism as Politics*, London: Zed.
——— 2010, 'From Metabolic Rift to Metabolic Value', *Organization & Environment*, 23(2): 205–19.
Sapir, Boris (ed.) 1974, *Lavrov: Gody emigratsii: Arkhivnye materially v dvukh tomakh* [*Lavrov: Years of Emigration: Letters and Documents in Two Volumes*], Boston, MA: D. Reidel.
Sarkar, Saral 1999, *Eco-Socialism or Eco-Capitalism*, London: Zed Books.
——— 2012 'The Waning Relevance of Marxism', *Arbeitskreis Ökopolitic*, retrieved from: http://ak-oekopolitik.blogspot.com/2012/11/waning-relevance-of-marxism.html.
Sartre, Jean-Paul 1963 *The Search for a Method*, New York: Vintage.
——— 2004, *Critique of Dialectical Reason*, Vol. 1, London: Verso.
Say, Jean-Baptiste 1832, *A Treatise on Political Economy*, Philadelphia, PA: Grigg & Elliott.
Schafer, Paul M. 2006, 'Editor's Introduction', in *The First Writings of Karl Marx*, Brooklyn: Ig.
Schmidt, Alfred 1971, *The Concept of Nature in Marx*, London: New Left.
Schnaiberg, Alan 1980, *The Environment: From Surplus to Scarcity*, New York: Oxford University Press.
Schneider, Eric and Dorion Sagan 2005, *Into the Cool: Energy Flow, Thermodynamics, and Life*, Chicago, IL: University of Chicago Press.
Schneider, Stephen and Randi Londer 1984, *The Coevolution of Climate and Life*, San Francisco, CA: Sierra Club Books.
Schrödinger, Erwin 1944, *What is Life?*, Cambridge: Cambridge University Press.
Schumpeter, Joseph A. 1954, *History of Economic Analysis*, New York: Oxford University Press.
Secchi, Angelo 1875–7, *Le Soleil*, 2 Volumes, Second edition, Paris: Gauthier-Villars.
Serbyn, Roman 1982, 'In Defense of an Independent Ukrainian Socialist Movement: Three Letters from Serhii Podolinsky to Valerian Smirnov', *Journal of Ukrainian Studies*, 7(2): 3–32.
Sheasby, Walt 1999, 'Anti-Prometheus, Post-Marx: The Real and the Myth in Green Theory', *Organization and Environment*, 12(1): 5–44.
——— 2001, 'Marx at Karlsbad', *Capitalism Nature Socialism* 12(3): 91–7.
——— 2002, *Metabolism in Marx's Theory*, unpublished manuscript, Rio Hondo College, Whittier, CA.
Sheehan, Helena 1985, *Marxism and the Philosophy of Science: A Critical History*, Atlantic Highlands, NJ: Humanities Press.
Shenstone, William Ashwell 1901, *Justus von Liebig: His Life and Work*, New York: Macmillan.
Sidenbladh, Elis 1876, *La Suede: Expose Statistique*, Paris: K. Nilsson.
——— 1878, *Royaume de Suède; Exposition Universelle de 1878 à Paris*, Stockholm.

Sieber, Nikolai 2001, 'Marx's Theory of Value and Money', *Research in Political Economy*, 19: 17–45.

Siferle, Robert Peter 2001, *The Subterranean Forest*, Cambridge: White Horse.

Sismondi, Simonde de 1991, *New Principles of Political Economy: Of Wealth in its Relation to Population*, New Brunswick, NJ: Transaction Publishers.

Smil, Vaclav, 1991, *General Energetics*, New York: John Wiley and Sons.

——— 2008, *Energy in Nature and Society*, Cambridge, MA: MIT Press.

Smith, Adam 1937, *An Inquiry into the Nature and Causes of the Wealth of Nations*, New York: Modern Library.

Smith, Crosbie, 1998, *The Science of Energy: A Cultural History of Energy and Physics in Victorian Britain*, London: The Athlone Press.

Smith, David Norman 2001, 'The Spectral Reality of Value: Sieber, Marx and Commodity Fetishism', *Research in Political Economy*, 19: 47–66.

Smith, Mark J. 1998, *Ecologism*, Minneapolis, MN: University of Minnesota.

Smith, Neil 1984, *Uneven Development: Nature, Capital and the Production of Space*, Oxford: Blackwell.

Soddy, Frederick 1922, *Cartesian Economics*, London: Hendersons.

Solow, Robert M. 1974, 'The Economics of Resources or the Resources of Economics', *The American Economic Review*, 64(2): 1–14.

——— 1993, 'The Economics of Resources or the Resources of Economics', in *Economics of the Environment: Selected Readings*, edited by Robert Dorfman and Nancy S. Dorfman, New York: W.W. Norton.

Soper, Kate 1996, 'Greening Prometheus: Marxism and Ecology', in *The Greening of Marxism*, edited by Ted Benton, New York: Guilford.

Spencer, Herbert 1880, *First Principles*, New York: A.L. Burt.

Stanley, John L. 2002, *Mainlining Marx*, New Brunswick, NJ: Transaction Publishing.

Steffen, Will et al. 2015, 'Planetary Boundaries: Guiding Human Development on a Changing Planet', *Science*, 347(6223): 736–46.

Steiner, George 1975, *After Babel*, New York: Oxford University Press.

Sternberger, Dolf 1977, *Panorama of the Nineteenth Century*, New York: Urizen.

Sterry Hunt, Thomas 1891, *Chemical and Geological Essays*, New York: Scientific Publishing Company.

Steuart, Sir James 1966, *An Inquiry into the Principles of Political Economy*, Volume One, Chicago, IL: University of Chicago Press.

Stokes, Kenneth M. 1994, *Man and the Biosphere: Toward a Coevolutionary Political Economy*, Armonk, NY: M.E. Sharpe.

Strasser, Susan 1999, *Waste and Want: A Social History of Trash*, New York: Henry Holt and Company.

Sweezy, Paul M. 1953, *The Present as History: Reviews on Capitalism and Socialism*, New York: Monthly Review

——— 1973, 'Cars and Cities', *Monthly Review*, 24(11): 1–18.

——— 1977, 'Comment', in *The Political Economy of the New Left*, by Assar Lindbeck, New York: New York University Press.

——— 1980, 'Japan in Perspective', *Monthly Review*, 31(9) (February): 1–14.

——— 1989, 'Capitalism and the Environment', *Monthly Review*, 41(2).

Tait, Peter Guthrie 1864, 'Energy', *North British Review*, February-May: 337–68.

Tanuro, Daniel 2003, *Green Capitalism: Why It Can't Work*, London: Merlin.

——— 2006, 'Marx's Concept of Social Metabolism and Ecosocialist Responses to Climate Change', Speech at the event 'Ecosocialism or Barbarism', Socialist Reistance, London, 2 December, retrieved from: http://www.europe-solidaire.org/spip.php?article4508.

——— 2007, 'The Devil Makes the Saucepans, But Not the Lids', *International Viewpoint*, 17 March, retrieved from: http://www.internationalviewpoint.org/spip.php?article1233.

——— 2008, 'Humanity, Society and Ecology', *Climate and Capitalism*, 18 October, retrieved from: http://climateandcapitalism.com/2008/10/21/humanity-society-and-ecology-global-warming-and-the-ecosocialist-alternative.

——— 2009, 'Climate Crisis: 21st Century Socialists Must be Ecosocialists', in *The Global Fight for Climate Justice*, edited by Ian Angus and Simon Butler, London: Resistance Books.

——— 2010, 'Marxism, Energy, and Ecology', *Capitalism Nature Socialism*, 21(4): 89–101.

——— 2011, 'The Futility of Green Capitalism', *International Viewpoint*, retrieved from: http://www.internationalviewpoint.org/spip.php?auteur54.

——— 2012, 'A Plea for the Ecological Reconstruction of Marxism', *International Viewpoint*, 3 December, retrieved from: http://www.internationalviewpoint.org/spip.php?article2815.

——— 2013, *Green Capitalism*, London: Merlin Press.

Taylor, Charles 1975, *Hegel*, Cambridge: Cambridge University Press.

Thompson, E.P. 1963, *The Making of the English Working Class*, New York: Vintage.

——— 1991, *Customs in Common*, London: Merlin Press.

Thomson, William 1848, 'On an Absolute Thermometric Scale founded on Carnot's Theory of the Motive Power of Heat, and Calculated from Regnault's Observations', *Proceedings of the Cambridge Philosophical Society*, 1: 66–71.

——— 1862, 'On the Secular Cooling of the Earth', *Proceedings of the Royal Society of Edinburgh*, 4: 610–11.

——— 1891, *Popular Lectures and Addresses*, Vol. 1, New York: Macmillan.

Toulmin, Stephen 1982, *The Return to Cosmology*, Berkeley, CA: University of California Press.

Tsuru, Shigeto 1976, *Towards a New Political Economy*, Tokyo: Kodansha.

——— 1994, *Economic Theory and Capitalist Society*, Brookfield, VT: Edward Elgar.

Turgot, Anne-Robert-Jacques 1898, *Reflections on the Formation and Distribution of Riches*, New York: Macmillan.

Tyndall, John, 1863, *Heat Considered as a Law of Motion*, London: Longman, Green & Co.

Uranovsky, Y.M. 1935, 'Marxism and Natural Science', in *Marxism and Modern Thought*, by Nikolai Bukharin et al., New York: Harcourt Brace and Company.

Ursul, A.D. (ed.) 1983, *Philosophy and the Ecological Problems of Civilization*, Moscow: Progress Publishers.

Verdet, Emile 1868–72, *Oeuvres de Émile*, Volumes VII–VIII, Paris: Imprimerie Impériale.

Vilenkin, Alex 2006, *Many Worlds in One: The Search for other Universes*, New York: Hill and Wang.

Voden, Alexis n.d., 'Talks with Engels', in *Reminiscences of Marx and Engels*, edited by Institute of Marxism-Leninism, Moscow: Foreign Languages.

Vogel, Steven 1996, *Against Nature*, New York: SUNY Press.

Wall, Derek 2010, *The Rise of the Green Left*, London: Pluto.

Wallis, Victor 2014, 'Ecosocialist Struggles', *Capitalism Nature Socialism* 25(1): 40–52.

Washburn, Sherwood L. and Ruth Moore 1974, *Ape into Man*, Boston, MA: Little, Brown.

Weart, Spencer 2003, *The Discovery of Global Warming*, Cambridge, MA: Harvard University Press.

Weber, Max 2003, *General Economic History*, Mineola, New York: Dover.

Weeks, John 1979, 'The Process of Accumulation and the "Profit Squeeze" Hypothesis', *Science and Society*, 43: 259–80.

Weiner, Douglas R. 1988, *Models of Nature: Ecology, Conservation, and Cultural Revolution in Soviet Russia*, Bloomington, IN: Indiana University Press.

——— 1999, *A Little Corner of Freedom: Russian Nature Protection from Stalin to Gorbachëv*, Berkeley, CA: University of California Press.

Wendling, Amy 2007, 'On Alienation and Machine Production: Capitalist Embodiment in Karl Marx', *Beiträge zur Marx-Engels-Forschung: Neue Folge 2007*: 253–67.

——— 2009, *Karl Marx on Technology and Alienation*, London: Palgrave Macmillan.

West, Cornel 1991, *The Ethical Dimensions of Marxist Thought*, New York: Monthly Review.

Weston, Del 2014, *The Political Economy of Global Warming: The Terminal Crisis*, London: Routledge.

Wetter, Gustav Andreas 1958, *Dialectical Materialism: A Historical and Systematic Survey of Philosophy in the Soviet Union*, New York: Routledge and Kegan Paul.

White, James D. 1996, *Karl Marx and the Intellectual Origins of Dialectical Materialism*, New York: St. Martin's Press.

White, James D. 2001, 'Nikolai Sieber and Karl Marx', *Research in Political Economy*, 19: 3–14.

Wilkinson, R.G. 1973, *Poverty and Progress: An Ecological Model of Economic Development*, London: Methuen and Co.

Williams, Chris 2010, *Ecology and Socialism*, Chicago, IL: Haymarket.

Williams, Raymond 1989, *Resources of Hope*, London: Verso.

Williamson, A.G. 1993, 'The Second Law of Thermodynamics and the Economic Process', *Ecological Economics*, 7(1): 70–1.

Willis, Robert 1851, *A System of Apparatus for the Use of Lecturers and Experimenters in Mechanical Philosophy*, London: John Weale.

Winder, Nick, Brian S. McIntosh and Paul Jeffrey 2005, 'The Origin, Diagnostic Attributes and Practical Application of Co-evolutionary Theory', *Ecological Economics*, 54(4): 347–61.

Wishart, Ryan 2012, 'Coal River's Last Mountain', *Organization and Environment*, 25(4): 470–85.

Wishart, Ryan, R. Jamil Jonna and Jordan Fox Besek 2014, 'The Metabolic Rift: A Selected Bibliography', retrieved from: http://monthlyreview.org/commentary/metabolic-rift/.

Wittman, Hannah 2009, 'Reworking the Metabolic Rift', *Journal of Peasant Studies*, 36(4): 805–26.

Worster, Donald 1994, *Nature's Economy: A History of Ecological Ideas*, Cambridge: Cambridge University Press.

Wrigley, E.A. 2010, *Energy and the English Industrial Revolution*, Cambridge: Cambridge University Press.

Yergin, Daniel 2011, *The Quest*, New York: Penguin.

Zarembka, Paul 2000, 'Accumulation of Capital, its Definition', *Research in Political Economy*, 18: 183–241.

Zimmerman, Michael J. 2010, 'Intrinsic vs. Extrinsic Value', *The Stanford Encyclopedia of Philosophy*, edited by Edward N. Zalta, retrieved from: http://plato.stanford.edu/archives/win2010/entries/value-intrinsic-extrinsic/.

Index

CPSIA information can be obtained
at www.ICGtesting.com
Printed in the USA
JSHW030329070223
37361JS00005B/5